Problems and Solutions in Euclidean Geometry

PROBLEMS AND SOLUTIONS IN EUCLIDEAN GEOMETRY

M. N. AREF
Alexandria University

WILLIAM WERNICK
City College, New York

DOVER PUBLICATIONS, INC
MINEOLA, NEW YORK

Bibliographical Note

This Dover edition, first published in 2010, is a reissue of a work originally published by Dover Publications, Inc, in 1968.

International Standard Book Number

ISBN-13: 978-0-486-47720-6
ISBN-10: 0-486-47720-7

Manufactured in the United States by Courier Corporation
47720701
www.doverpublications.com

Dedicated to my father
who has been always of lifelong inspiration
and encouragement to me

M.N.A.

CONTENTS

PREFACE

This book is intended as a second course in Euclidean geometry. Its purpose is to give the reader facility in applying the theorems of Euclid to the solution of geometrical problems. Each chapter begins with a brief account of Euclid's theorems and corollaries for simplicity of reference, then states and proves a number of important propositions. Chapters close with a section of miscellaneous problems of increasing complexity, selected from an immense mass of material for their usefulness and interest. Hints of solutions for a large number of problems are also included.

Since this is not intended for the beginner in geometry, such familiar concepts as point, line, ray, segment, angle, and polygon are used freely without explicit definition. For the purpose of clarity rather than rigor the general term *line* is used to designate sometimes a ray, sometimes a segment, sometimes the length of a segment; the meaning intended will be clear from the context.

Definitions of less familiar geometrical concepts, such as those of modern and space geometry, are added for clarity, and since the use of symbols might prove an additional difficulty to some readers, geometrical notation is introduced gradually in each chapter.

January, 1968

M. N. Aref

William Wernick

SYMBOLS

EMPLOYED IN THIS BOOK

$\angle$	angle
$\odot$	circle
▱	parallelogram
⏢	quadrilateral
▭	rectangle
□	square
$\triangle$	triangle

$a = b$	a equals b
$a > b$	a is greater than b
$a < b$	a is less than b
$a \parallel b$	a is parallel to b
$a \perp b$	a is perpendicular to b
a / b	a divided by b
$a : b$	the ratio of a to b
AB^2	the square of the distance from A to B
$\because$	because
$\therefore$	therefore

CHAPTER 1

TRIANGLES AND POLYGONS

Theorems and Corollaries

Lines and Angles

1.1. *If a straight line meets another straight line, the sum of the two adjacent angles is two right angles.*

Corollary 1. *If any number of straight lines are drawn from a given point, the sum of the consecutive angles so formed is four right angles.*

Corollary 2. *If a straight line meets another straight line, the bisectors of the two adjacent angles are at right angles to one another.*

1.2. *If the sum of two adjacent angles is two right angles, their non-coincident arms are in the same straight line.*

1.3. *If two straight lines intersect, the vertically opposite angles are equal.*

1.4. *If a straight line cuts two other straight lines so as to make the alternate angles equal, the two straight lines are parallel.*

1.5. *If a straight line cuts two other straight lines so as to make:* (*i*) *two corresponding angles equal; or* (*ii*) *the interior angles, on the same side of the line, supplementary, the two straight lines are parallel.*

1.6. *If a straight line intersects two parallel straight lines, it makes:* (*i*) *alternate angles equal;* (*ii*) *corresponding angles equal;* (*iii*) *two interior angles on the same side of the line supplementary.*

Corollary. *Two angles whose respective arms are either parallel or perpendicular to one another are either equal or supplementary.*

1.7. *Straight lines which are parallel to the same straight line are parallel to one another.*

Triangles and Their Congruence

1.8. *If one side of a triangle is produced,* (*i*) *the exterior angle is equal to the sum of the interior non-adjacent angles;* (*ii*) *the sum of the three angles of a triangle is two right angles.*

Corollary 1. *If two angles of one triangle are respectively equal to two angles of another triangle, the third angles are equal and the triangles are equiangular.*

Corollary 2. *If one side of a triangle is produced, the exterior angle is greater than either of the interior non-adjacent angles.*

COROLLARY 3. *The sum of any two angles of a triangle is less than two right angles.*

1.9. *If all the sides of a polygon of n sides are produced in order, the sum of the exterior angles is four right angles.*

COROLLARY. *The sum of all the interior angles of a polygon of n sides is* $(2n - 4)$ *right angles.*

1.10. *Two triangles are congruent if two sides and the included angle of one triangle are respectively equal to two sides and the included angle of the other.*

1.11. *Two triangles are congruent if two angles and a side of one triangle are respectively equal to two angles and the corresponding side of the other.*

1.12. *If two sides of a triangle are equal, the angles opposite to these sides are equal.*

COROLLARY 1. *The bisector of the vertex angle of an isosceles triangle, (i) bisects the base; (ii) is perpendicular to the base.*

COROLLARY 2. *An equilaterial triangle is also equiangular.*

1.13. *If two angles of a triangle are equal, the sides which subtend these angles are equal.*

COROLLARY. *An equiangular triangle is also equilateral.*

1.14. *Two triangles are congruent if the three sides of one triangle are respectively equal to the three sides of the other.*

1.15. *Two right-angled triangles are congruent if the hypotenuse and a side of one are respectively equal to the hypotenuse and a side of the other.*

INEQUALITIES

1.16. *If two sides of a triangle are unequal, the greater side has the greater angle opposite to it.*

1.17. *If two angles of a triangle are unequal, the greater angle has the greater side opposite to it.*

1.18. *Any two sides of a triangle are together greater than the third.*

1.19. *If two triangles have two sides of the one respectively equal to two sides of the other and the included angles unequal, then the third side of that with the greater angle is greater than the third side of the other.*

1.20. *If two triangles have two sides of the one respectively equal to two sides of the other, and the third sides unequal, then the angle contained by the sides of that with the greater base is greater than the corresponding angle of the other.*

1.21. *Of all straight lines that can be drawn to a given straight line from a given external point, (i) the perpendicular is least; (ii) straight lines which make equal angles with the perpendicular are equal; (iii) one making a greater angle with the perpendicular is greater than one making a lesser angle.*

COROLLARY. *Two and only two straight lines can be drawn to a given straight line from a given external point, which are equal to one another.*

QUADRILATERALS AND OVER FOUR-SIDED POLYGONS

1.22. *The opposite sides and angles of a parallelogram are equal, each diagonal bisects the parallelogram, and the diagonals bisect one another.*

COROLLARY 1. *The distance between a pair of parallel straight lines is everywhere the same.*

COROLLARY 2. *The diagonals of a rhombus bisect each other at right angles.*

COROLLARY 3. *A square is equilateral.*

1.23. *A quadrilateral is a parallelogram if (i) one pair of opposite sides are equal and parallel; (ii) both pairs of opposite sides are equal or parallel; (iii) both pairs of opposite angles are equal; (iv) the diagonals bisect one another.*

1.24. *Two parallelograms are congruent if two sides and the included angle of one are equal respectively to two sides and the included angle of the other.*

COROLLARY. *Two rectangles having equal bases and equal altitudes are congruent.*

1.25. *If three or more parallel straight lines intercept equal segments on one transversal, they intercept equal segments on every transversal.*

COROLLARY. *A line parallel to a base of a trapezoid and bisecting a leg bisects the other leg also.*

1.26. *If a line is drawn from the mid-point of one side of a triangle parallel to the second side, it bisects the third side. This line is called a mid-line of a triangle.*

COROLLARY 1. *Conversely, a mid-line of a triangle is parallel to the third side and is equal to half its magnitude.*

COROLLARY 2. *In any triangle, a mid-line between two sides and the median to the third side bisect each other.*

1.27. *In a right triangle, the median from the right vertex to the hypotenuse is equal to half the hypotenuse.*

1.28. *If one angle of a right triangle is* 30°, *the side opposite this angle is equal to half the hypotenuse.*

COROLLARY. *Conversely, if one side of a right triangle is half the hypotenuse, the angle opposite to it is* 30°.

1.29. *The median of a trapezoid is parallel to the parallel bases and is equal to half their sum.*

COROLLARY. *The line joining the mid-points of the diagonals of a trapezoid is parallel to the parallel bases and is equal to half their difference.*

1.30. *In an isosceles trapezoid, the base angles and the diagonals are equal to one another.*

INTRODUCTION TO CONCURRENCY

1.31. *The perpendicular bisectors of the sides of a triangle are concurrent in a point equidistant from the vertices of the triangle which is the center of the circumscribed circle and called the circumcenter of the triangle.*

1.32. *The bisectors of the angles of a triangle are concurrent in a point equidistant from the sides of the triangle which is the center of the inscribed circle and called the incenter of the triangle.*

COROLLARY 1. *The bisector of any interior angle and the external bisectors of the other two exterior angles are concurrent in a point outside the triangle which is equidistant from the sides (or produced) of the triangle and called an excenter of the triangle.*

COROLLARY 2. *There are four points equidistant from the three sides of a triangle: one inside the triangle, which is the incenter, and three outside it, which are the excenters.*

1.33. *The altitudes of a triangle are concurrent in a point called the orthocenter of the triangle.*

1.34. *The medians of a triangle are concurrent in a point $\frac{2}{3}$ the distance from each vertex to the mid-point of the opposite side. This point is called the centroid of the triangle.*

Solved Problems

1.1. *ABC is a triangle having $BC = 2\,AB$. Bisect BC in D and BD in E. Prove that AD bisects $\angle CAE$.*

CONSTRUCTION: Draw $DF \parallel AC$ to meet AB in F (*Fig.* 1.)

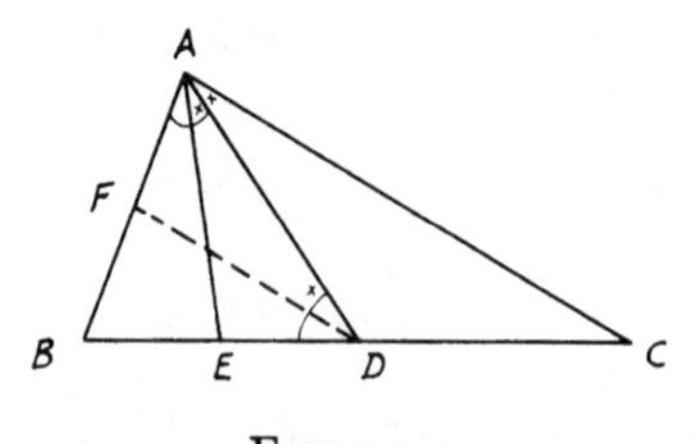

FIGURE 1

Proof: $\because$ D is the mid-point of BC and $DF \parallel AC$, $\therefore$ F is the mid-point of AB (Th. 1.26). Also, $AB = BD = \frac{1}{2}BC$. $\therefore$ $BF = BE$. $\therefore$ $\triangle$s ABE and DBF are congruent. $\therefore$ $\angle EAF = \angle EDF$, but $\angle BAD = \angle BDA$ (since $BA = BD$, Th. 1.12). $\therefore$ Subtraction gives $\angle EAD$

$= \angle FDA$. But $\angle FDA = \angle DAC$ (since, $DF \parallel AC$). $\therefore$ $\angle EAD = \angle DAC$.

1.2. *ABC is a triangle. D and E are any two points on AB and AC. The bisectors of the angles ABE and ACD meet in F. Show that* $\angle BDC + \angle BEC = 2 \angle BFC$.

CONSTRUCTION: Join AF and produce it to meet BC in G (*Fig.* 2).

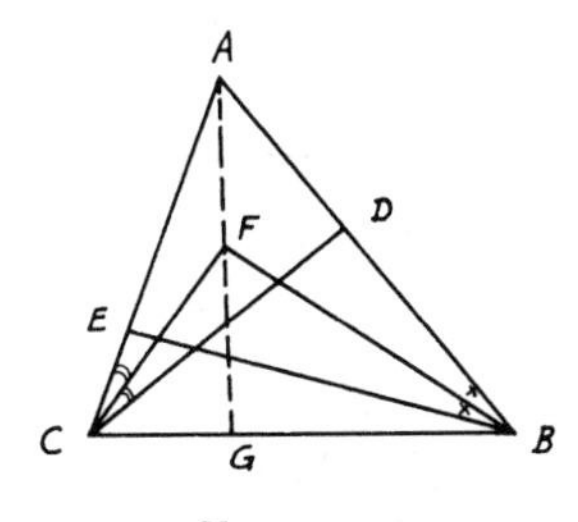

FIGURE 2

Proof: $\angle BDC$ is exterior to $\triangle ADC$. $\therefore$ $\angle BDC = \angle A + \angle ACD$ (Th. 1.8). Also, $\angle BEC$ is exterior to $\triangle AEB$. $\therefore$ $\angle BEC = \angle A + \angle ABE$; hence $\angle BDC + \angle BEC = 2\angle A + \angle ACD + \angle ABE$ (1). Similarly, $\angle$s BFG, CFG are exterior to $\triangle$s AFB, AFC. $\therefore$ $\angle BFG + \angle CFG = \angle BFC = \angle A + \frac{1}{2} \angle ABE + \frac{1}{2} \angle ACD$ (2). Therefore, from (1) and (2), $\angle BDC + \angle BEC = 2\angle BFC$.

1.3. *ABC is a right-angled triangle at A and AB > AC. Bisect BC in D draw DE perpendicular to the hypotenuse BC to meet the bisector of the right angle A in E. Prove that (i) AD = DE; (ii)* $\angle DAE = \frac{1}{2} (\angle C - \angle B)$.

CONSTRUCTION: Draw $AF \perp BC$ (*Fig.* 3).

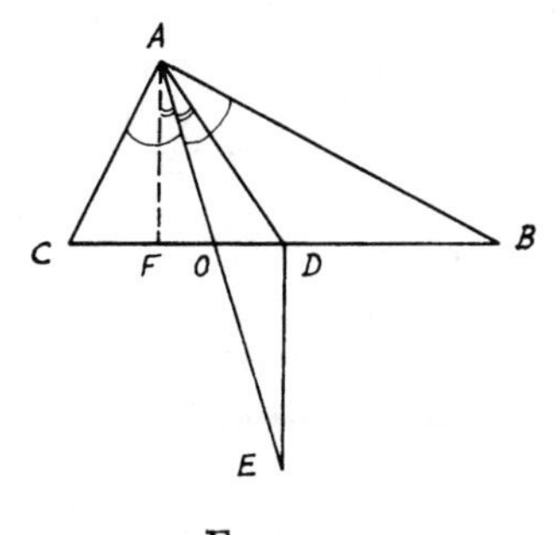

FIGURE 3

Proof: (i) $\because$ D is the mid-point of BC. $\therefore$ $AD = DB = DC$ (Th. 1.27). $\therefore$ $\angle DAB = \angle DBA$. But, $\angle DBA$ or $FBA = \angle FAC$ (since they are complementary to $\angle FAB$). $\therefore$ $\angle DAB = \angle FAC$. But since AO bisects $\angle A$, $\therefore \angle OAB = \angle OAC$ and subtraction gives $\angle OAD = \angle OAF$. $\because$ $DE \parallel AF$ (both $\perp BC$), $\therefore$ $\angle OAF = \angle OED$, and hence $\angle OAD = \angle OED$. $\therefore$ $AD = DE$ (Th. 1.13).

(ii) As shown above, $\angle DAB = \angle DBA$ or B. Since $\angle FAB = \angle C$ (both complement $\angle FAC$), $\therefore (\angle FAB - \angle DAB) = (\angle C - \angle B) = 2\angle DAO$ or $\angle DAE = \frac{1}{2}(\angle C - \angle B)$.

1.4. *The sides AB, BC, and AC of a triangle are bisected in F, G and H respectively. If BE is drawn perpendicular to AC, prove that* $\angle FEG = \angle FHG$ *(Fig. 4).*

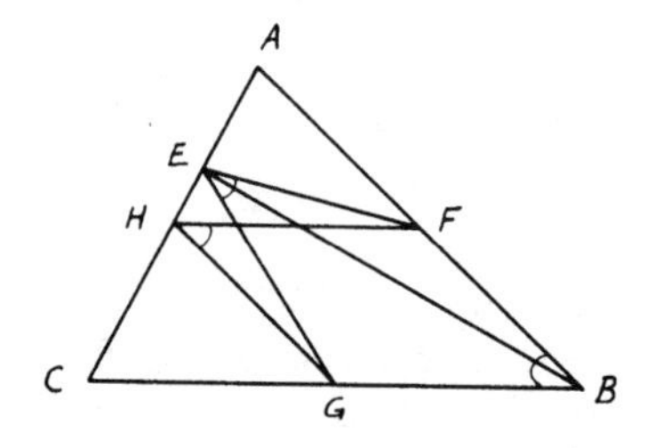

FIGURE 4

Proof: $\because$ G and H are the mid-points of BC and AC. $\therefore$ $HG \parallel AB$ and = its half (Cor. 1, Th. 1.26), or $HG \parallel FB$ and equal to it. Hence $FBGH$ is a ▱. $\therefore$ $\angle FHG = \angle B$. But $\triangle AEB$ is right angled at E and F is the mid-point of AB. $\therefore$ $EF = FB$. $\therefore$ $\angle FEB = \angle FBE$ and, similarly, $\angle GEB = \angle GBE$. Therefore, by adding, $\angle FEG = \angle B = \angle FHG$.

1.5. *In an isosceles triangle ABC, the vertex angle A is $\frac{2}{7}$ right angle. Let D be any point on the base BC and take a point E on CA so that CE = CD. Join ED and produce it to meet AB produced in F (Fig. 5). If the bisector of the* $\angle EDC$ *meets AC in R, show that (i) AE = EF; (ii) DR = DC.*

Proof: (i) $\angle A = \frac{2}{7}$ right angle. $\because$ The sum of the 3 $\angle$s of $\triangle ABC$ = 2 right angles (Th. 1.8), half the difference is $\angle B = \angle C = \frac{6}{7}$ right angle. But, in $\triangle CED$, $CE = CD$. $\therefore$ $\angle CED = \angle CDE = \frac{1}{2}(2 - \frac{6}{7})$ right angle $= \frac{4}{7}$ right angle. $\because$ $\angle CED$ is exterior to the $\triangle AEF$, $\therefore$ $\angle CED = \angle A + \angle AFE$. But $\angle A = \frac{2}{7}$ right angle. $\therefore$ $\angle AFE = \frac{2}{7}$ right angle or $\angle A = \angle AFE$. $\therefore$ $AE = EF$ (Th. 1.13).

(ii) $\because$ $\angle CDE = \frac{4}{7}$ right angle, $\therefore$ $\angle CDR = \frac{2}{7}$ right angle. But $\angle C = \frac{6}{7}$ right angle, hence in the $\triangle CDR$, the difference from 2 right angles would be $\angle CRD = \frac{6}{7}$ right angle. $\therefore \angle CRD = \angle C$. $\therefore DR = DC$ (Th. 1.13).

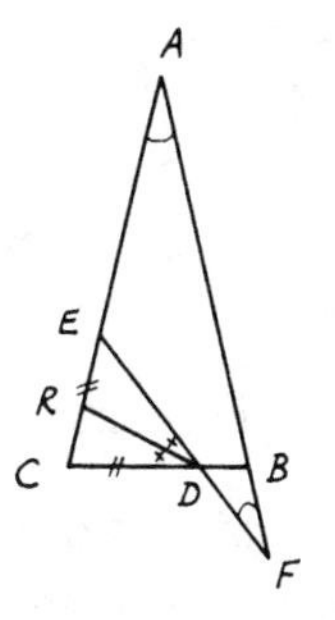

FIGURE 5

1.6. *The vertex C is a right angle in the triangle ABC. If the points D and E are taken on the hypotenuse, so that BC = BD and AC = AE, show that DE equals the sum of the perpendiculars from D and E on AC and BC respectively.*

CONSTRUCTION: Draw $CN \perp AB$ and join CD and CE (*Fig.* 6).

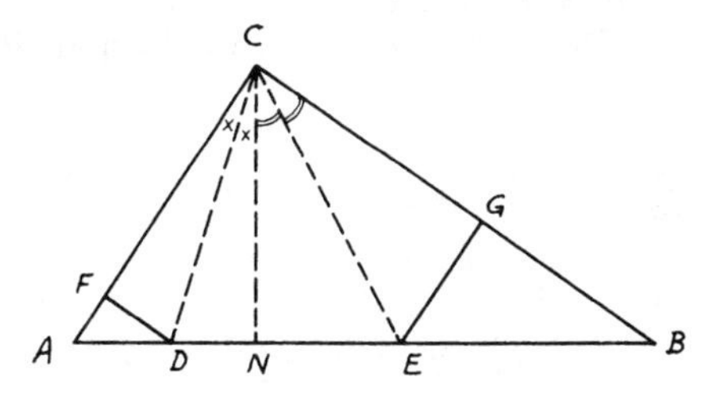

FIGURE 6

Proof: $\because BC = BD$. $\therefore \angle BCD = \angle BDC$ (Th. 1.12). But $\angle BCD = \angle BCN + \angle NCD$ and $\angle BDC = \angle DCA + \angle A$ (Th. 1.8). Hence $\angle BCN + \angle NCD = \angle DCA + \angle A$ (1) and $\because CN \perp AB$, $\therefore \angle BCN = \angle A$ (since both complement $\angle NCA$) (2). From (1) and (2), $\therefore \angle NCD = \angle DCA$. But $\angle CND = \angle CFD$ = right angle. $\therefore \triangle$s CND, CFD are congruent. $\therefore DN = DF$. Similarly, $EN = EG$ and, by adding, $\therefore DE = DF + EG$.

1.7. *The side AB in the rectangle ABCD is twice the side BC. A point P is taken on the side AB so that* $BP = \frac{1}{4} AB$. *Show that BD is perpendicular to CP.*

CONSTRUCTION: Bisect AB in Q and draw $QR \perp AB$ to meet BD in R (*Fig.* 7).

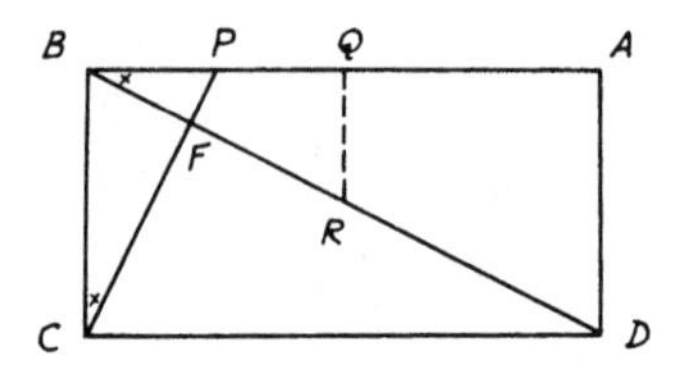

FIGURE 7

Proof: $\because AB = 2\,BC$, $\therefore BC = BQ$. In the $\triangle ABD$, Q is the mid-point of AB and $QR \parallel AD$. $\therefore$ R is the mid-point of BD and $QR = \frac{1}{2}AD$ (Th. 1.26, Cor. 1). Since $BP = \frac{1}{4}AB = \frac{1}{2}AD$, $\therefore$ $BP = QR$. Therefore, $\triangle$s BQR and CBP are congruent. $\therefore$ $\angle QBR = \angle BCP$. But since $\angle QBR + \angle RBC =$ right angle, $\therefore$ $\angle BCP + \angle RBC =$ right angle, yielding $\angle BFC =$ right angle [Th. 1.8(ii)]. Hence $BD \perp CP$.

1.8. *ABC is a triangle. BF, CG are any two lines drawn from the extremities of the base BC to meet AC and AB in F and G respectively and intersect in H. Show that AF + AG > HF + HG.*

CONSTRUCTION: Draw HD and $HE \parallel$ to AB and AC respectively (*Fig.* 8).

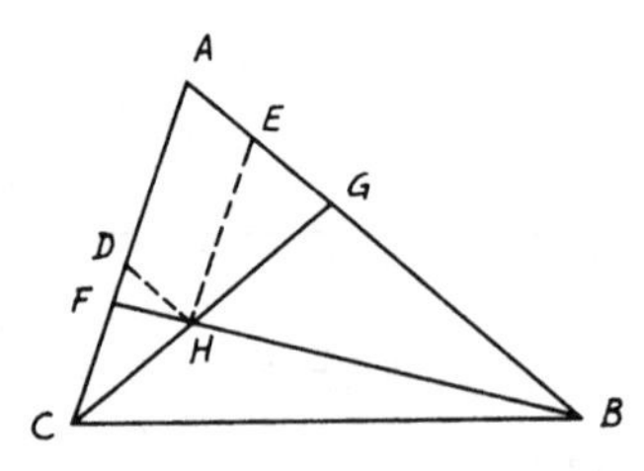

FIGURE 8

Proof: The figure $ADHE$ is a parallelogram [Th. 1.23(ii)]. $\therefore$ $AD = HE$ and $AE = DH$. In the $\triangle DHF$, $DH + DF > HF$ (Th. 1.18). $\therefore$ $AE + DF > HF$ (1). Similarly in $\triangle EGH$, $EH + EG > HG$ or $AD + EG > HG$ (2). Adding (1) and (2) gives $AE + EG + AD + DF > HF + HG$. $\therefore$ $AF + AG > HF + HG$.

1.9. *ABCD is a square. The bisector of the $\angle DBA$ meets the diagonal AC in F. If CK is drawn perpendicular to BF, intersecting BD in L and* *produced to meet AB in R, prove that* $AR = 2\,SL$, *where S is the intersection of the diagonals.*

CONSTRUCTION: From S draw $ST \parallel CR$ to meet BF in J and AB in T (*Fig.* 9).

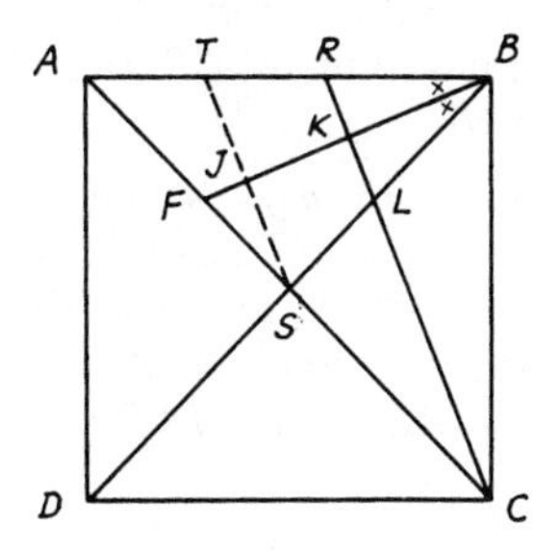

FIGURE 9

Proof: $\because CR \perp BF$, $\therefore ST \perp BF$. $\triangle$s BLK, BRK are congruent (Th. 1.11). $\therefore BR = BL$. Again $\triangle$s BTJ, BSJ are congruent. $\therefore BT = BS$. Hence $TR = SL$ and, since the two diagonals of a square bisect one another (Th. 1.22), $\therefore S$ is the mid-point of AC. But ST is $\parallel CR$ in the $\triangle ACR$. $\therefore T$ is the mid-point of AR (Th. 1.26). $\therefore AR = 2TR$. $\therefore AR = 2SL$.

1.10. *The side AB in the triangle ABC is greater than AC, and D is the mid-point of BC. From C draw two perpendiculars to the bisectors of the internal and external vertical angles at A to meet them in F and G respectively. Prove that* (*i*) $DF = \frac{1}{2}(AB - AC)$; (*ii*) $DG = \frac{1}{2}(AB + AC)$.

CONSTRUCTION: Produce CF, CG to meet AB, BA produced in H and M (*Fig.* 10).

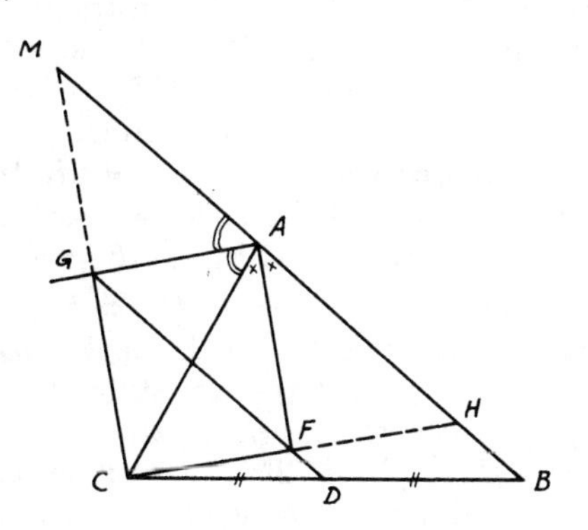

FIGURE 10

Proof: (i) $\because$ CF is $\perp AF$ and AF bisects the $\angle A$, then the $\triangle$s AFC, AFH are congruent. $\therefore$ $CF = FH$ and $AC = AH$. In the $\triangle CBH$, D, F are the mid-points of BC and CH. $\therefore$ $DF = \frac{1}{2} BH = \frac{1}{2}(AB - AH) = \frac{1}{2}(AB - AC)$.

(ii) Similarly, G is the mid-point of CM, and $AC = AM$. In the $\triangle CBM$, D, G are the mid-points of CB and CM. $\therefore$ $DG = \frac{1}{2} BM = \frac{1}{2}(AB + AM) = \frac{1}{2}(AB + AC)$.

1.11. *ABC is a triangle and AD is any line drawn from A to the base BC. From B and C, two perpendiculars BE and CF are drawn to AD or AD produced. If R is the mid-point of BC, prove that RE = RF.*

CONSTRUCTION: Produce CF and BE so that $CF = FG$ and $BE = EH$, then join AG, BG, AH, and CH (*Fig.* 11).

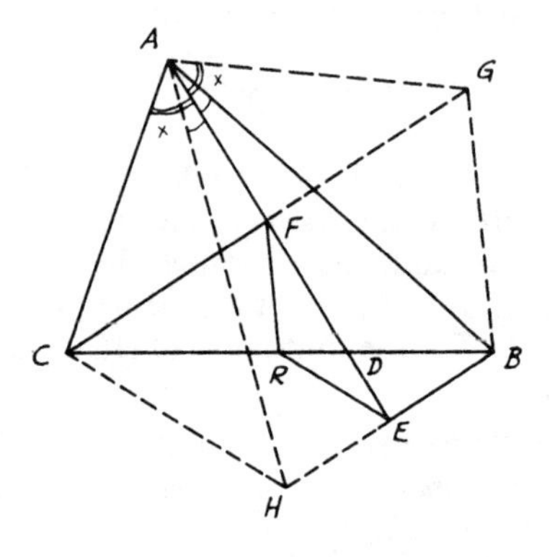

FIGURE 11

Proof: $\because$ $BE \perp AE$ and $BE = EH$ (construct). Then $\triangle$s ABE, AHE are congruent (Th. 1.10). $\therefore$ $AB = AH$ and $\angle BAE = \angle HAE$. Similarly, from the congruent $\triangle$s ACF, AGF: $AC = AG$ and $\angle CAF = \angle GAF$. Therefore by subtracting, $\angle BAG = \angle CAH$. Hence $\triangle$s AGB, ACH are congruent. $\therefore$ $BG = CH$. In the $\triangle CBG$, R, F are the mid-points of CB and CG. $\therefore$ $RF = \frac{1}{2} BG$. Similarly, in the $\triangle CBH$, $RE = \frac{1}{2} CH$. Since $BG = CH$, $\therefore$ $RE = RF$.

1.12. *Any finite displacement of a segment AB can be considered as though a rotation about a point called the pole. If the length of the segment and the angle which it rotates are given, describe the method to determine the pole. What would be the conditions of the angles between the rays around the pole?*

ANALYSIS: Let A_1B_1, A_2B_2 be the two positions of the given segment which is rotated a given angle (*Fig.* 12). Construct the two perpendicular bisectors CP_{12}, DP_{12} to A_1A_2, B_1B_2 respectively to intersect at the required pole P_{12}. Since $\triangle$s A_1CP_{12}, A_2CP_{12} are congruent (Th. 1.10), $\therefore$ $P_{12}A_1 = P_{12}A_2$ and $\angle A_1P_{12}C = \angle A_2P_{12}C = \frac{1}{2}$ given $\angle A_1P_{12}A_2$. Similarly, $P_{12}B_1 = P_{12}B_2$ and $\angle B_1P_{12}D = \angle B_2P_{12}D = \frac{1}{2}$ given $\angle B_1P_{12}B_2$, but $\angle A_1P_{12}A_2 = \angle B_1P_{12}B_2 =$ given. $\therefore$ $\angle A_1P_{12}B_1 = \angle A_2P_{12}B_2$. Hence $\triangle$s $A_1P_{12}B_1$, $A_2P_{12}B_2$

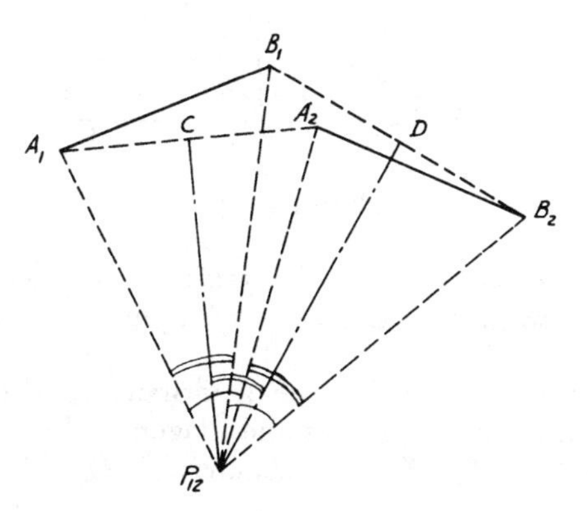

FIGURE 12

are congruent; i.e., A_1B_1 is rotated around the pole P_{12} to a new position A_2B_2 through the given $\angle$s $A_1P_{12}A_2$ or $B_1P_{12}B_2$.

SYNTHESIS: Join A_1A_2, B_1B_2 and construct their $\perp$ bisectors CP_{12}, DP_{12} to meet at the required pole P_{12}. Notice that $\angle A_1P_{12}B_1 = \angle A_2P_{12}B_2 = \angle CP_{12}D$. This is called *Chasle's theorem.*

1.13. *ABC is a triangle in which the vertical angle C is* 60°. *If the bisectors of the base angles A, B meet BC, AC in P, Q respectively, show that* $AB = AQ + BP$.

CONSTRUCTION: Take point R on AB so that $AR = AQ$. Join RP and RQ. Let AP, BQ intersect in E. Join ER and PQ. Let QR intersect AE in D (*Fig.* 13).

Proof: Since $\angle C = 60°$, $\therefore$ $\angle A + \angle B = 120°$ (Th. 1.8). Hence $\angle EAB + \angle EBA = 60°$. $\therefore$ $\angle AEB = 120°$. $\therefore$ $\angle PEB = \angle QEA = 60°$. Since $AR = AQ$ (construct) and AD bisects $\angle A$, $\therefore$ $\triangle$s ADQ, ADR are congruent (Th. 1.10). $\therefore$ $DQ = DR$ and $AD \perp QR$. Hence $QE = ER$ and $\angle QED = \angle RED = 60°$. $\therefore$ $\angle REB = \angle PEB = 60°$. $\because$ EB bisects $\angle B$, $\therefore$ $\triangle$s EBP, EBR are congruent. $\therefore$ $RB = BP$.

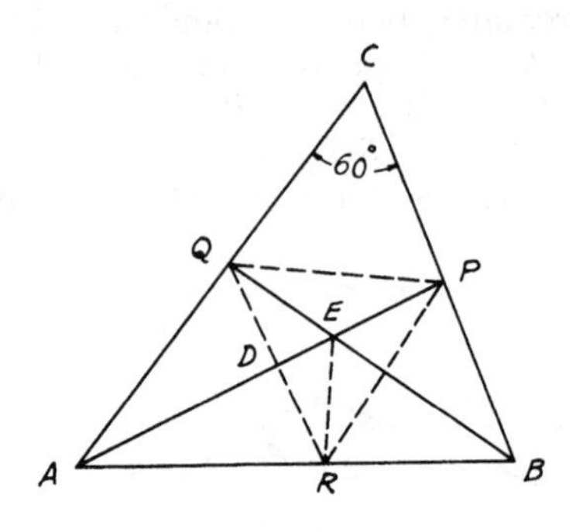

FIGURE 13

1.14. *In any quadrilateral the two lines joining the mid-points of each pair of opposite sides meet at one point, with the line joining the mid-points of its diagonals.*

CONSTRUCTION: Let $ABCD$ be the quadrilateral and E, F, G, H, K, and L the mid-points of its sides and diagonals. Join EF, FG, GH, HE, EK, KG, GL, LE, FK, KH, HL, and LF (*Fig.* 14).

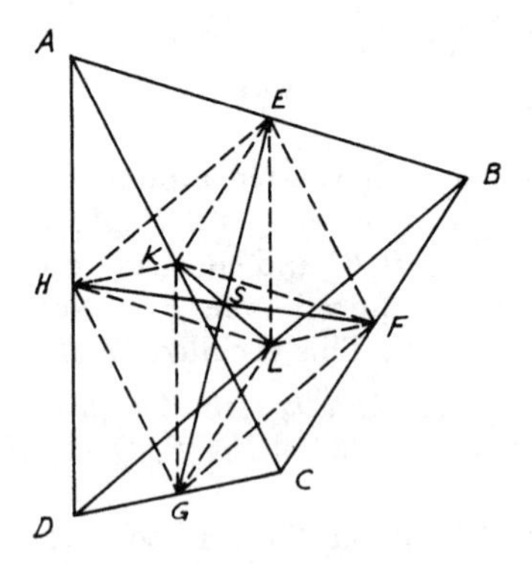

FIGURE 14

Proof: In the $\triangle ABC$, E and F are the mid-points of AB and BC. $\therefore$ $EF \parallel AC$ and equals its half (Th, 1.26. Cor. 1). Also, in the $\triangle DAC$, $HG \parallel AC$ and equals its half. Hence $EF =$ and $\parallel HG$. $\therefore$ $EFGH$ is a parallelogram (Th. 1.23). Therefore, its diagonals EG and FH bisect one another (Th. 1.22). Similarly, $FKHL$ and $EKGL$ are parallelograms also. $\therefore$ FH and LK in the first ▱ bisect one another, and EG and LK in the second ▱ bisect one another also. Since the point of intersection of EG, FH, and LK is their mid-point, which cannot be more than one point, then EG and FH intersect at one point with the line LK.

1.15. *ABC is a triangle. If the bisectors of the two exterior angles B and C of the triangle meet at D and DE is the perpendicular from D on AB produced, prove that* $AE = \frac{1}{2}$ *the perimeter of the* $\triangle ABC$.

CONSTRUCTION: Draw DF and DG $\perp$s BC and AC produced; then join DA (*Fig.* 15).

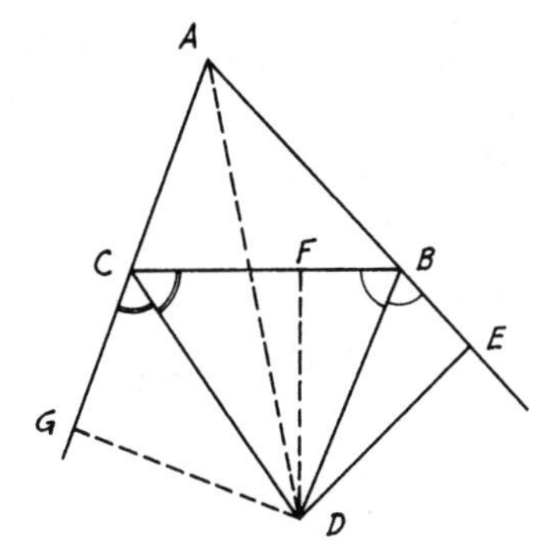

FIGURE 15

Proof: $\triangle$s DBE and DBF are congruent (Th. 1.11). $\therefore BE = BF$ and $DE = DF$. Similarly, $\triangle$s DCF and DCG are congruent also. $\therefore CF = CG$ and $DG = DF$. Hence $DE = DG$. Therefore, $\triangle$s ADE and ADG are congruent (Th. 1.15). $\therefore AE = AG$. But $AE = AB + BE = AB + BF$ and $AG = AC + CF$. $\therefore AE = \frac{1}{2}(AE + AG) = \frac{1}{2}(AB + BF + AC + CF) = \frac{1}{2}(AB + BC + AC)$.

1.16. *AB is a straight line. D and E are any two points on the same side of AB. Find a point F on AB so that (i) the sum of FD and FE is a minimum; (ii) the difference between FD and FE is a minimum. When is this impossible?*

CONSTRUCTION: (i) Draw $EN \perp AB$; produce EN to M so that $EN = NM$. Join MD to cut AB in the required point F (*Fig.* 16).

Proof: Take another point F' on AB and join EF, EF', $F'D$, and $F'M$. $\because$ $\triangle$s ENF and MNF are congruent (Th. 1.10). $\therefore EF = FM$. $\therefore MD = FD + FE$. $\because$ D and M are two fixed points, the line joining them is a minimum. This is clear from F', since $EF' = F'M$ (from the congruence of $\triangle$s $EF'N$ and $MF'N$) but $MF' + F'D > MD$ in the $\triangle DF'M$ (Th. 1.18). $\therefore EF' + F'D > EF + FD$. Hence F is the required point and $(FD + FE)$ is a minimum.

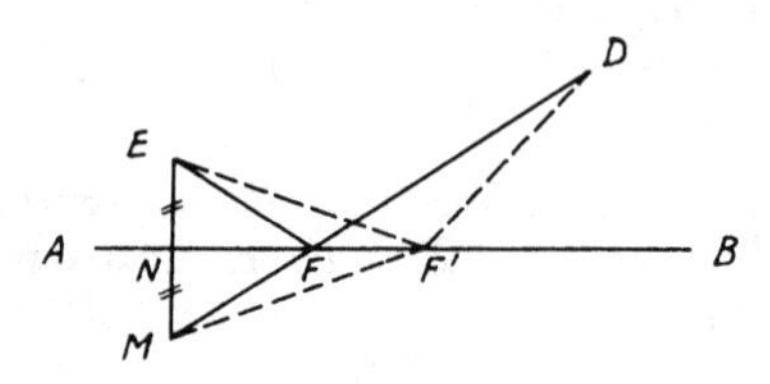

FIGURE 16

CONSTRUCTION: (ii) Join ED. Bisect ED in N, then drawn NF $\perp$ ED to meet AB in the required point F. Join FD and FE (*Fig.* 17).

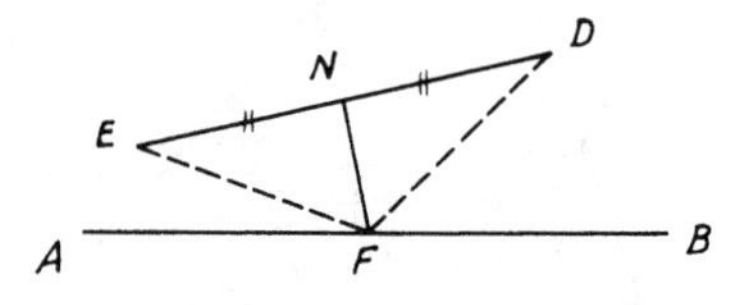

FIGURE 17

Proof: $\triangle$s FDN and FEN are congruent (Th. 1.10). $\therefore$ $FD = FE$. Hence their difference will be zero and it is a minimum. This case will be impossible to solve when ED is $\perp$ AB, where NF will be $\parallel$ AB and therefore F cannot be determined.

1.17. *Any straight line is drawn from A in the parallelogram ABCD. BE, CF, DG are* $\perp$*s from the other vertices to this line. Show that if the line lies outside the parallelogram, then* $CF = BE + DG$, *and when the line cuts the parallelogram, then CF = the difference between BE and DG.*

CONSTRUCTION: Join the diagonals AC and BD which intersect in L. Draw LM $\perp$ to the line through A (*Fig.* 18).

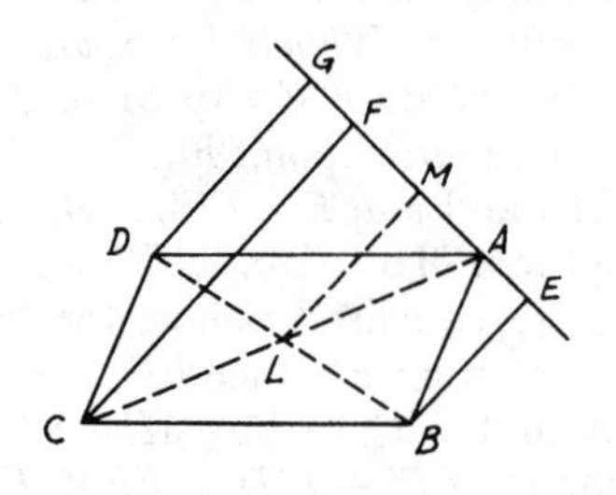

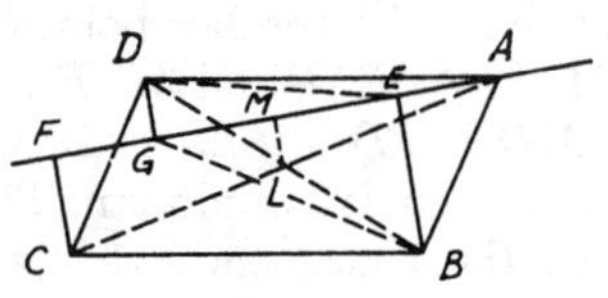

(i) FIGURE 18 (ii)

Proof: (i) The line lies outside the ▱ $ABCD$. The diagonals bisect one another (Th. 1.22). $\therefore$ L is the mid-point of AC and BD. In the trapezoid $BDGE$, L is the mid-point of BD and LM is $\parallel$ to BE and DG ($\perp$s to the line AG). Hence M is the mid-point of GE and $LM = \frac{1}{2}(BE + DG)$ (Th. 1.29). Again in the $\triangle ACF$, L is the mid-point of AC and LM is $\parallel$ CF (both $\perp AG$). $\therefore$ M is the mid-point of AF and $LM = \frac{1}{2} CF$ (Th. 1.26, Cor. 1). $\therefore$ $CF = BE + DG$.

(ii) The line passes through the ▱ $ABCD$. In the trapezoid $BEDG$, L is the mid-point of BD and LM is $\parallel$ BE and DG (as shown above). $\therefore$ M is the mid-point of the diagonal GE. $\therefore$ $LM = \frac{1}{2}(BE - DG)$ (Th. 1.29, Cor.). But, as shown above, also $LM = \frac{1}{2} CF$. $\therefore$ $CF = BE - DG$.

1.18. *ABC is a triangle. AD and AE are two perpendiculars drawn from A on the bisectors of the base angles of the triangle B and C respectively. Prove that DE is $\parallel$ BC.*

CONSTRUCTION: Produce AD and AE to meet BC in F and G (*Fig.* 19).

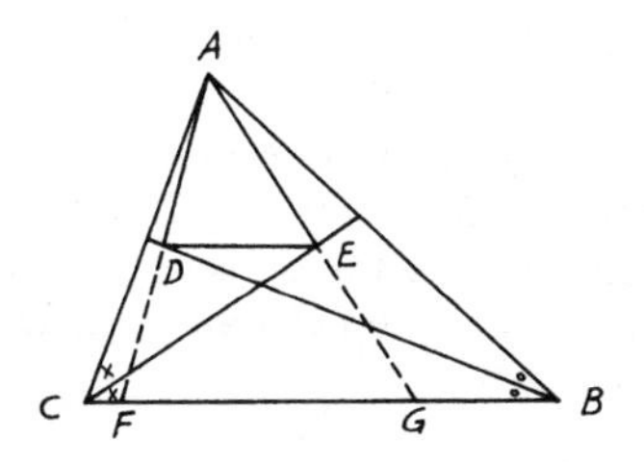

FIGURE 19

Proof: $\triangle$s ADB and FDB are congruent, since AD is $\perp DB$ (Th. 1.11). $\therefore$ $AD = DF$. Similarly, $\triangle$s AEC and GEC are congruent (Th. 1.11). $\therefore$ $AE = EG$. In the $\triangle AFG$, D and E are proved to be the mid-points of AF and AG. $\therefore$ DE is $\parallel$ FG or BC (Th. 1.26, Cor. 1).

1.19. *P and Q are two points on either side of the parallel lines AB and CD so that AB lies between P and CD. Two points L and M are taken on AB and CD. Find another two points X and Y on AB and CD so that $(PX + XY + YQ)$ is a minimum and XY is $\parallel$ LM.*

CONSTRUCTION: From Q draw QR equal and $\parallel$ to LM; then join PR to cut AB in X. Draw XY $\parallel$ to LM to meet CD in Y. Hence X and Y are the two required points (*Fig.* 20).

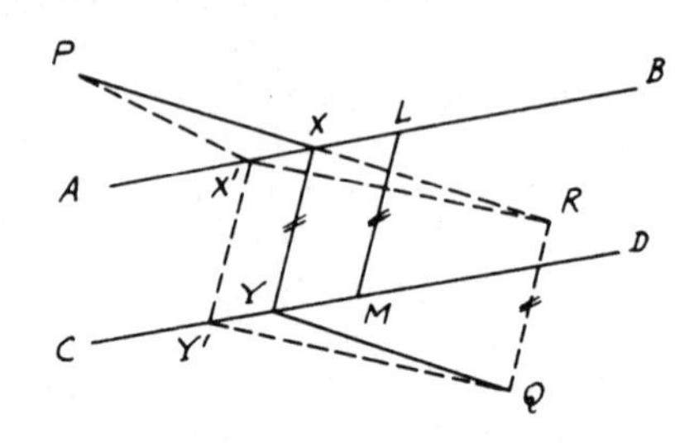

FIGURE 20

Proof: $\because$ RQ is equal and $\parallel$ to LM and $XY \parallel LM$, $\therefore$ $RQ =, \parallel XY$ (Th. 1.7). $\therefore$ $XYQR$ is a parallelogram. $\therefore$ $YQ = XR$ and $\because$ $XY = LM$, hence the sum of $(PX + XY + YQ) = PX + LM = XR = PR + LM$. To prove this is a minimum, let us take $X'Y'$ any other $\parallel$ to LM and join $X'P$, $X'R$ and $Y'Q$. $X'Y' = LM$ (in $\square X'Y'ML$). Also $X'Y'QR$ is a $\square$. $\therefore$ $X'R = Y'Q$. But in the $\triangle X'PR$, $(PX' + X'R) > PR$ (Th. 1.18). $\because$ $X'Y' = LM$. $\therefore$ $(PX' + X'Y' + Y'Q) > PR + LM$ or $(PX' + X'Y' + Y'Q) > (PX + XY + YQ)$. $\therefore$ X and Y are the two required points.

1.20. *Show that the sum of the two perpendiculars drawn from any point in the base of an isosceles triangle on both sides is constant.*

CONSTRUCTION: Let ABC be the isosceles $\triangle$ and D any point on its base BC. DE, DF are the $\perp$s on AB and AC. Draw $BG \perp AC$, and $DL \perp BG$ (*Fig.* 21).

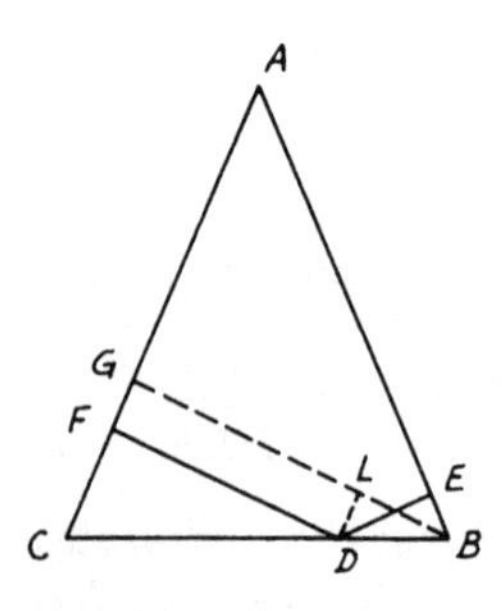

FIGURE 21

Proof: The figure $DLGF$ is a rectangle (construct). $\therefore$ $DF = LG$. Again, $DL \parallel FG$ (both $\perp BG$). $\therefore$ $\angle BDL = \angle C$. $\because$ $\angle C = \angle B$ (since $AC = AB$, Th. 1.12), $\therefore$ $\angle BDL = \angle B$. Also $\angle BED$

$= \angle DLB =$ right angle. Hence $\triangle$s BED and BLD are congruent (Th. 1.11). $\therefore DE = BL$. Therefore, $DE + DF = BL + LG = BG$ $=$ fixed quantity (since BG is $\perp$ from B on AC and both are fixed).

1.21. *Draw a straight line parallel to the base BC of the triangle ABC and meeting AB and AC (or produced) in D, E so that DE will be equal to (i) the sum of BD and CE; (ii) their difference.*

CONSTRUCTION: Bisect the $\angle B$ to meet the bisectors of the $\angle C$ internally and externally in P and Q. Draw from P and Q the parallels D_1PE_1 and QE_2D_2 to BC to meet AB and AC respectively in D_1, D_2 and E_1, E_2. Then E_1D_1 and E_2D_2 are the two required lines (*Fig.* 22).

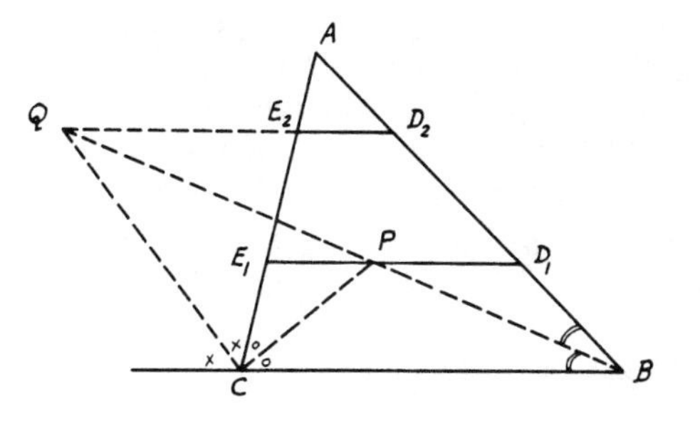

FIGURE 22

Proof: (i) $\because E_1D_1$ is $\parallel BC$. $\therefore \angle CBP = \angle BPD_1$. But $\angle CBP$ $= \angle PBD_1$ (construct). $\therefore \angle BPD_1 = \angle PBD_1$. $\therefore BD_1 = D_1P$ (Th. 1.13). Similarly, $CE_1 = E_1P$. Hence $BD_1 + CE_1 = D_1P$ $+ E_1P = D_1E_1$. $\therefore E_1D_1$ is the required first line.

(ii) In a similar way, $\because QE_2D_2$ is $\parallel BC$, $\therefore \angle CBQ = \angle BQD_2$. But $\angle CBQ = \angle QBD_2$. $\therefore \angle BQD_2 = \angle QBD_2$. $\therefore BD_2 = QD_2$. Likewise, $CE_2 = QE_2$. $\therefore BD_2 - CE_2 = QD_2 - QE_2 = D_2E_2$. $\therefore$ D_2E_2 is the required second line.

1.22. *ABCD is a parallelogram. From D a perpendicular DR is drawn to AC. BN is drawn $\parallel$ to AC to meet DR produced in N. Join AN to intersect BC in P. If DRN cuts BC in Q, prove that (i) P is the mid-point of BQ; (ii) AR = BN + RC.*

CONSTRUCTION: Draw $BS \perp AC$, then join CN (*Fig.* 23).

Proof: (i) $\triangle$s ABS, CDR are congruent (Th. 1.11). $\therefore AS = CR$ and $BS = DR$. Since the figure $BSRN$ is a rectangle (BS and NR are both $\perp AC$), $\therefore BS = NR$. $\therefore DR = NR$. Hence $\triangle$s ADR, ANR are congruent (Th. 1.10). $\therefore \angle DAR = \angle RAN$. But $\angle DAR = \angle RCB$ $= \angle CBN$ and $\angle RAN = \angle PNB$ (since BN is $\parallel AC$). $\therefore \angle CBN$

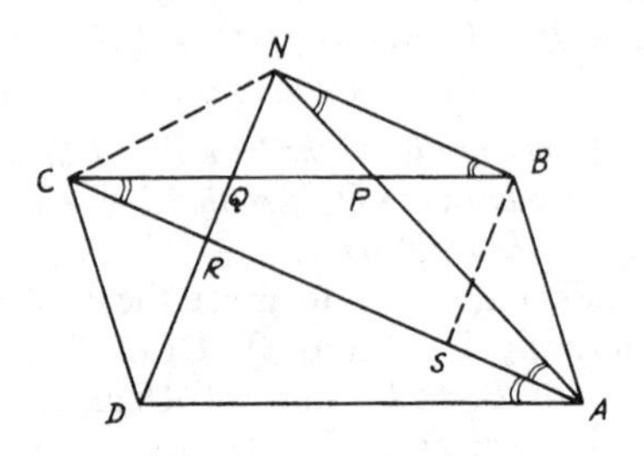

FIGURE 23

$= \angle PNB$. $\therefore PB = PN$. $\because \angle BNQ =$ right angle, $\therefore$ in the $\triangle BNQ$, $\angle PNQ = \angle PQN$ (complementary to equal $\angle$s). $\therefore PQ = PN$. Hence $PB = PQ = PN$ or P is the mid-point of BQ.

(ii) Since $BSRN$ is a rectangle, $\therefore BN = SR$. But $AS = CR$ (as shown earlier). $\therefore AR = AS + SR = CR + BN$.

1.23. *ABC is an isosceles triangle in which the vertical angle A* = 120°. *If the base BC is trisected in D and E, prove that ADE is an equilateral triangle.*

CONSTRUCTION: Draw AF and DG $\perp$s to BC and AB (*Fig.* 24).

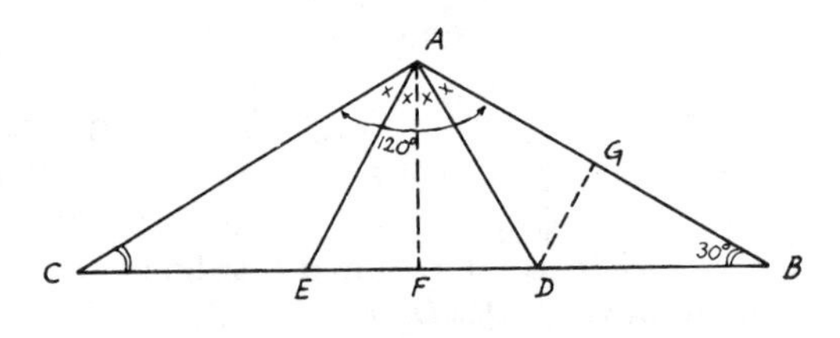

FIGURE 24

Proof: $\because \angle A = 120°$ in $\triangle ABC$, $\therefore \angle B + \angle C = 60°$ (Th. 1.8). But $\angle B = \angle C$ since $AB = AC$ (Th. 1.12). $\therefore \angle B = \angle C = 30°$. $\triangle BDG$ is right angled at G and $\angle B = 30°$. $\therefore DG = \frac{1}{2} BD$ (Th. 1.28). $\because AF$ is $\perp BC$ and $AB = AC$, $\therefore BF = CF$ (Th. 1.12, Cor. 1). Since $BD = DE = CE$ (given), $\therefore DF = FE$ or $DF = \frac{1}{2} DE = \frac{1}{2} BD$. $\therefore DG = DF$. $\triangle$s ADG, ADF are congruent (Th. 1.15). $\therefore \angle DAG = \angle DAF$. Similarly, $\angle EAC = \angle EAF$. But since $\triangle$s ADB, AEC are congruent (Th. 1.10), $\angle DAB = \angle EAC$ and $AD = AE$. Hence $\angle DAB = \angle DAF = \angle EAF = \angle EAC = 30°$. Therefore, $\angle DAE = 60°$, but, since $AD = AE$, then $\triangle ADE$ is an equilateral triangle.

1.24. *ABC and CBD are two angles each equal to 60°. A point O is taken inside the angle ABC and the perpendiculars OP, OQ, and OR are drawn from O to BA, BC, and BD respectively. Show that $OR = OP + OQ$.*

CONSTRUCTION: From O draw the line FOG to make an angle 60° with AB or BC; then drop the two perpendiculars BL and GK from B and G on FOG and AB respectively (*Fig.* 25).

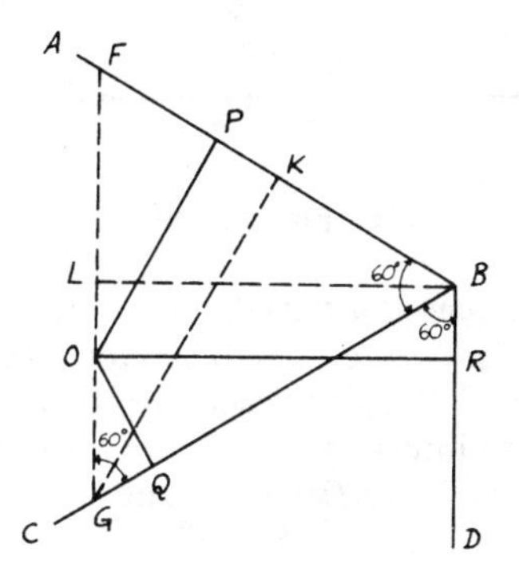

FIGURE 25

Proof: Since FOG makes with AB and BC angles = 60°, ∴ $\triangle FBG$ is equilateral. According to Problem 1.20, $OP + OQ = GK = BL$ (as the altitudes of an equilateral triangle are equal). ∵ $\angle FGB = \angle GBD = 60°$, ∴ BD is $\parallel FG$. ∴ $BROL$ is a rectangle. ∴ $LB = OR$. ∴ $OR = OP + OQ$.

1.25. *Construct a right-angled triangle, given the hypotenuse and the difference between the base angles.*

CONSTRUCTION: The hypotenuse BC and the difference of the base angles $\angle E'AD'$ are given. Draw $\angle EAD = \angle E'AD'$ and take $AD = \frac{1}{2} BC$. Then drop $DE \perp AE$ and produce DE from both sides to B and C so that $DB = DC = \frac{1}{2} BC$, the given hypotenuse. Then ABC is the required triangle (*Fig.* 26).

Proof: According to Problem 1.3, the angle subtended by the altitude and median from the right-angled vertex to the hypotenuse equals the difference of the base angles. Also AD is half the hypotenuse (Th. 1.27). In the $\triangle ABC$, $DA = DB = DC = \frac{1}{2}$ given hypotenuse. ∴ $\angle DAB = \angle DBA$ and $\angle DAC = \angle DCA$. ∴ $\angle DAB + \angle DAC = \angle BAC$ = right angle (half the sum of the angles of a $\triangle$, Th. 1.8). Again, $\angle EAD = (\angle C - \angle B)$ = given angle. ∴ ABC is the required $\triangle$.

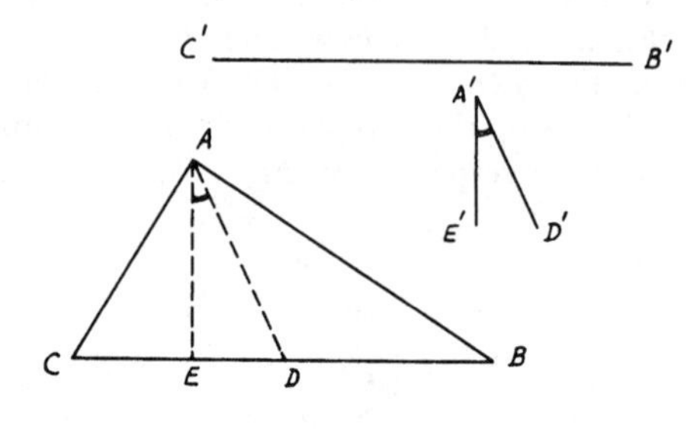

FIGURE 26

1.26. *If the three medians of a triangle are known, construct the triangle.*

CONSTRUCTION: Draw the $\triangle CDG$ having CD, DG, and CG equal to $C'D'$, $A'F'$ and $B'E'$ the given medians. Draw the medians CJ and GK in the $\triangle CDG$ to intersect in F. From F draw $FA \parallel$ and equal to DG or the given $F'A'$. Join AD and produce it to B, so that $AD = DB$. Then ABC is the required $\triangle$ (*Fig.* 27).

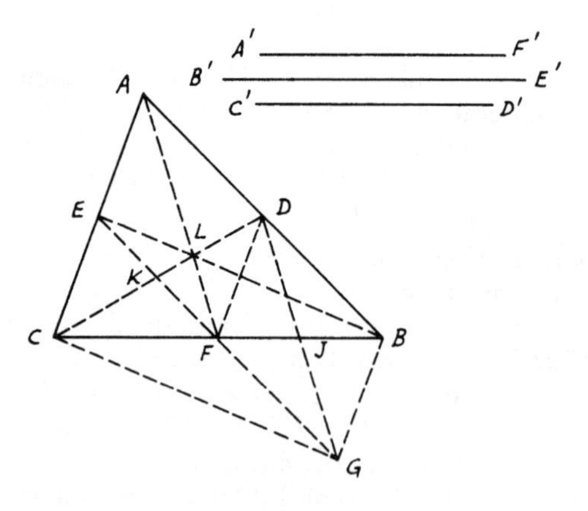

FIGURE 27

Proof: Join DF. In the $\triangle CDG$, F is the point of concurrency of its medians. Hence $FJ = \frac{1}{2} CF$ and $FK = \frac{1}{2} FG$ (Th. 1.34). Since FA is $\parallel$ and equal to DG, $\therefore$ $ADGF$ is a parallelogram (Th. 1.23). $\therefore$ FG is $\parallel$ and equal to AD and DB (Th. 1.22). Again, $DBGF$ is also a parallelogram. Therefore, DG and FB bisect one another (Th. 1.22). $\therefore$ FJ

$= JB = \frac{1}{2} FB$. $\therefore$ $CF = FB$ or F is the mid-point of BC. But L is the intersection of the medians CD and AF. $\therefore$ BE is also the third median (Th. 1.34). $\because$ DF is $\parallel$ and $= CE$ (Th. 1.26, Cor. 1) and also is $\parallel$ and $= BG$, $\therefore$ BG is $\parallel$ and equal to CE. Hence $BGCE$ is another parallelogram. $\therefore$ BFC is one diagonal and $BE = CG$. Therefore, the medians of the $\triangle ABC$ are equal to the sides of $\triangle CDG$ or equal to the given medians. Hence ABC is the required $\triangle$.

1.27. *ABC is a triangle. On AB and AC as sides, two squares ABDE and ACFG are drawn outside the triangle. Show that CD, BF, and the perpendicular from A on BC meet in one point.*

CONSTRUCTION: Draw $BP \perp CD$. BP produced meets HA (the perpendicular from A on BC) produced in R. Join RC cutting BF in Q (*Fig.* 28).

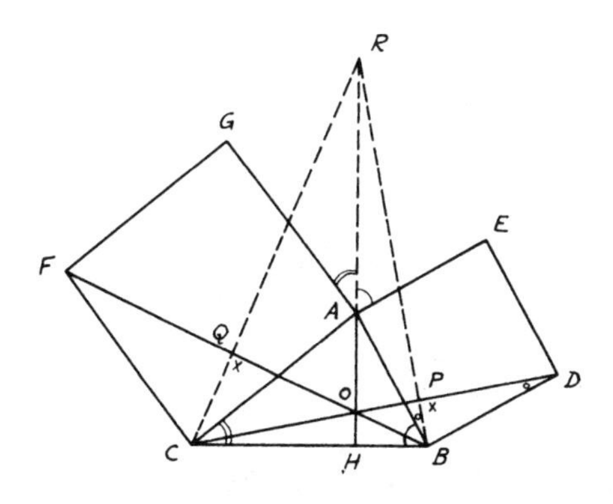

FIGURE 28

Proof: $\because$ BP is $\perp$ CD, $\therefore$ $\angle ABP = \angle BDP$ (since both are complementary to $\angle PBD$). $\because$ $\angle BAE$ = right angle, $\therefore$ $\angle RAE + \angle BAH$ = right angle (Th. 1.1). But since AH is $\perp$ BC, $\therefore$ in the $\triangle AHB$, $\angle BAH + \angle ABH$ = right angle. $\therefore$ $\angle RAE = \angle ABH$, since, in the square $ABDE$, $\angle BAE = \angle ABD$ = right angle. By adding, $\angle RAE + \angle BAE = \angle ABH + \angle ABD$ or $\angle RAB = \angle CBD$. Since also $AB = BD$, $\therefore$ $\triangle$s ABR and BDC are congruent (Th. 1.11). $\therefore$ $AR = BC$. Again, $\angle RAG = \angle ACH$ (since both are complementary to $\angle CAH$), and, since $AC = CF$ in the square $ACFG$, then the $\triangle$s ARC and CBF are congruent (Th. 1.10). $\therefore$ $\angle ACR = \angle CFB$. $\because$ $\angle ACR + \angle RCF = \angle ACF$ = right angle in the square AF, $\angle CFB + \angle RCF$ or $\angle QCF$ = right angle. $\therefore$ In the $\triangle CQF$, $\angle CQF$ = right angle or CQ is $\perp$ BF. Now, in the $\triangle BRC$,

RH, CP, and BQ are the altitudes to the sides. Therefore, they are concurrent (Th. 1.33); i.e., AH, CD and BF meet in one point.

1.28. *If any point D is taken on the base BC of an isosceles triangle ABC and DEF is drawn perpendicular to the base BC and meets AB and AC or produced in E and F, show that (DE + DF) is a constant quantity and equals twice the perpendicular from A to BC.*

CONSTRUCTION: Draw AP and AQ $\perp$s BC and DEF respectively. From B draw $BF' \perp BC$ to meet CA produced in F' (*Fig.* 29).

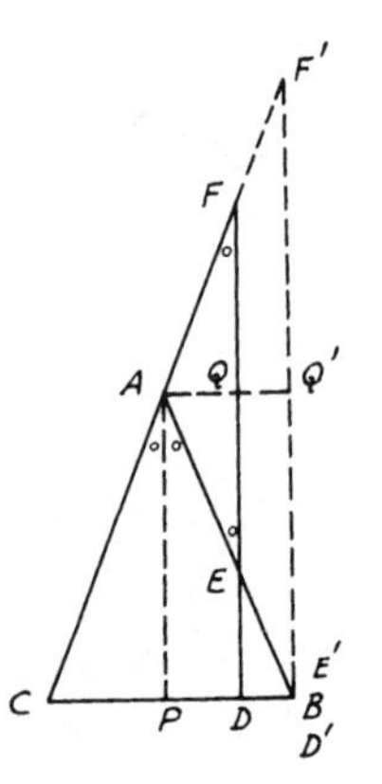

FIGURE 29

Proof: AP is $\perp$ BC in the isosceles $\triangle ABC$. $\therefore$ AP bisects $\angle A$ (converse Th. 1.12, Cor. 1). But DEF is $\parallel$ AP (both are $\perp$ BC). $\therefore$ $\angle BAP = \angle AEF$ and $\angle CAP = \angle AFE$ (Th. 1.6). Since $\angle BAP = \angle CAP$, $\therefore$ $\angle AEF = \angle AFE$. $\therefore$ $AE = AF$ (in the $\triangle AEF$, Th. 1.13). But AQ is $\perp$ EF (construct). $\therefore$ the $\triangle$s AEQ and AFQ are congruent (Th. 1.11). $\therefore$ $EQ = QF$. Hence $(DE + DF) = 2(DE + EQ) = 2DQ$. Since $AQDP$ is a rectangle, $\therefore$ $DQ = AP$. Therefore, $(DE + DF) = 2\ AP$ = constant. The extreme case is when D and E approach the extremities of the base until they coincide with B, where obviously $D'F' = 2\ D'Q' = 2\ AP$ = constant.

1.29. *Any line is drawn through O, the point of concurrence of the medians of a triangle ABC. From A, B, and C three perpendiculars AP, BQ, and CR are drawn to this line. Show that AP = BQ + CR.*

CONSTRUCTION: Let AD and CE be two medians of the $\triangle ABC$. From D and G, the mid-points of BC, AO drop DF and GN $\perp$s to the straight line ROQ (*Fig.* 30).

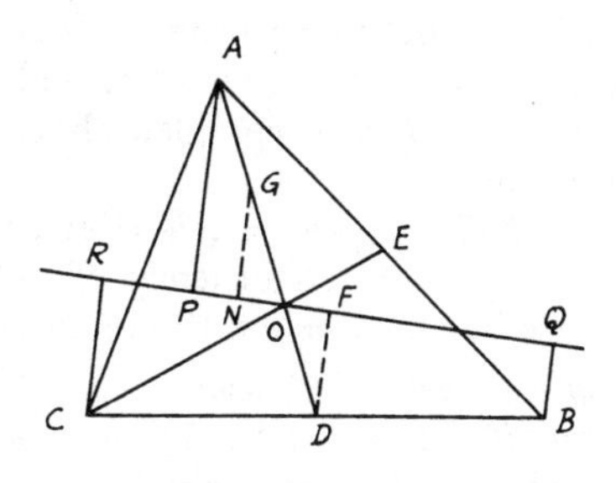

FIGURE 30

Proof: The figure $BQRC$ is a trapezoid and D is the mid-point of CB. Since DF is $\parallel$ BQ and CR ($\perp$s on line ROQ), $\therefore$ F is the mid-point of RQ (Th. 1.25) and $DF = \frac{1}{2}(BQ + CR)$ (Th. 1.26, Cor. 3). Now, since O is the point of concurrency of the medians of the $\triangle ABC$, then $AO = \frac{2}{3} AD$ (Th. 1.34). $\therefore$ $AO = 2\,OD$. $\therefore$ $GO = OD$. But DF, GN are $\perp$s to ROQ. Hence the two opposite $\triangle$s ODF and OGN are congruent (Th. 1.11). $\therefore$ $DF = GN$. Again, in the $\triangle AOP$, G is the mid-point of AO and GN is $\parallel$ AP ($\perp$s to ROQ). $\therefore$ N is the mid-point of OP and $GN = \frac{1}{2} AP$ (Th. 1.26, Cor. 1). Therefore, $AP = BQ + CR$.

1.30. *Construct an isosceles trapezoid having given the lengths of its two parallel sides and a diagonal.*

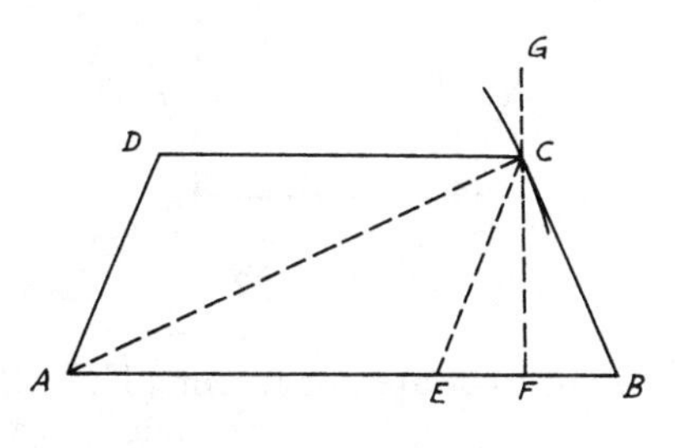

FIGURE 31

ANALYSIS: Suppose $ABCD$ is the required trapezoid (*Fig.* 31). If through C, CE is drawn $\parallel$ AD, $AECD$ is a $\square$, $\therefore$ $AE = DC$ = given length. Also $CE = AD = CB$. $\therefore$ C lies on the straight line which bisects EB at right angles. But C also lies on the circle with center A and radius equal to the given diagonal. Hence,

SYNTHESIS: From AB the greater of the two given sides, cut off AE equal to the less; bisect EB in F; from F draw $FG \perp EB$. With center A and radius equal to the given diagonal, describe a circle cutting FG in C. Join CE, CB, and complete the ▱ ED. $ABCD$ will be the required trapezoid.

Proof: Since DE is a ▱, $\therefore$ $DC = AE =$ given length.

Also $DA = EC = CB$ (from congruency of $\triangle$s BFC, EFC, Th. 1.10) and AB, AC are by construction the required lengths.

1.31. *On the sides of a triangle ABC, squares ABDE, ACFG, BCJK are constructed externally to it. BF, AJ, CD, AK are joined. If FC, DB are produced to meet AJ, AK in H, I respectively and X, Y are the intersections of BF, CD with AC, AB respectively, show that X, H are equidistant from C and Y, I are also equidistant from B. Prove also that the perpendiculars from A, B, C on GE, DK, FJ respectively intersect in the centroid of the triangle ABC and that GE, DK, FJ are respectively double the medians from A, B, C.*

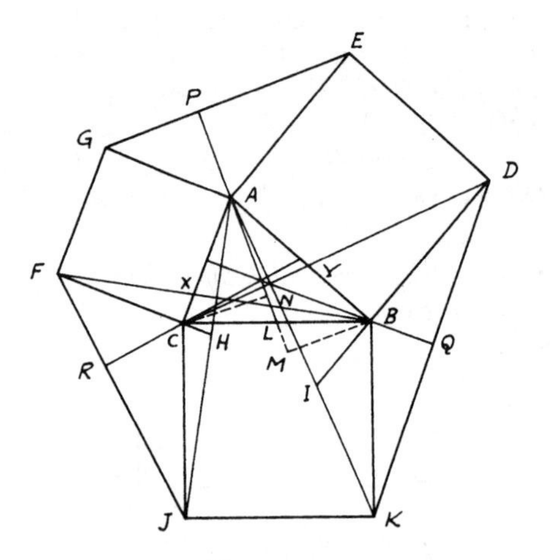

FIGURE 32

Proof: (i) $\triangle$s ACJ, FCB are congruent (Th. 1.10) (*Fig.* 32). $\therefore$ $\angle CAJ = \angle CFB$. Since $\angle CHJ =$ right angle $+ \angle CAH$ and $\angle CXB =$ right angle $+ \angle CFB$, $\therefore$ $\angle CHJ = \angle CXB$. Hence $\triangle$s ACH, FCX are congruent (Th. 1.11). $\therefore$ $CX = CH$. Similarly, $BY = BI$.

(ii) Let AP, BQ, CR be the $\perp$s from A, B, C on GE, DK, FJ respectively. Produce PA to meet BC in L and draw BM, CN $\perp$s PAL. Then $\triangle$s APE, BMA are congruent (Th. 1.11). $\therefore$ $AP = BM$. Similarly, $\triangle$s APG, CNA are congruent. $\therefore$ $AP = CN = BM$. Also $\triangle$s BML, CNL are congruent. $\therefore$ $BL = CL$. Hence PAL is a median in $\triangle ABC$. Similarly, QB, RC produced are the other two medians of $\triangle ABC$. Therefore, they meet at the centroid. Again, from the congruence of $\triangle$s APE, BMA, $\therefore PE = MA$. Also, $PG = NA$. $\therefore$ $GE = MA + NA = 2(AN + LN) = 2\,AL$.

1.32. *The point of concurrence S of the perpendiculars drawn from the middle points of the sides of a triangle ABC, the orthocenter O, and the centroid G are collinear and* $OG = 2\,SG$.

CONSTRUCTION: Let D, E be the mid-points of BC, AB. Bisect AO, CO, AG, GO in H, K, L, M. Join DE, HK, LM (*Fig.* 33).

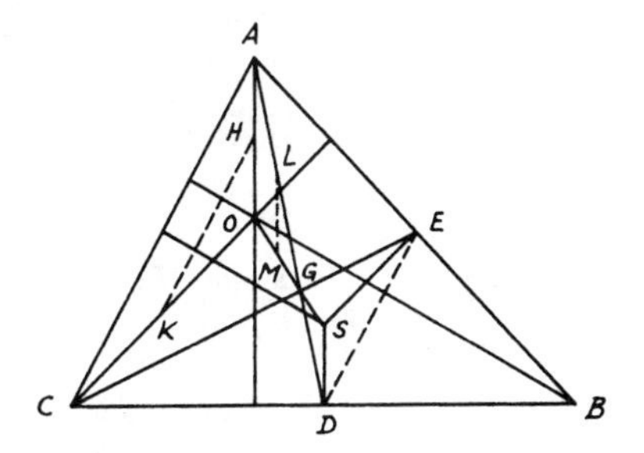

FIGURE 33

Proof: HK is $\parallel$ and $= ED$ (each being $\parallel AC$ and $\frac{1}{2} AC$ in $\triangle$s AOC, ABC, Th. 1.26). $\triangle$s ESD, KOH have their corresponding angles equal. Since their sides are respectively $\parallel$, $\therefore$ they are congruent and $SD = HO$ (Th. 1.11). $\because LM = \frac{1}{2} AO = HO$, $\therefore$ in $\triangle AGO$, since L, M are the mid-points of AG, GO, $\therefore SD = LM = \frac{1}{2} AO$. But G is the centroid. $\therefore GD = \frac{1}{2} AG = LG$ (Th. 1.34) and $\angle SDG = \angle GLM$ (since SD is $\parallel LM$, being $\parallel AO$). $\therefore$ $\triangle$s SDG, MLG are congruent (Th. 1.10). $\therefore$ $\angle SGD = \angle MGL$. But they are vertically opposite at G. $\therefore$ SG is in line with GM or GO (converse, Th. 1.3). Also, $SG = MG$ or $OG = 2\,SG$. The line OGS referred to is the *Euler line*.

Miscellaneous Exercises

1. Two triangles are congruent if two sides and the enclosed median in one triangle are respectively equal to two sides and the enclosed median of the other.
2. The two sides AB, AC in the triangle ABC are produced to D, E respectively so that $BD = BC = CE$. If BE and CD intersect in F, show that $\angle BFD = \text{right angle} - \frac{1}{2}\angle A$.

3. From any point D on the base BC of an isosceles triangle ABC, a perpendicular is drawn to cut BA and CA or produced in M and N. Prove that AMN is an isosceles triangle.
4. ABC is a triangle in which AB is greater than AC. If D is the middle point of BC, then the angle CAD will be greater than the angle BAD.
5. The straight line drawn from the middle point of the base of a triangle at right angles to the base will meet the greater of the two sides, not the less.
6. ABC is a right-angled triangle at A. The altitude AD is drawn to the hypotenuse BC. DA and CB are produced to P and Q respectively so that $AP = AB$ and $BQ = AC$. Show that $CP = AQ$.
7. A square is described on the hypotenuse BC of a right-angled triangle ABC on the opposite side to the triangle. If M is the intersection of the diagonals of the square and LMN is drawn perpendicular to MA to meet AB, AC produced in L, N respectively, then $BL = AC$, $CN = AB$.
8. Show that the sum of the altitudes of a triangle is less than the sum of its three sides.
9. On BC as a base, an equilateral triangle ABC and an isosceles triangle DBC are drawn on the same side of BC such that $\angle D$ = half $\angle A$. Prove that $AD = BC$.
10. P is any point inside or outside the triangle ABC. AP, BP, CP are produced to R, S, T respectively so that $AP = PR$, $BP = PS$, $CP = PT$. Show that the triangles RST and ABC are equiangular.
11. The interior and exterior angles at C of a triangle ABC are bisected by CD, CF to meet AB and BA produced in D, F respectively. From D a line DR is drawn parallel to BC to meet AC in R. Show that FR produced bisects BC. (Produce DR to meet CF in S.)
12. ABC is an isosceles triangle in which $AB = AC$. On AB a point G is taken and on AC produced the distance CH is taken so that $BG = CH$. Prove that $GH > BC$.
13. Construct a triangle having given the base, the difference of the base angles, and the difference of the other two sides. (Use Problem 1.10).
14. Show that the sum of the three medians in a triangle is less than its perimeter and greater than $\frac{3}{4}$ the perimeter.
15. ABC is an isosceles triangle and D any point in the base BC; show that perpendiculars to BC through the middle points of BD and DC meet AB, AC in points H, K respectively so that $BH = AK$ and $AH = CK$. (Join DH, DK.)
16. The side AB of an equilateral triangle ABC is produced to D so that $BD = 2\,AB$. A perpendicular DF is drawn from D to CB produced. Show that FAC is a right angle.
17. On the two arms of a right angle with vertex at A, AB is taken $= AD$ and also $AC = AE$, so that B, C are on the same area of $\angle A$. Prove that the perpendicular from A to CD when produced bisects BE.

18. ABC is an obtuse angle and $AB = 2\ BC$. From G, the middle point of AB, a perpendicular is drawn to it and from C another perpendicular is drawn to CB to meet the first one in H. Show that $\angle AHG = \frac{1}{3}\ \angle AHC$.
19. Show that the sum of the three bisectors of the angles of a triangle is greater than half its perimeter.
20. Construct a triangle, having given the base, the vertical angle, and (a) the sum; (b) the difference of the sides.
21. ABC is a triangle. On AB a point M is taken so that $AM = \frac{1}{3}\ AB$. BC is bisected in N and AN and CM intersect in R. Show that R is the middle of AN and that $MR = \frac{1}{4}\ MC$. (Draw $NP \parallel MC$.)
22. A triangle ABC is turned about the vertex A into the position $AB'C'$; if AC bisects BB', prove that AB' (produced, if necessary) will bisect CC'.
23. $ABCDEF$ is a hexagon. Prove that its perimeter $> \frac{2}{3}\ (AD + BE + CF)$.
24. Any point D is taken in AB one of the equal sides of an isosceles triangle ABC; DEF is drawn meeting AC produced in F and being bisected by BC in E. Show that $CF = BD$.
25. $ABCD$ is a quadrilateral in which $AB = CD$ and $\angle C > \angle B$. Prove that $DB > AC$ and $\angle A > \angle D$.
26. Prove that the interior angle of a regular pentagon is three times the exterior angle of a regular decagon. (Use Cor., Th. 1.9.)
27. The vertical angle A in an isosceles triangle ABC is half a right angle. From A, the altitude AD is drawn to the base BC. If the perpendicular from C to AB cuts AD in P and meets AB in Q, show that $PQ = BQ$.
28. A, B, C are three points on a straight line. On AB and AC squares $ABDE$, $ACFG$ are described so as to lie on the same side of the straight line. Show that the straight line through A at right angles to GB bisects EC.
29. Construct a triangle, having given one of its sides and the point of concurrence of its medians.
30. The sum of the distances (perpendiculars) of the vertices of a triangle on any straight line is equal to the sum of the distances of the mid-points of the sides of the triangle on the same line. (Use Problems 1.17 and 1.29.)
31. The vertical angle A of an isosceles triangle ABC is $\frac{1}{3}$ of each of the base angles. Two points M, N are taken on AB, AC respectively so that $BM = BC = CN$. If BN and CM intersect in D, show that $\angle MDN = \angle B$. (See Problem 1.5.)
32. If any two points F, G are taken inside an acute angle BAC, find two points P on AB and Q on AC, so that the sum of $FP + PQ + QG$ will be a minimum.
33. If a triangle and a quadrilateral are drawn on the same base and the quadrilateral is completely inside the triangle, show that the perimeter of the triangle $>$ the perimeter of the quadrilateral.

34. P is any point in AB, one of the shorter sides of a given rectangle $ABCD$. Show how to construct a rhombus having one of its angular points at P, and with its other angular points one on each of the other sides of the rectangle.

35. $ABCDE$ is an irregular pentagon. Prove that if each pair of its sides, when produced, meet in five points, the sum of the five resulting angles will be equal to $\frac{1}{3}$ the sum of the angles of the pentagon.

36. ABC is a triangle in which the angle $B = 120°$. On AC at the opposite side of the triangle, an equilateral triangle ACD is described. Show that DB bisects $\angle B$ and equals $(AB + BC)$.

37. $ABCD$ is a parallelogram. If the two sides AB and AD are bisected in E, F respectively, show that CE, CF, when joined, will cut the diagonal BD into three equal parts.

38. A, B, C are three given points not on the same straight line. Draw a line to pass through A so that if two perpendiculars are drawn to it from B and C, then the one from C will be double that from B.

39. ABC is a right-angled triangle at B. On AB and BC two squares $ABDE$, $BCGH$ are described outside the triangle. From E, G two perpendiculars EL, GK are drawn to AC produced. Show that $AC = EL + GK$.

40. D is the middle point of the base BC of a triangle ABC. Prove that if the vertical angle A is acute, then $6\,AD >$ the perimeter of the triangle.

41. The bisectors of the angles of any quadrilateral form a second quadrilateral, the opposite angles of which are supplementary. If the first quadrilateral is a parallelogram, the second is a rectangle the diagonals of which are parallel to the sides of the parallelogram and equal to the difference of its adjacent sides. If the first quadrilateral is a rectangle, the second is a square.

42. The straight line AB is trisected in C so that AC is double BC and parallel lines through A, B, C (all on the same side of AB) meet a given line in L, M, N. Prove that AL with twice BM is equal to three times CN.

43. Show that the distance (perpendicular) of the centroid of a triangle from a straight line is equal to the arithmetic mean of the distances of its vertices from this line. (Use Problem 1.17 and Exercises 30 and 42.)

44. Construct a triangle having given the positions of the middle points of its three sides.

45. ABC is a triangle in which $AC > AB$ and D is the middle point of the base BC. From D a straight line DFG is drawn to cut AC in F and BA produced in G so that $\angle AFG = \angle AGF$. Prove that AF is equal to half the difference between AC and AB. (Draw $AM \parallel DFG$ and drop $BM \perp AM$. Produce BM to meet AC in N. Join MD. Use Problem 1.10.)

46. Show that the sum of the perpendiculars from any point inside an equilateral triangle on its sides is constant and equal to any altitude in the triangle. (From the point draw a line $\parallel$ to any side and use Problem 1.20.)

47. ABC is an equilateral triangle. If the two angles B, C are bisected by BD, CD and from D two parallels DR, DQ are drawn to AB, AC to meet BC in R, Q, show that R, Q are the points of trisection of BC.

48. AD, BF, CG are the three medians of a triangle ABC. AR, BR are drawn parallel to BF, AC and meet in R. Show that R, G, F are collinear and that RC bisects DG.

49. In a triangle ABC, AB is greater than AC. Find a point P on BC such that $AB - AP = AP - AC$. [Apply Problem 1.10, part (ii).]

50. Construct a triangle having given the base, one of the base angles, and the difference between the other two sides.

51. ABC is an isosceles triangle having $\angle A = 45°$. If AD, CF, the two altitudes from A, C on BC, AB respectively, meet in G, prove that $AC - FG = CF$.

52. The base angle in an isosceles triangle ABC is three times the vertical angle A. D is any point on the base BC. On CA the distance CF is taken $=$ to CD. If FD is joined and produced to meet AB produced in G, prove that the bisector of the external angle $C \parallel FDG$ and that $FD + FG > AB$. (Use Problem 1.5.)

53. ABC is a right-angled triangle at A. Produce BA and CA to X, Y so that $AX = AC$, $AY = AC$. If XY is bisected in M, show that MA produced is perpendicular to BC.

54. Construct a triangle having given its perimeter and two base angles.

55. On AB and BC of a triangle ABC, two squares $ABDE$, $BCJK$ are constructed outside the triangle. Prove that $CD \perp AK$.

56. Prove that any angle of a triangle is either acute, right, or obtuse, according to whether the median from it to the opposite side is greater than or equal or less than half this side. State and prove the opposite of this problem.

57. From a given point P draw three straight lines PA, PB, PC of given lengths such that A, B, C will be on the same line and $AB = BC$.

58. ABC is a triangle and $AC > AB$. From A two straight lines AD, AG are drawn to meet the base in D, G so that $\angle DAC = \angle B$ and $\angle GAB = \angle C$. Show that $DC > BG$.

59. Prove by the methods of Chapter 1 that if D, E, F are the feet of the perpendiculars from A, B, C respectively on the opposite sides of an acute-angled triangle ABC, then AD, BE, CF bisect the angles of the pedal triangle DEF.

60. In Exercise 59, if DG, DH are drawn perpendicular to AC, AB and G, H are joined, prove that GH is equal to half the perimeter of the pedal triangle DEF. (Produce DG, DH to meet EF in L, M.)

61. Construct a triangle, having given the feet of the perpendiculars on the sides from the opposite vertices.

62. ABC is an isosceles triangle in which the vertical angle $A = 20°$. From B, a straight line BD is drawn to subtend a $60°$ angle with BC and meet AC in D. From C another straight line CF is drawn to subtend a $50°$ angle with BC and meet AB in F. DF is joined and produced to meet CB produced in G. Show that $BD = BG$. (Draw $DR \parallel BC$ to meet AB in R. Join RC cutting DB in M, and join MF.)

63. Prove that if pairs of opposite sides of a quadrilateral are produced to meet in two points, then the bisectors of the two angles at these two points will subtend an angle which is equal to half the sum of one pair of opposite angles in the quadrilateral.

64. AB, AC are two straight lines intersecting in A. From D, any point taken on AC, a straight line DG is drawn parallel to the bisector of the angle A. If F is any point on DG, show that the difference between the perpendiculars drawn from F on AB, AC is fixed.

65. The side AB in a triangle ABC is greater than AC. If the two altitudes BD, CF are drawn, show that $BD > CF$.

66. Draw a straight line to subtend equal angles with two given intersecting lines and bear equal distances from two given points. (From the mid-point of the line joining the two points, draw $\perp$ the bisector of the angle between the straight lines.)

67. Construct a right-angled triangle, having given the hypotenuse and the altitude from the right vertex on the hypotenuse.

68. In any triangle the sum of two medians is greater than the third median, and the median bisecting the greater side is less than the median bisecting the smaller side.

69. Show that the bisector of any angle of a triangle is less than half the sum of the two surrounding sides.

70. $ABCD$ is a quadrilateral and XY is any straight line. Show that if M is the intersection of the lines joining the mid-points of the opposite side, AB, CD and BC, AD, then the sum of the distances of A, B, C, D from XY will be equal to four times the distance of M from XY.

71. A point is moving on a hypotenuse of a given right-angled triangle. Find the location of this point such that the line joining the feet of the perpendiculars from it on the sides of the triangle is minimum.

72. $ABCD$ is a parallelogram. On BC another parallelogram $BCA'D'$ is described so that AB, BD' will be adjacent sides. A third parallelogram $ABD'C'$ is constructed. Show that AA', CC', DD' are concurrent.

73. ABC is a right-angled triangle at A. D is the middle point of the hypotenuse BC. Join AD and produce it to E, so that $AD = DE$. If the perpendicular from E on BC meets the bisectors of the angles B, C or produced in F, G, show that (a) $EB = EF$; (b) $EC = EG$.

74. Construct a right-angled triangle, having given (a) one of its sides and the sum of the hypotenuse and second side; (b) one of its sides and the difference between the hypotenuse and the second side.

75. $ABCD$, $EBFD$ are a square and a rectangle having the same diagonal BD. If A, E are on the same side of BD, then AG is drawn perpendicular to AE to meet BE (or produced) in G. Prove that $BG = ED$.

76. ABC is a triangle and AE, BM, CN are its three medians. Produce AE to D, so that $AE = ED$ and BC from both sides to F, G, so that $FB = BC = CG$. Show that the perimeter of each of the triangles ADF, ADG is equal to twice the sum of the medians of the triangle ABC.

77. The point of intersection of the diagonals of the square described on the hypotenuse of a right-angled triangle is equally distant from the sides containing the right angle.

78. O is a point inside an equilateral triangle ABC. OA, OB, OC are joined and on OB is described, on the side remote from A, an equilateral triangle OBD. Prove that CD is equal to OA.

79. Construct a quadrilateral having given the lengths of its sides and of the straight line joining the middle points of two of its opposite sides.

80. In a triangle ABC, AC is its greatest side. AB is produced to B' so that AB' is equal to AC; CB is produced to B'' so that CB'' is equal to CA: CB', CB'' meet in D. Show that if $AB > BC$, then $AD > CD$.

81. $ABCD$ is a parallelogram having $AB = 2\,BC$. The side BC is produced from both sides to E and F so that $BE = BC = CF$. Show that AF is $\perp DE$. (Join LG and prove that $AGLD$ is a rhombus.)

82. Show that the centers of all the parallelograms which can be inscribed in a given quadrilateral so as to have their sides parallel to the diagonals of the quadrilateral lie in a straight line.

83. D, E, F are the middle points of the sides BC, CA, AB of a triangle ABC. FG is drawn parallel to BE meeting DE produced in G. Prove that the sides of the triangle CFG are equal to the medians of the triangle ABC.

84. On the halves of the base of an equilateral triangle, equilateral triangles are described remote from the vertex. If their vertices are joined to the vertex of the original triangle, show that the base is trisected by these lines. (From the vertices draw $\perp$s to the base.)

85. Which of the triangles that have the same vertical angle has the least perimeter, if two of its sides coincide with the arms of the given angle? Construct such a triangle if the perimeter is given. (Let A be the given angle. On one of its arms, take $AD = \frac{1}{2}$ the given perimeter, then complete the isosceles $\triangle ADE$. The bisectors of $\angle$s A, D meet in G. Draw $BGC \parallel DE$ to meet AD, AE in B, C. Therefore, the isosceles $\triangle ABC$ is the required triangle with the least given perimeter, since $BG + BA = BD + BA =$ half the given perimeter, while G is the mid-point of BC. To show that $\triangle ABC$ has the least perimeter, take BB' on $AB = CC'$ on CE. Join $B'C'$, then $B'C' > BC$ according to Exercise 12. But, since $AB' + AC' = AB + AC$, then perimeter of $\triangle AB'C' >$ perimeter of $\triangle ABC$.)

86. If the bisectors of two angles of a triangle are equal, then the triangle is isosceles.

87. If two straight lines are drawn inside a rectangle parallel to one of the diagonals and at equal distances from it, then the perimeter of the parallelogram formed by joining the nearer extremities of these two lines will be constant.

88. Describe the quadrilateral $ABCD$, given the lengths of the sides in order and the angle between the two opposite sides AB, CD. (Through B draw BE parallel and equal to CD.)

89. If the two sides of a quadrilateral are equal, these sides being either adjacent or opposite, the line joining the middle points of the other sides makes equal angles with the equal sides.

90. Let O be the middle point of AB, the common hypotenuse of two right-angled triangles ACB and ADB. From C, D draw straight lines at right angles to OC, OD respectively to intersect at P. Show that $PC = PD$.

91. The perpendicular from any vertex of a regular polygon, having an even number of sides, to the straight line joining any two other vertices passes through a fourth vertex of the polygon.

92. D is the middle point of the base BC of an isosceles triangle ABC and E the foot of the perpendicular from D on AC. Show that the line joining the middle point of DE to A is perpendicular to BE.

93. In Problem 1.31, if AS, BT, CV, the medians from A to GE, B to DK, C to FJ in the triangles AGE, BDK, CFJ, are produced inside the triangle ABC, they will meet at the orthocenter of the triangle ABC and that CB, CA, AB will be double AS, BT, CV respectively. (Apply a similar procedure to that of Problem 1.31 as illustrated.)

94. The exterior angles of the triangle ABC are bisected by straight lines forming a triangle LMN; L, M, N being respectively opposite A, B, C. If P, Q, R be the orthocenters of the triangles LBC, MCA, NAB respectively, show that the triangle PQR has its sides equal and parallel to those of ABC. [Let the bisectors of the interior angles of $\triangle ABC$ meet in O (Th. 1.32). Therefore, OA, OB, OC are $\perp$s MN, NL, LM (Th. 1.1, Cor. 2). Consequently, the figures $ARBO$, $AQCO$ are ▱s. $\therefore$ RB, AO, QC are equal and $\parallel$. $\therefore$ $RBCQ$ is a ▱. $\therefore$ RQ, BC are equal and $\parallel$. Similarly with PQ, AB and RP, AC.]

95. Show that the perpendiculars from the middle points of the sides of any triangle to the opposite sides of its pedal triangle are concurrent. (Join any two vertices of the pedal triangle to the mid-point of the opposite side of the original triangle and prove these lines are equal.)

CHAPTER 2

AREAS, SQUARES, AND RECTANGLES

Theorems and Corollaries

AREAS OF POLYGONS

2.35. *Parallelograms on the same base and between the same parallels, or of the same altitude, are equal in area.*

COROLLARY. *Parallelograms on equal bases and between the same parallels, or of the same altitude, are equal in area.*

2.36. *The area of a parallelogram is equal to that of a rectangle whose adjacent sides are equal to the base and altitude of the parallelogram respectively.*

2.37. *If a triangle and a parallelogram are on the same base and between the same parallels, or of the same altitude, the area of the triangle is equal to half that of the parallelogram.*

COROLLARY. *If a triangle and a parallelogram are on equal bases, and between the same parallels or of the same altitude, the area of the triangle is equal to half that of the parallelogram.*

2.38. *The area of a triangle is equal to half that of a rectangle whose adjacent sides are respectively equal to the base and altitude of the triangle.*

2.39. *Triangles on the same, or on equal bases, and between the same parallels or of the same altitude, are equal in area.*

2.40. *Triangles of equal area, which are on equal bases in the same straight line and on the same side of it, are between the same parallels.*

COROLLARY. *Triangles of equal area, on the same base, and on the same side of it are between the same parallels.*

2.41. *Triangles of equal area, on the same, or on equal bases, are of the same altitude.*

2.42. *In every parallelogram, each diagonal bisects its area into two equal triangles.*

Squares and Rectangles Related to Lines and Triangles

2.43. *The squares on equal straight lines are equal in area, and conversely equal squares are on equal straight lines.*

Theorems 2.44 to 2.50. *The following identities have been proved true geometrically.*

2.44. $x(a + b + c) = xa + xb + xc.$

2.45. $(a + b)^2 = a(a + b) + b(a + b).$

2.46. $a(a + b) = a^2 + ab.$

2.47. $(a + b)^2 = a^2 + 2ab + b^2.$

2.48. $(a + b)(a - b) = a^2 - b^2.$

2.49. $(a - b)^2 = a^2 - 2ab + b^2.$

2.50. $(a + b)^2 - (a - b)^2 = 4ab.$

2.51. *In any right-angled triangle, the square on the hypotenuse is equal to the sum of the squares on the sides containing the right angle. This is called Pythagoras's theorem.*

Corollary 1. *If N is any point in a straight line AB, or in AB produced, and P any point on the perpendicular from N to AB, then the difference of the squares on AP, BP is equal to the difference of the squares on AN, BN.*

Corollary 2. *The converse of Corollary 1 is true.*

2.52. *If the square on one side of a triangle is equal to the sum of the squares on the other sides, then the angle contained by these sides is a right angle.*

2.53. *In an obtuse-angled triangle, the square on the side opposite the obtuse angle is greater than the sum of the squares on the sides containing the obtuse angle, by twice the rectangle contained by either of these sides and the projection, on this side produced, of the other side adjacent to the obtuse angle.*

2.54. *In any triangle, the square on the side opposite an acute angle is less than the sum of the squares on the sides containing the acute angle, by twice the rectangle contained by either of these sides and the projection on it of the other side adjacent to the acute angle.*

2.55. *In any triangle, the sum of the squares on two sides is equal to twice the square on half the base, together with twice the square on the median which bisects the base.*

Corollary 1. *If a straight line BC is bisected at D and A is any point in BC, or BC produced, then the sum of the squares on AB and AC is equal to twice the sum of the squares on BD and AD.*

Corollary 2. *The sum of the squares on the sides of a parallelogram is equal to the sum of the squares on its diagonals.*

Solved Problems

2.1. *If two equal triangles are on the same base and on opposite sides of it, the straight line joining their vertices is bisected by their common base, produced if necessary. Conversely, if the straight line joining the vertices of two triangles on the same base and on opposite sides of it be bisected by their common base or base produced, then the two triangles are equal in area.*

(i) Let ABC, BCD be the equal triangles. Join AD meeting BC in E. AE will be equal to ED.

CONSTRUCTION: Draw $BE \parallel AC$ and $CF \parallel AB$. Join FD, AF (*Fig.* 34).

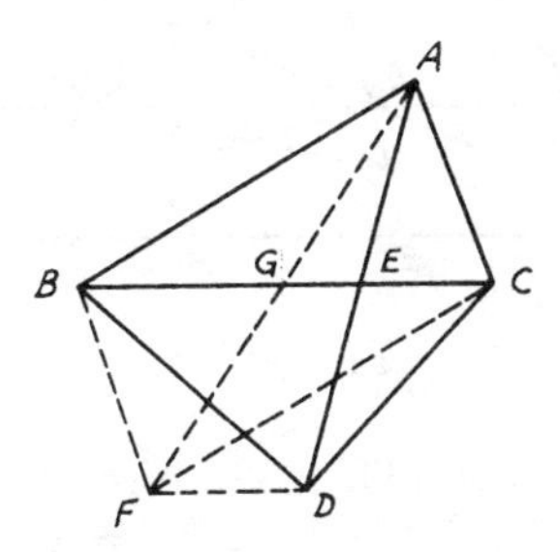

FIGURE 34

Proof: $ACFB$ is a ▱. $\therefore AG = GF$. $\because \triangle BFC = \triangle ABC$ (Th. 1.22) and $\triangle BDC = \triangle ABC$ (hypothesis), $\therefore \triangle BFC = \triangle BDC$. $\therefore FD$ is $\parallel BC$ (Th. 2.40). But G is the middle point of AF, $\therefore E$ is the middle point of AD in the $\triangle AFD$ (Th. 1.26).

(ii) Let AD joining the vertices of the $\triangle$s ABC, DBC be bisected by BC in E.

CONSTRUCTION: Bisect BC in G. Join AG and produce it to F so that $GF = AG$. Join BF, FC, FD.

Proof: $\because AG = GF$ and $BG = GC$, $\therefore ABFC$ is a ▱ (Th. 1.23). $\therefore \triangle FBC = \triangle ABC$. Again, since $AE = ED$ (hypothesis) and $AG = GF$, $\therefore FD$ is $\parallel GE$ (Th. 1.26, Cor. 1). $\therefore \triangle BDC = \triangle FBC = \triangle ABC$. Hence, in a similar way, it can be shown that:

1. In a triangle ABC, the median from A bisects all straight lines parallel to BC and terminated by AB and AC or by these produced.
2. If the base of a triangle be divided into any number of equal parts by straight lines drawn from the vertex, any straight line parallel to the base, which is terminated by the sides or sides produced, will be divided by these straight lines into the same number of equal parts.

2.2. *Two triangles are on the same base and between the same parallels. Prove that the sides or sides produced intercept equal segments on any straight line parallel to the base. Let ABC, DBC be the triangles. Draw $GEHF$ parallel to BC meeting AB, AC, DB, DC produced in E, F, G, H respectively. EF will be equal to GH.*

CONSTRUCTION: Through F draw $FLK \parallel AB$; through G draw $GNM \parallel CD$. Produce BC to L, N and AD to M, K. Join BF, GC (*Fig.* 35).

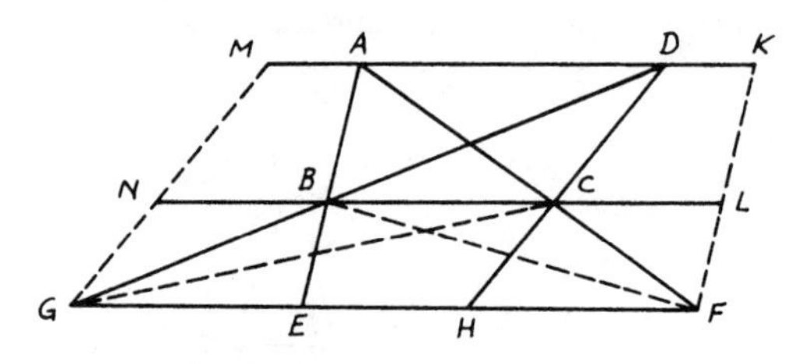

FIGURE 35

Proof: $\triangle ABC = \triangle DBC$ (Th. 2.39). Also, $\triangle BCF = \triangle BGC$. $\therefore$ by adding, $\triangle ABF = \triangle DGC$. Since $\triangle ABF = \frac{1}{2}\square AL$ (Th. 2.37) and $\triangle DGC = \frac{1}{2}\square CM$, $\therefore \square AL = \square CM$ and since they are between the same $\parallel$s NL, MK, $\therefore$ they are on equal bases (converse, Th. 2.35). $\therefore$ $BL = CN$. But $EF = BL$ and $GH = CN$. $\therefore$ $EF = GH$. Similarly, it can be proved that if two triangles are on equal bases and between the same parallels, the sides or sides produced intercept equal segments on any straight line parallel to the base.

2.3. *Bisect a triangle by a straight line drawn from a given point on one of its sides.*

Let ABC be the given $\triangle$, P the given point on AC.

CONSTRUCTION: Bisect AC in D. Join PB and draw $DE \parallel PB$. Join PE, BD. PE will be the required line (*Fig.* 36).

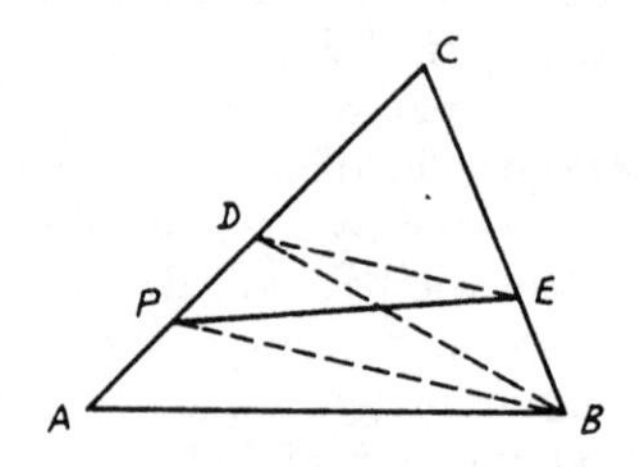

FIGURE 36

Proof: ∵ $AD = DC$, ∴ $\triangle ADB = \triangle DBC$ (Th. 2.39). ∴ $\triangle ADB$ is $\frac{1}{2}\triangle ABC$. ∵ $DE \parallel PB$, ∴ $\triangle PEB = \triangle PDB$. Adding $\triangle PAB$ to each, ∴ fig. $PABE = \triangle ADB = \frac{1}{2}\triangle ABC$. Hence PE bisects $\triangle ABC$.

2.4. *Trisect a given quadrilateral by means of two straight lines drawn from (i) one of its vertices; (ii) a given point on one of its sides.*

CONSTRUCTION: (i) Let $ABCD$ be the given quadrilateral. Convert ⏢$ABCD$ into an equal $\triangle DAE$, through the required vertex D, by drawing $CE \parallel DB$ and joining DE. Trisect AE in F, G. Draw FP, $GQ \parallel DB$. Hence DP, DQ are the two required lines (*Fig.* 37).

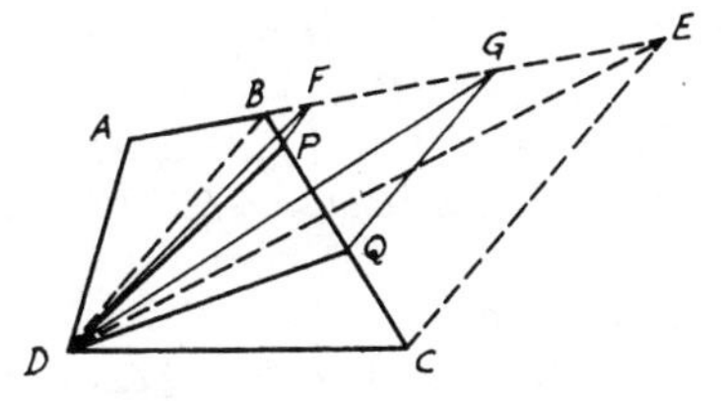

FIGURE 37

Proof: ∵ $AF = \frac{1}{3}AE$, ∴ $\triangle DAF = \frac{1}{3}\triangle DAE$. ∵ $\triangle DBF = \triangle DBP$ ($FP \parallel DB$), ∴ $\triangle DAF =$ ⏢$DABP = \frac{1}{3}\triangle DAE = \frac{1}{3}$⏢$ABCD$.

Similarly, $\triangle$s DPQ, DQC can be easily shown to be each $= \frac{1}{3}$⏢$ABCD$.

CONSTRUCTION: (ii) Convert ⏢ into an equal $\triangle D'EF$ through a given point D' on DC. Trisect EF in P, G. Draw $GQ \parallel D'B$. Hence $D'P$, $D'Q$ are the two lines (*Fig.* 38).

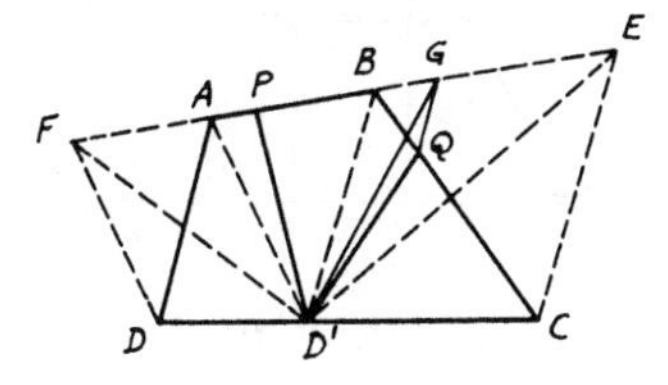

FIGURE 38

2.5. *ABCD is a parallelogram. P, Q are taken on AB, DP so that $AP = \frac{1}{3}AB$, $DQ = \frac{1}{3}DP$. Find the area of $\triangle QBC$ in terms of ▱ABCD.*

CONSTRUCTION: Draw $EQF \parallel AD$ and join AQ, BD (*Fig.* 39).

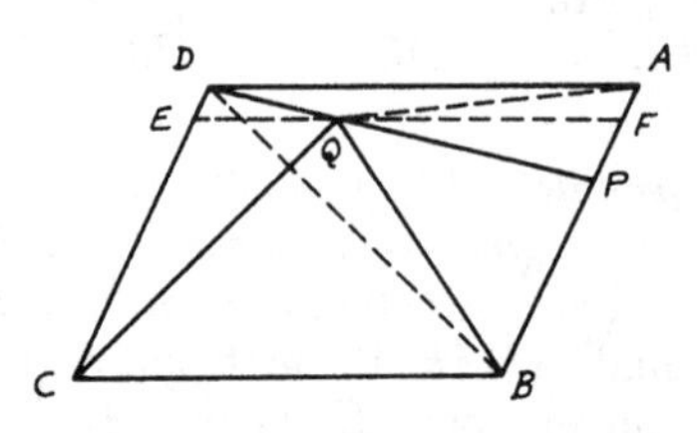

FIGURE 39

Proof: $\because DQ = \frac{1}{3} DP$, $\therefore \triangle ADQ = \frac{1}{3} \triangle ADP$ (Th. 2.39, Cor.). Again, $AP = \frac{1}{3} AB$. $\therefore \triangle ADP = \frac{1}{3} \triangle ADB$. Since $\triangle ADB = \frac{1}{2} \square ABCD$, $\therefore \triangle ADQ = \frac{1}{3} \times \frac{1}{3} \times \frac{1}{2} \square ABCD = \frac{1}{18} \square ABCD$. But $\square AFED = 2 \triangle ADQ$. $\therefore \square AFED = \frac{1}{9} \square ABCD$. Hence $\square FBCE = \frac{8}{9} \square ABCD$.

Since $EF \parallel AD$ and BC and Q lies on EF, $\therefore \triangle QBC = \frac{1}{2} \square FBCE$ (Th. 2.37). Therefore $\triangle QBC = \frac{1}{2} \times \frac{8}{9} = \frac{4}{9} \square ABCD$.

2.6. *A square is drawn inside a triangle so that one of its sides coincides with the base of the triangle and the other two corners lie on the other two sides of the triangle. Show that twice the area of the triangle is equal to the rectangle contained by one side of the square and the sum of base of the triangle and the altitude on this base.*

CONSTRUCTION: Let $DEFG$ be the inscribed square. Draw $AN \perp BC$ to cut GD in M. Then join CD (*Fig.* 40).

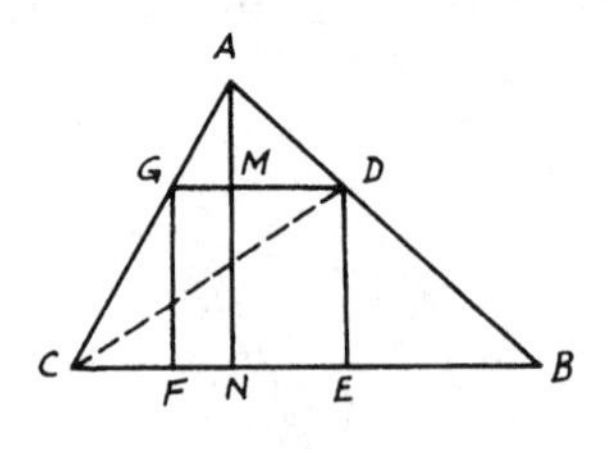

FIGURE 40

Proof: $2 \triangle DBC = CB \times DE$. Also, $2 \triangle GDC = GD \times GF$ (since $GF = \perp$ from C to GD) and $2 \triangle ADG = GD \times AM$. $\because DE = GD$, adding yields $2 (\triangle DBC + \triangle GDC + \triangle ADG) = DE (CB + GF + AM) = DE (CB + AN)$ or $2 \triangle ABC = DE (CB + AN)$.

2.7. *If E is the point of intersection of the diagonals in the parallelogram*

ABCD and P is any point in the triangle ABE, prove that $\triangle PDC = \triangle ABP + \triangle PBD + \triangle PAC$.

CONSTRUCTION: Draw $QPR \parallel AB$ to meet AD, BC in Q, R respectively (*Fig.* 41).

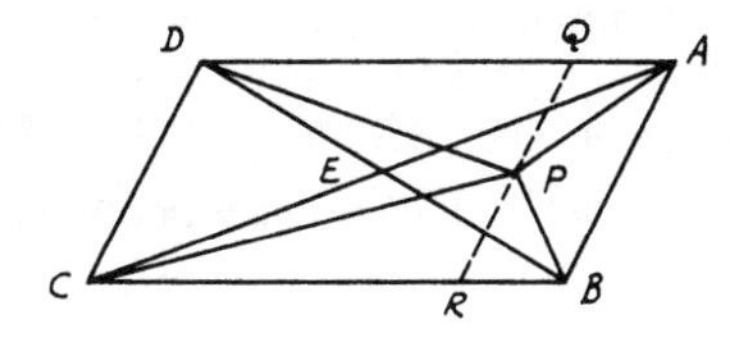

FIGURE 41

Proof: $\triangle ABD = \triangle ABP + \triangle APD + \triangle PBD = \frac{1}{2}\square ABCD$. Also, $\triangle ABC = \triangle ABP + \triangle BPC + \triangle PAC = \frac{1}{2}\square ABCD$. By adding, $2\triangle ABP + \triangle APD + \triangle BPC + \triangle PBD + \triangle PAC = \square ABCD$ (1). But $\triangle ABP = \frac{1}{2}\square AQRB$ and $\triangle PDC = \frac{1}{2}\square DQRC$. $\therefore$ $\triangle ABP + \triangle PDC = \frac{1}{2}\square ABCD$ (2). Therefore, the remainder $\triangle APD + \triangle BPC = \frac{1}{2}\square ABCD$ (3). Subtracting (3) from (1) gives $2\triangle ABP + \triangle PBD + \triangle PAC = \frac{1}{2}\square ABCD$ (4). Now, by equating (2) and (4), $\triangle PDC = \triangle ABP + \triangle PBD + \triangle PAC$.

2.8. *ABC is a right-angled triangle at A. Two squares ABFG and ACKL are described on AB and AC outside the triangle. If BK, CF cut AC, AB in M, N, show that AM = AN.*

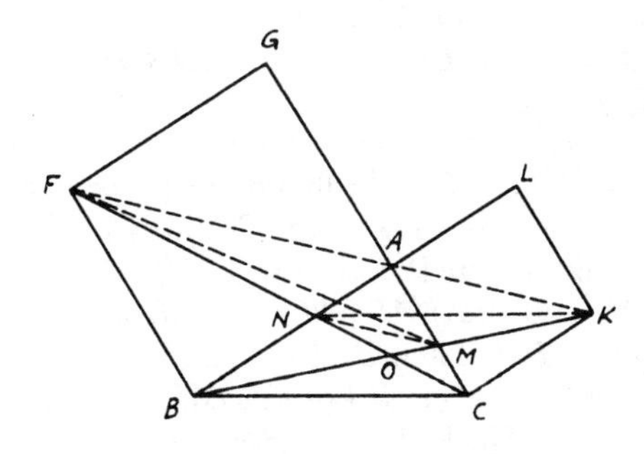

FIGURE 42

ANALYSIS: Join MN, FM, KN (*Fig.* 42). Assume AM is equal to AN; $\therefore$ $\angle AMN = \angle ANM = \frac{1}{2}$ a right angle $= \angle CAK$. $\therefore$ MN is $\parallel$ FK (FK is a straight line since $\angle BAC =$ right angle). $\therefore$ $\triangle FMN$

$= \triangle KMN$ (Th. 2.39). By adding $\triangle OMN$, $\triangle FOM = \triangle KON$. But this is the case, for $\triangle FMB = \triangle FCB$, since $FB \parallel AC$ (Th. 2.39). Take away $\triangle FOB$; $\therefore$ $\triangle FMO = \triangle OBC$. Similarly, $\triangle KNO = \triangle OBC$; $\therefore$ $\triangle FMO = \triangle KNO$. Hence,

SYNTHESIS: $\because$ $\triangle FMB = \triangle FCB$, take $\triangle FOB$ from each; $\therefore$ $\triangle FMO = \triangle OBC$. Similarly, $\triangle KNO = \triangle OBC$, $\therefore$ $\triangle FMO = \triangle KNO$. $\therefore$ $\triangle FNM = \triangle KNM$. $\therefore$ NM is $\parallel$ FK. $\therefore$ $\angle AMN = \angle MAK = \frac{1}{2}$ right angle $= \angle ANM$. $\therefore$ $AM = AN$.

2.9. *Construct a parallelogram that will be equal in area and perimeter to a given triangle.*

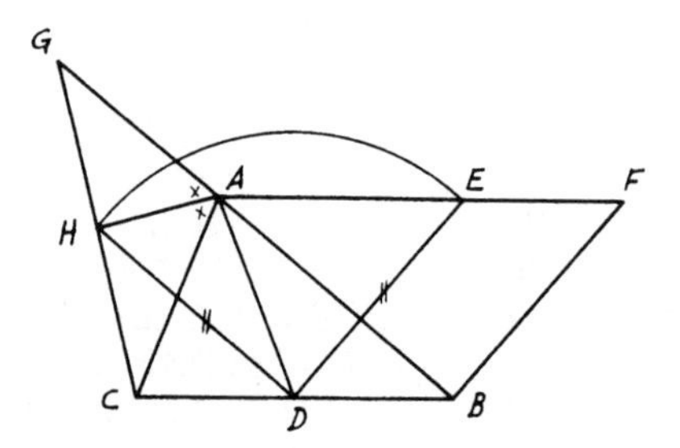

FIGURE 43

ANALYSIS: Let ABC be the given triangle. Assume that $BDEF$ is the required $\square$ (*Fig.* 43). Since $\square BDEF$ will be equal in area to the given $\triangle ABC$, then it should lie on half the base BC and between BC and the parallel to it through A. Hence D is the mid-point of BC. $\therefore$ $CB = 2\,DB = DB + EF$. Since also $\square BDEF$ will be equal in perimeter to $\triangle ABC$, $\therefore$ $AB + AC = 2\,DE = DE + FB$; i.e., $DE = \frac{1}{2}(AB + AC)$. This is true if DE will be equal to the line joining the mid-point of BC to the foot of the perpendicular from C or B on the external bisector of $\angle A$ (see Problem 1.10). Thus,

SYNTHESIS: Bisect BC in D and draw $CH \perp$ the external bisector of $\angle A$. Join DH and draw $AF \parallel BC$. With D as center and DH as radius, take $DE = DH$. Then $BDEF$ is the required $\square$.

Proof: Produce CH to meet BA produced in G. Join AD. $\triangle$s ACH, AGH are congruent (Th. 1.11). $\therefore$ $CH = HG$ and $AC = AG$. In the $\triangle BCG$, $DH = \frac{1}{2}BG = \frac{1}{2}(AB + AC) = DE$. $\therefore$ $AB + AC = DE + BF$. Since $BC = BD + EF$, $\therefore$ perimeter of $\square BDEF$ = perimeter of $\triangle ABC$. $\because$ $AF \parallel BC$, $\therefore$ $\square BDEF = \triangle ABC = 2\,\triangle ABD$.

2.10. *ABC is a triangle, and D, E are any two points on AB, AC respectively. AB, AC are produced to G, H, so that BG = AD and CH = AE. If BH,*

CG intersect in L and DE, GH are joined, show that $\triangle LGH = \triangle ADE + \triangle LBC$.

CONSTRUCTION: Join *CD, HD* (*Fig.* 44).

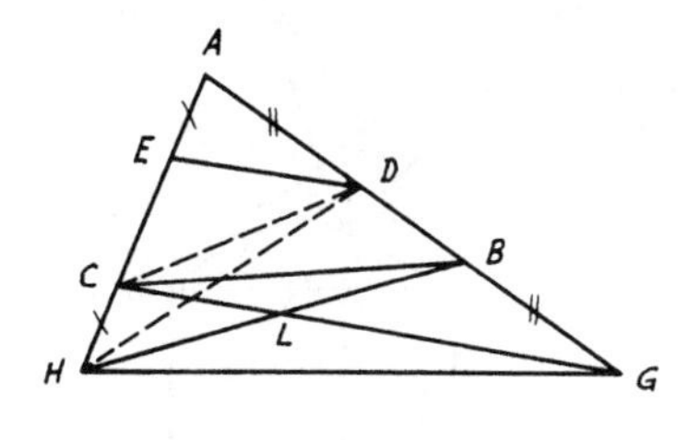

FIGURE 44

Proof: $\triangle ADE = \triangle CDH$ (have equal bases and altitudes, Th. 2.39). But, $\triangle ADH = \triangle ADC + \triangle CDH = \triangle ADC + \triangle ADE$. Since also $\triangle ADH = \triangle HBG$ and $\triangle ADC = \triangle CBG$ (Th. 2.39), $\therefore$ $\triangle HBG = \triangle CBG + \triangle ADE$. By subtracting the common $\triangle LBG$ from $\triangle$s *HBG, CBG*, $\therefore$ $\triangle LGH = \triangle ADE + \triangle LBC$.

2.11. *ABCD is a parallelogram. From A and C, two parallel lines AE, CF are drawn to meet BC, AD in E, F. If a line is drawn from E parallel to AC to meet AB in P, show that* $PF \parallel BD$.

CONSTRUCTION: Join *PC, PD, BF* (*Fig.* 45).

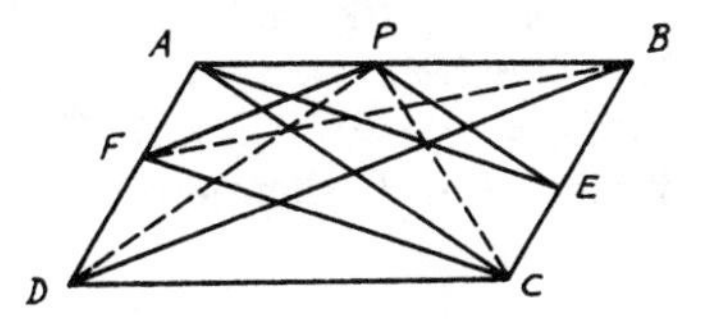

FIGURE 45

Proof: $\triangle AFB = \triangle AFC$ (on same base and between same parallels, Th. 2.39). $\because$ $AE \parallel FC$, $\therefore$ *AECF* is a ▱. $\therefore$ $\triangle AFC = \triangle ACE$ (Th. 2.42). $\therefore$ $\triangle AFB = \triangle ACE$. Since $\triangle ACE = \triangle APC$ (because $PE \parallel AC$) and $\triangle APC = \triangle APD$ (since $AP \parallel DC$), $\therefore$ $\triangle AFB = \triangle APD$. By subtracting common $\triangle AFP$ from both, $\therefore$ $\triangle FPB = \triangle FPD$. Since these two $\triangle$s are on the same base *FP*, $\therefore$ $FP \parallel BD$ (Th. 2.40).

2.12. *A point D is taken inside a triangle ABC. On the sides of the triangle ABC, the rectangles BCEF, CAGH, ABKL are drawn outside the triangle such that the area of each rectangle = twice the area of* $\triangle ABC$. *Prove that the sum of the areas of* $\triangle$*s DEF, DGH, DKL equals four times* $\triangle ABC$.

CONSTRUCTION: Draw $DR \perp EF$. Join AD, BD, CD (*Fig.* 46).

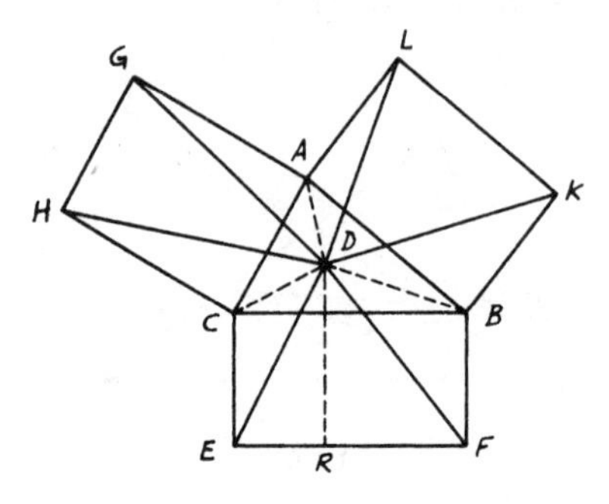

FIGURE 46

Proof: $\triangle DBF = \frac{1}{2} \square BR$ (Th. 2.38). Also, $\triangle DCE = \frac{1}{2} \square CR$. By adding, $\triangle DBF + \triangle DCE = \frac{1}{2}$ $BCEF$. Since $\triangle DEF =$ fig. $DBFEC - (\triangle DBF + \triangle DCE)$, $\therefore$ $\triangle DEF = \triangle DBC + \square BCEF - \frac{1}{2} \square BCEF = \triangle DBC + \frac{1}{2} \square BCEF$. Similarly, $\triangle DGH = \triangle DCA + \frac{1}{2} \square CAGH$ and $\triangle DKL = \triangle DAB + \frac{1}{2} \square ABKL$. Since these rectangles are each $= 2 \triangle ABC$ (hypothesis), adding gives $\triangle DEF + \triangle DGH + \triangle DKL = \triangle ABC + \frac{1}{2} (\square BCEF + \square CAGH + \square ABKL) = 4 \triangle ABC$.

2.13. *ABC is a triangle having the angle C a right angle. Equilateral triangles ADB, AEC are described externally to the triangle ABC and CD is drawn. Show that* $\triangle ACD = \triangle AEC + \frac{1}{2} \triangle ABC$.

CONSTRUCTION: Bisect AB in F, then join EF, CF, EB. Let EF cut AC in G (*Fig.* 47).

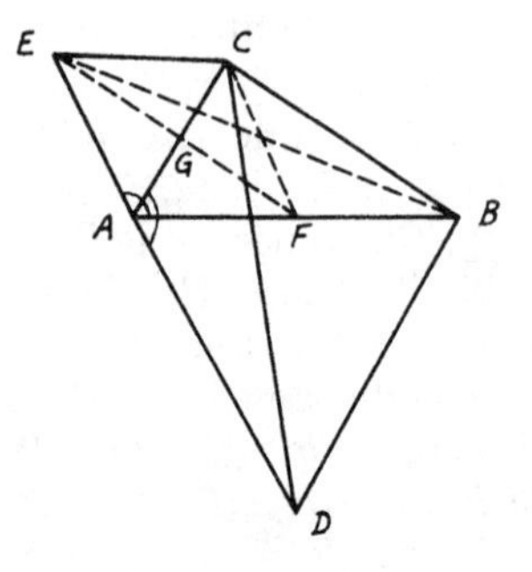

FIGURE 47

Proof: $\because \angle BAD = \angle CAE = 60°$ (angles of equilateral $\triangle$s), then adding $\angle CAB$ to each gives $\angle CAD = \angle BAE$.

Now, $\triangle$s ACD, AEB are congruent (Th. 1.10). $\therefore$ They are equal in area; i.e., $\triangle ACD = \triangle AEB$. But $\triangle ABC$ is right-angled at C, and F is the mid-point of AB. $\therefore CF = AF$ (Th. 1.27). $\therefore$ $\triangle$s ECF, EAF are congruent (Th. 1.14). $\therefore \angle CEF = \angle AEF$. Consequently, $\triangle$s GEC, GEA are congruent (Th. 1.10). $\therefore \angle CGE = \angle AGE =$ right angle. Hence $EF \perp AC$ or $\parallel CB$. $\because \triangle AEB = \triangle AEF + \triangle FEB = \triangle AEF + \triangle FEC$ (since $EF \parallel CB$), $\therefore \triangle AEB = \square AFCE = \triangle AEC + \triangle ACF = \triangle AEC + \frac{1}{2}\triangle ABC$. Since $\triangle ACD = \triangle AEB$ (as shown above), $\therefore \triangle ACD = \triangle AEC + \frac{1}{2}\triangle ABC$.

2.14. *ABC is a triangle right-angled at A. AD is drawn perpendicular to BC. If two squares BE, CF are described on AB, AC each on the same side of its base as the triangle ABC, show that the triangle ABC is equal to the triangle DEF together with the square on AD.*

CONSTRUCTION: Draw FG, $EH \perp$s AD. Join BE, CF (*Fig.* 48).

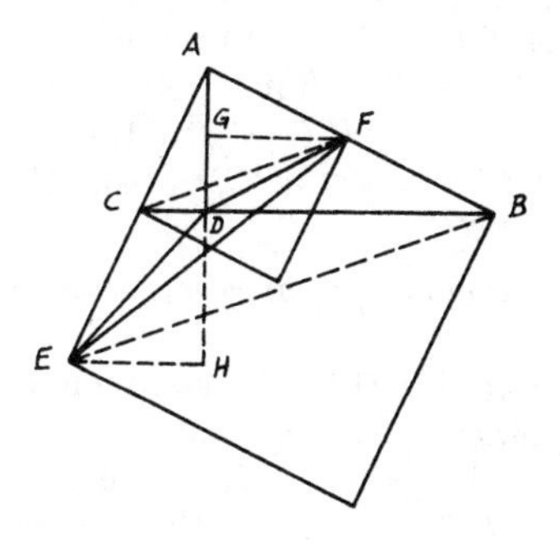

FIGURE 48

Proof: $\because FG \parallel BC$, $\therefore \angle AFG = \angle ABC$. But $\angle ABC = \angle CAD$ (both are complementary to $\angle BAD$). $\because AF = AC$ (in the square CF), $\therefore$ $\triangle$s AGF, CDA are congruent (Th. 1.11). $\therefore FG = AD$. Similarly, $\triangle$s ADB, EHA are congruent. $\therefore AD = EH = FG$. Now, $\triangle AEF = \triangle DEF + \triangle ADF + \triangle ADE = \triangle DEF + \frac{1}{2} AD \cdot FG + \frac{1}{2} AD \cdot EH = \triangle DEF + AD^2$.

2.15. *ABCD is a quadrilateral and G, H are the middle points of AC, DB. If CB, DA are produced to meet in E, show that* $\triangle EGH = \frac{1}{4}\square ABCD$.

CONSTRUCTION: Join GB, GD, HA, HC (*Fig.* 49).

Proof: $\because G$ is the mid-point of AC, $\therefore \triangle EGC = \frac{1}{2}\triangle AEC$ (Th. 2.39). Also, $\triangle DGC = \frac{1}{2}\triangle DCA$. $\therefore \triangle EGC + \triangle DGC = \frac{1}{2}(\triangle ACE + \triangle DCA) = \frac{1}{2}\triangle EDC$ (1). Similarly, $\triangle BGC + \triangle DGC =$

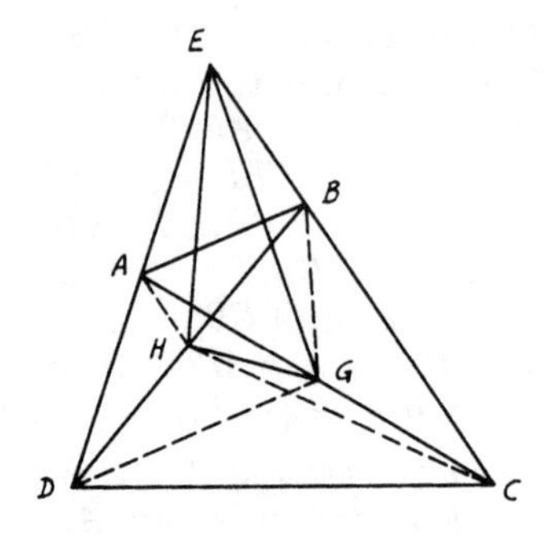

FIGURE 49

$\frac{1}{2}\square ABCD$ (2). Hence, subtracting (2) from (1), $\triangle EBG = \frac{1}{2}\triangle ABE$. Similarly, $\triangle EHD + \triangle HDC = \frac{1}{2}\triangle EDC$ and $\triangle AHD + \triangle HDC = \frac{1}{2}$ $ABCD$. By subtracting, $\triangle AHE = \frac{1}{2}\triangle ABE$. Again, $\triangle EHC = \frac{1}{2}\triangle EDC$. But $\triangle AHE = \frac{1}{2}\triangle ABE = \triangle EBG$. $\therefore$ $\triangle EGH + \triangle BGC + \triangle HGC = \frac{1}{2}\square ABCD$ (3) and $\triangle BGC + \triangle HGC = \frac{1}{2}\square ABCH = \frac{1}{4}\square ABCD$ (4). Hence from (3) and (4), $\triangle EGH = \frac{1}{4}\square ABCD$.

2.16. *The middle points of the three diagonals of a complete quadrilateral are collinear.*

CONSTRUCTION: Let $ABCD$ be the complete quadrilateral and G, H, J the middle points of its diagonals AC, BD, EF. Join EG, EH, FG, FH. Draw EL, FK $\perp$s GH, GH produced. Assuming that GH produced cuts EF in J, the problem is then reduced to prove that J is the mid-point of EF (*Fig.* 50).

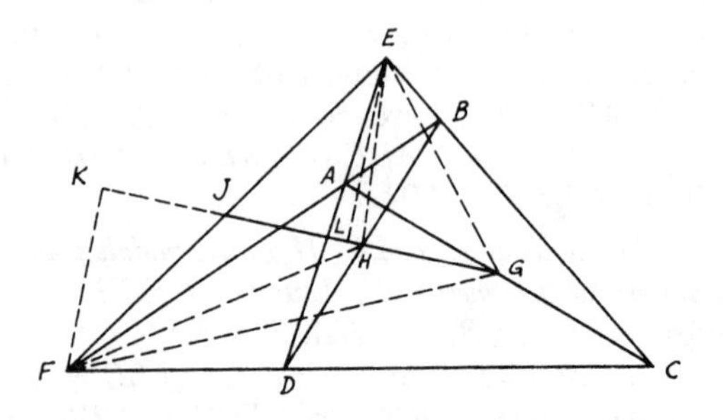

FIGURE 50

Proof: According to Problem 2.15, $\triangle EGH = \frac{1}{4} \square ABCD$. Similarly, $\triangle FGH = \frac{1}{4} \square ABCD$. $\therefore$ $\triangle EGH = \triangle FGH$. Since these $\triangle$s are on the same base GH, $\therefore$ their altitudes on GH should be equal; i.e., $EL = FK$ (Th. 2.41). Hence $\triangle$s ELJ, FKJ are congruent (Th. 1.11). $\therefore$ $EJ = JF$ or J is the mid-point of EF. Therefore, G, H, J are collinear.

2.17. *ABCD is a square and E is the intersection point of the diagonals. If N is any point on AE, show that (i)* $AB^2 - BN^2 = AN \cdot NC$; *(ii)* $AN^2 + NC^2 = 2\, BN^2$ *(Fig. 51).*

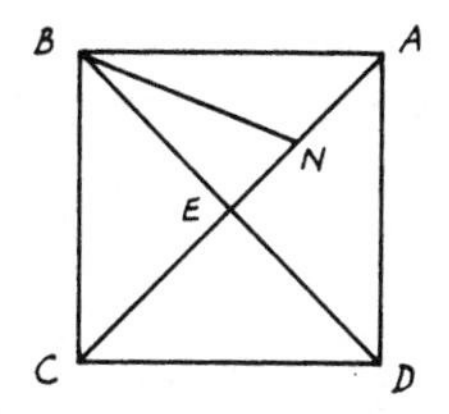

FIGURE 51

Proof: (i) The diagonals of a square are $\perp$ and bisect one another. $\therefore$ $AE = EC$ and $\perp BE$. $AN \cdot NC = AN(NE + EC) = AN(AN + 2\, NE) = AN^2 + 2\, AN \cdot NE$. In $\triangle ABN$, $AB^2 = BN^2 + AN^2 + 2\, AN \cdot NE$ (Th. 2.53). Hence $AB^2 - BN^2 = AN \cdot NC$.

COROLLARY 2.17.1. This is also true for any-angled isosceles triangle ABC having $\angle B$ other than a right angle.

(ii) $AN^2 = (AE - NE)^2$ and $NC^2 = (CE + NE)^2 = (AE + NE)^2$. Adding gives $AN^2 + NC^2 = 2\,(AE^2 + NE^2) = 2\,(BE^2 + NE^2) = 2\, BN^2$.

2.18. *In any trapezoid, the sum of the squares on the diagonals is equal to the sum of the squares on the non-parallel sides plus twice the rectangle contained by the parallel sides.*

CONSTRUCTION: Draw the $\perp$s CF, DG on AB produced (*Fig.* 52).

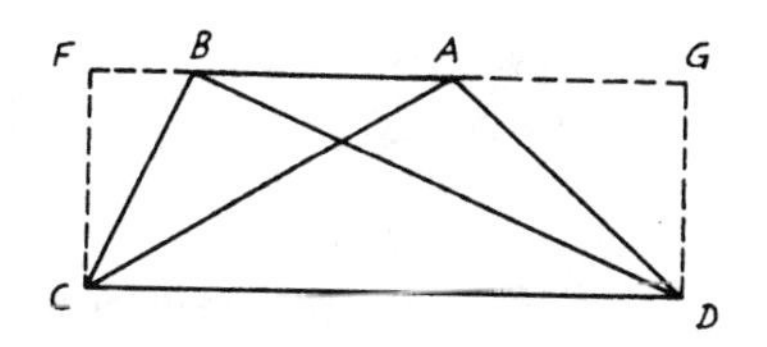

FIGURE 52

Proof: In $\triangle$s ABD, ABC, $BD^2 = AD^2 + AB^2 + 2\,AB \cdot AG$, $AC^2 = BC^2 + AB^2 + 2\,AB \cdot BF$ (Th. 2.53). Adding yields $BD^2 + AC^2 = AD^2 + BC^2 + 2\,AB\,(AB + AG + BF)$ or $BD^2 + AC^2 = AD^2 + BC^2 + 2\,AB \cdot CD$.

2.19. *ABC is a right-angled triangle at A. Show that, if AD is the perpendicular from A to the hypotenuse and denoting AB, AC, BC, AD by c, b, a, e, then* $1/e^2 = (1/c^2) + (1/b^2)$ *(Fig. 53).*

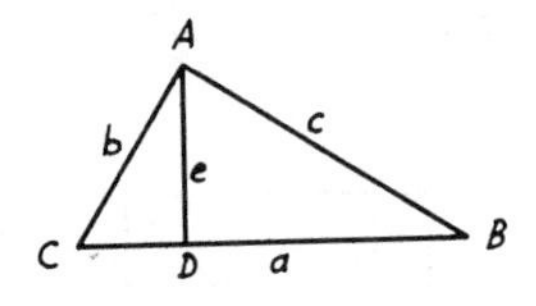

FIGURE 53

Proof: $ae = bc =$ twice $\triangle ABC$. $\therefore$ $e = bc/a$ or $e^2 = b^2c^2/a^2$. $\therefore$ $1/e^2 = a^2/b^2c^2$. $\because$ $a^2 = b^2 + c^2$ (Th. 2.51), therefore, $1/e^2 = (b^2 + c^2)/b^2c^2 = (1/c^2) + (1/b^2)$.

2.20. *Any point P is taken inside or outside triangle ABC. From P perpendiculars PE, PD, PF are drawn to the sides BC, CA, AB respectively. Show that* $AF^2 + BE^2 + CD^2 = BF^2 + CE^2 + AD^2$. *Enunciate and prove the converse of this theorem.*

CONSTRUCTION: Join PA, PB, PC (*Fig.* 54).

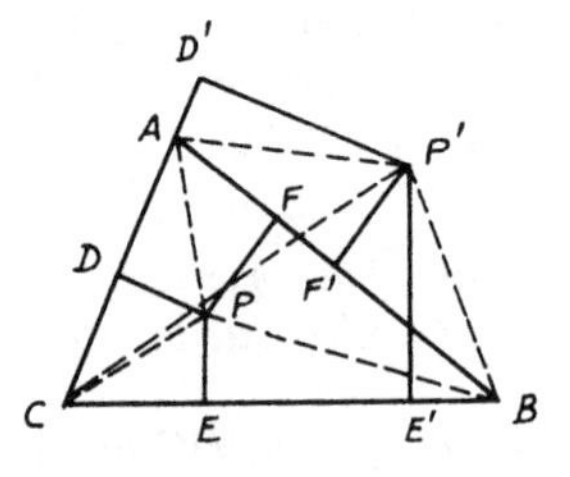

FIGURE 54

Proof: $AF^2 - BF^2 = AP^2 - BP^2$ (Th. 2.51 Cor. 1). Also, $BE^2 - CE^2 = BP^2 - CP^2$ and $CD^2 - AD^2 = CP^2 - AP^2$. Hence adding yields $(AF^2 - BF^2) + (BE^2 - CE^2) + (CD^2 - AD^2) = 0$ or $AF^2 + BE^2 + CD^2 = BF^2 + CE^2 + AD^2$. Now, the converse will be: If the above expression in a triangle ABC is true, then the perpendiculars through D, E, F are concurrent. This is evident, since if the $\perp$s from E, F meet in P, then PD should be $\perp AC$. From the above $(AF^2 - BF^2) + (BE^2 - CE^2) = (AP^2 - BP^2) + (BP^2 - CP^2) = (AP^2 - CP^2)$. Hence the remainders are equal or $(CD^2 - AD^2) = (CP^2 - AP^2)$. According to (Th. 2.51, Cor. 2), this is true when DP is $\perp AC$. Therefore, the $\perp$s from such points D, E, F are concurrent.

2.21. *ABC is an isosceles triangle having* $AB = AC$. *Find (i) point D on BC so that, if DE is drawn* $\perp BC$ *to meet AB in E, then* $AD^2 + DE^2 = AB^2$; *(ii) point F on CB produced so that, if FG is* $\perp BC$ *to meet AB produced in G, then* $AF^2 - FG^2 = AB^2$.

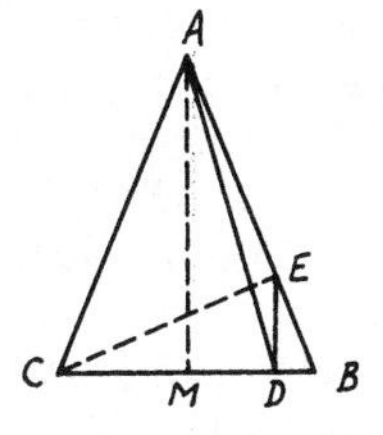

FIGURE 55

ANALYSIS: (i) Suppose D is the required point (*Fig.* 55). Then, $AD^2 + DE^2 = AB^2$. According to Problem 2.17(i), $AB^2 = AD^2 + BD \cdot DC$. Hence $DE^2 = BD \cdot DC$. But this is only true if $\angle CEB$ = right angle. Therefore,

SYNTHESIS: Draw $CE \perp AB$ and from E draw $ED \perp BC$. Then D is the required point.

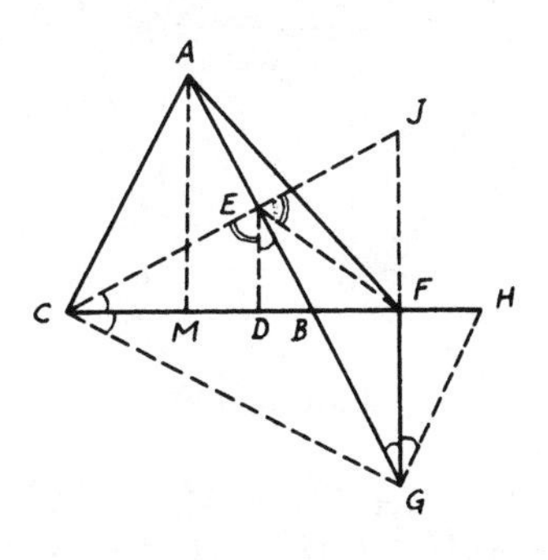

FIGURE 56

ANALYSIS: (ii) Suppose F is the required point (*Fig.* 56). Then $AF^2 - FG^2 = AB^2$. In the $\triangle ABF$, $AF^2 = AB^2 + BF^2 + 2\,BF \cdot$

BM (Th. 2.53) or $AF^2 - BF \cdot FC = AB^2$. Hence $FG^2 = BF \cdot FC$, which is true if BF is made equal to FH and $\triangle CGH$ right-angled at G. Thus $\triangle BGH$ is isosceles and $\angle BGF = \angle FGH$. But since $\angle BGF = \angle BED = \angle ECD$ and $\angle FGH = \angle FCG$, $\therefore$ $\angle ECD$ or $JCF = \angle FCG$. Hence $FG = FJ$. $\therefore$ $EF = FJ$ (since $\angle JEG$ is a right angle). $\therefore$ $\angle FEJ = \angle FJE = \angle DEC$ (since $FJ \parallel DE$). Therefore FE, DE are equally inclined to CJ, and thus,

SYNTHESIS: Draw $CE \perp AB$ and produce it to J; then draw EF making with EJ an $\angle FEJ = \angle DEC$ and meeting CB produced in F which is the required point.

2.22. *ABC is a triangle and three squares BCDE, ACFG, ABHK are constructed on its sides outside the triangle. Show that (i)* $AB^2 + AD^2 = AC^2 + AE^2$; *(ii)* $GK^2 + CB^2 = 2(AB^2 + AC^2)$; *(iii)* $AD^2 + BG^2 + CH^2 = AE^2 + BF^2 + CK^2$; *(iv)* $GK^2 + HE^2 + DF^2 = 3(AB^2 + BC^2 + AC^2)$; *(v) area of hexagon GKHEDF* $= 4\triangle ABC + (AB^2 + BC^2 + AC^2)$.

CONSTRUCTION: Draw $AMN \perp BC$, DE and $AP \perp GK$. Produce PA and draw CQ, BR $\perp$s to it (*Fig.* 57).

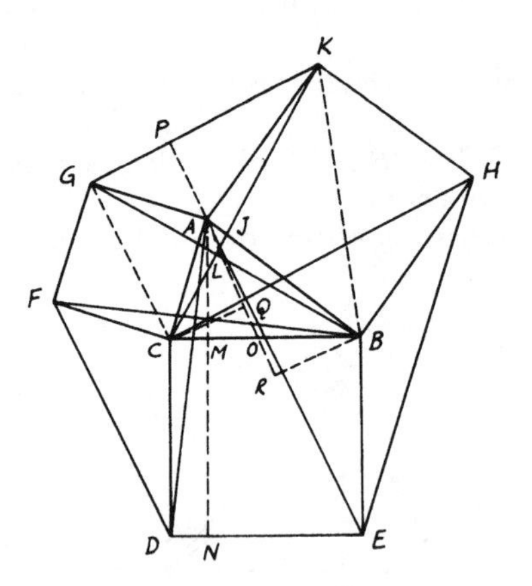

FIGURE 57

Proof: (i) $AB^2 + AD^2 = AM^2 + BM^2 + AN^2 + DN^2$ (Th. 2.51). Also, $AC^2 + AE^2 = AM^2 + CM^2 + AN^2 + NE^2$. Since $BM = NE$ and $DN = CM$, $\therefore$ $AB^2 + AD^2 = AC + AE^2$.

(ii) $\triangle$s BAG, KAC are congruent. $\therefore$ $\angle ABG = \angle AKC$. $\because$ $\angle AKC + \angle AJK =$ right angle, $\therefore$ $\angle AJK$ or $\angle BJL + \angle ABG =$ right angle. Hence, $BG \perp CK$. Now, in the quadrilateral $BCGK$, the diagonals are $\perp$ to one another. Hence $GK^2 + CB^2 = BK^2 + CG^2 = 2\,(AB^2 + AC^2)$.

(iii) According to (i), $AB^2 + AD^2 = AC^2 + AE^2$. Similarly, $BC^2 + BG^2 = AB^2 + BF^2$ and $AC^2 + CH^2 = BC^2 + CK^2$. Adding gives $AD^2 + BG^2 + CH^2 = AE^2 + BF^2 + CK^2$. Alternatively, $BG = CK$ from congruence of $\triangle$s BAG, KAC above and similarly it is easily shown that $AD = BF$ and $CH = AE$. Hence (iii) is true again.

(iv) As in (ii), $HE^2 + AC^2 = 2\,(BC^2 + AB^2)$ and $DF^2 + AB^2 = 2\,(AC^2 + BC^2)$. Therefore, by adding, $GK^2 + HE^2 + DF^2 = 3\,(AB^2 + BC^2 + AC^2)$.

(v) According to Problem 1.31(ii), $\triangle$s APG, CQA and $\triangle$s APK, BRA are congruent. $\because$ AQR is a median in $\triangle ABC$, $\therefore$ $\triangle$s CQO, BRO are congruent. $\therefore$ $\triangle CQA + \triangle BRA = \triangle ABC = \triangle APG + \triangle APK = \triangle AGK$. Similarly, $\triangle ABC = \triangle BHE = \triangle CDF$. Therefore, hexagon $GKHEDF = 4\,\triangle ABC + (AB^2 + BC^2 + AC^2)$.

2.23. *ABCD is a quadrilateral. E, F are the middle points of the diagonals AC, BD. If G is the mid-point of EF and P is any point outside the quadrilateral, show that* $AP^2 + BP^2 + CP^2 + DP^2 = AG^2 + BG^2 + CG^2 + DG^2 + 4\,PG^2$.

Construction: Join PE, PF (*Fig.* 58).

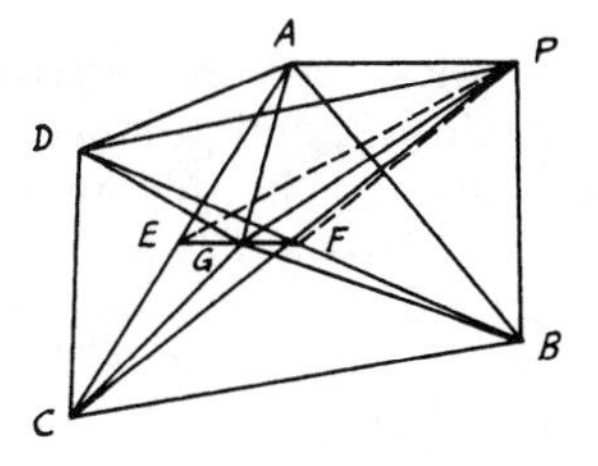

Figure 58

Proof: In the $\triangle APC$, $AP^2 + PC^2 = 2\,PE^2 + 2AE^2$ (Th. 2.55). Similarly, in the $\triangle PBD$, $PB^2 + PD^2 = 2PF^2 + 2BF^2$. Hence, by adding, $AP^2 + BP^2 + CP^2 + DP^2 = 2\,(PE^2 + PF^2 + AE^2 + BF^2)$ (1). In the $\triangle PEF$, $2\,(PE^2 + PF^2) = 4\,(PG^2 + GF^2) = 4\,PG^2 + 2GE^2 + 2GF^2$ (2). In $\triangle$s AGC, BGD, $AG^2 + BG^2 + CG^2 + DG^2 = 2\,(AE^2 + BF^2) + 2\,(GE^2 + GF^2)$ (3). Hence, from these equations, $AP^2 + BP^2 + CP^2 + DP^2 = AG^2 + BG^2 + CG^2 + DG^2 + 4PG^2$.

2.24. *Divide a straight line into two parts, so that the rectangle contained by the whole and one part may be equal to the square on the other part.*

CONSTRUCTION: (i) Let AB be the given straight line. It is required to divide it internally in K, so that $AK^2 = AB \cdot BK$. On AB describe square $ACDB$. Bisect AC in E. Join EB and produce CA to G making $EG = EB$. On AG describe square $AGHK$, then K is the required point. Produce HK to L [*Fig.* 59 (i)].

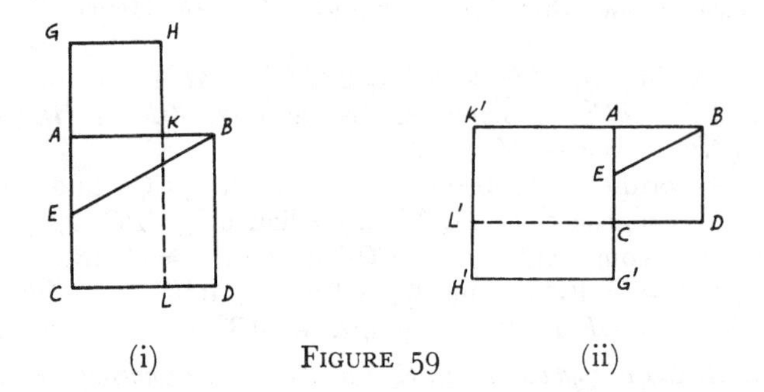

(i) FIGURE 59 (ii)

Proof: $\because$ CA is bisected in E and produced to G, $\therefore$ $EG^2 = CG \cdot GA + EA^2 = EB^2 = EA^2 + AB^2$, since $EG = EB$. Hence $CG \cdot GA = AB^2 =$ fig. AD. But fig. $GL = CG \cdot GA$. $\because$ $GA = GH$, $\therefore$ fig. GL. $=$ fig. AD. Take from each fig. AL. $\therefore$ fig. $GK =$ fig. KD or $AK^2 = AB \cdot BK$.

CONSTRUCTION: (ii) To divide the line externally, describe square $ABDC$, bisect AC in E. Join EB, and produce AC through C to G', making $EG' = EB$. On AG', on the side away from AD, describe square $AG'H'K'$. Produce DC to meet $H'K'$ in L' [*Fig.* 59(ii)]. Then it can be proved that $C'G' \cdot G'A = AB^2$, as before. Hence fig. $G'L' =$ fig. AD. Add to each fig. AL'. $\therefore$ fig. $G'K' =$ fig. $K'D$ and fig. $K'D = AB \cdot BK'$. $\because$ $AB = BD$, $\therefore$ $AB \cdot BK' = AK'^2$.

Algebraic equivalent: If in the previous problem AB and AK contain a and x units of length respectively, then $a(a - x) = x^2$; $\therefore$ $x^2 + ax - a^2 = 0$. Hence AK and AK' correspond to the roots of x in this quadratic equation.

2.25. *ABC and DEF are two triangles, DEF being the greater, which are so located that each pair of corresponding sides are parallel. From A, C two perpendiculars AG, CH are drawn to DF, from B, C another two perpendiculars BK, CL are drawn to EF, and from A, B a third pair of perpendiculars AM, BN are drawn to DE. Show that* $AK^2 + BH^2 + CM^2 = AL^2 + BG^2 + CN^2$.

CONSTRUCTION: From A, B, C draw $AP'P$, QBQ', $CR'R$ $\perp$s to each pair of opposite parallel sides of the $\triangle$s (*Fig.* 60).

Proof: $AK^2 + BH^2 + CM^2 = AP^2 + PK^2 + BQ^2 + QH^2 + CR^2$

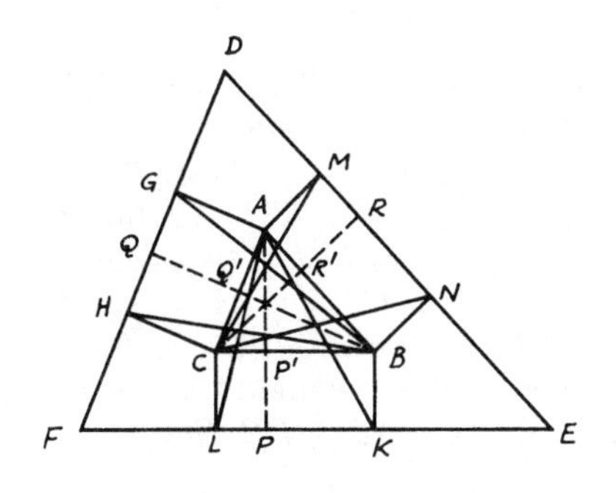

FIGURE 60

$+ RM^2$ (Th. 2.51) (1). Again, $AL^2 + BG^2 + CN^2 = AP^2 + PL^2 + BQ^2 + QG^2 + CR^2 + RN^2$ (2). But, since the sides are $\parallel$, the $\perp$s from A, B, C on the sides of $\triangle DEF$ are the altitudes of $\triangle ABC$. Hence they meet at one point O (Th. 1.33). $\therefore$ According to Problem 2.20, $BP'^2 + CQ'^2 + AR'^2 = CP'^2 + AQ'^2 + BR'^2$ or $PK^2 + QH^2 + RM^2 = PL^2 + QG^2 + RN^2$. Then subtracting from (1) and (2), $AK^2 + BH^2 + CM^2 = AL^2 + BG^2 + CN^2$.

2.26. *If in a quadrilateral the sum of the squares on one pair of opposite sides is equal to the sum of the squares on the other pair, the diagonals will be perpendicular to one another and the lines joining the middle points of opposite sides are equal.*

CONSTRUCTION: Let $ABCD$ be a quadrilateral in which $AB^2 + CD^2 = BC^2 + AD^2$, and let E, F, G, H be the middle points of AB, BC, CD, DA (*Fig.* 61). Let AC, BD meet in M and draw AR, CP $\perp$s BD. Join EF, FG, GH, HE. (P, R must fall on opposite sides of AC.)

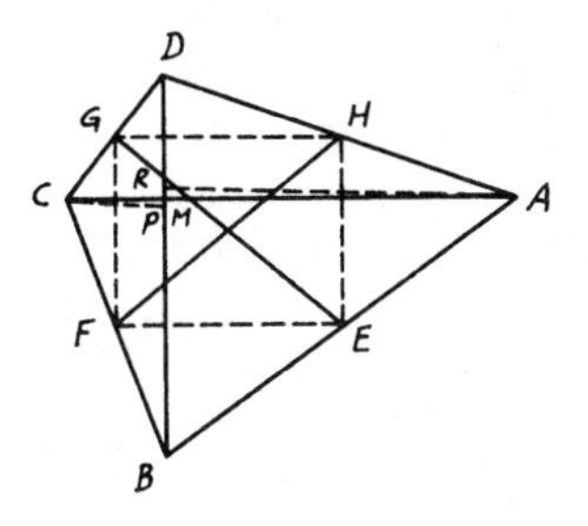

FIGURE 61

Proof: (i) In the $\triangle ABM$, $AB^2 = AM^2 + BM^2 + 2BM \cdot RM$ (Th. 2.53), and in the $\triangle CDM$, $CD^2 = CM^2 + DM^2 + 2DM \cdot PM$. Hence, by adding, $AB^2 + CD^2 = AM^2 + BM^2 + CM^2 + DM^2 + 2BM \cdot RM + 2DM \cdot PM$. Similarly, $AD^2 + BC^2 = AM^2 + BM^2 + CM^2 + DM^2 - 2BM \cdot PM - 2DM \cdot RM$ (Th. 2.54). $\because$ $AB^2 + CD^2 = AD^2 + BC^2$, $\therefore$ $2BM \cdot RM + 2DM \cdot PM = -2BM \cdot PM - 2DM \cdot RM$. $\therefore$ $(PM + RM)(BM + DM) = 0$ or $PR \cdot BD = 0$, which is impossible unless $PR = 0$. Hence P, R coincide with M, thus making AR, CP one line with $AC \perp DB$.

(ii) Since the lines joining the middle points EF, GH are $\parallel$ AC and HE, FG are $\parallel$ BD and since $AC \perp DB$, $EFGH$ is a rectangle and thus its diagonals EG, FH are equal.

2.27. *ABC is an isosceles triangle in which $AB = AC$. From C, a perpendicular CD is drawn to AB (Fig. 62). Denoting AB, AC, BC by b, b, a show that (i) $BC^2 = 2\,AB \cdot BD$; (ii) $CD = (a/2b)\sqrt{4\,b^2 - a^2}$.*

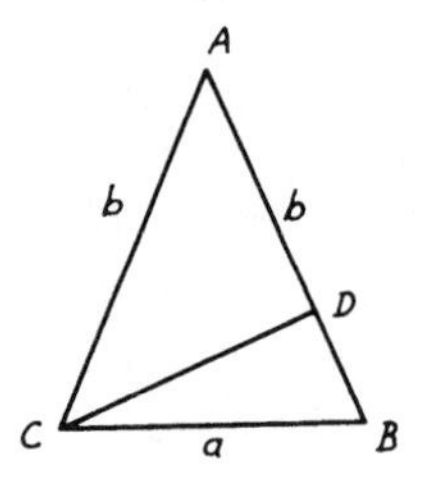

FIGURE 62

Proof: (i) $BC^2 = AB^2 + AC^2 - 2AB \cdot AD$ (Th. 2.54) $= 2AB^2 - 2AB \cdot AD = 2AB(AB - AD) = 2AB \cdot BD$.

(ii) $CD^2 = a^2 - DB^2$. $\because$ $a^2 = 2b \cdot BD$, $\therefore$ $BD = a^2/2b$. $\therefore$ $CD^2 = a^2 - (a^2/2b)^2$. $\because$ $CD = \sqrt{(4a^2b^2 - a^4)/4b^2} = (a/2b)\sqrt{4b^2 - a^2}$.

2.28. *ABC is a triangle and O any point. The parallelograms $AOBC'$, $BOCA'$, $COAB'$ are completed. Show that the lines AA', BB', CC' are concurrent and that the sum of the squares on their lengths is equal to the sum of the squares on the sides of the triangle ABC and on the distances of O from its vertices.*

CONSTRUCTION: Join $A'C'$, $B'C'$, $A'B'$ (*Fig.* 63).

Proof: Since AB', BA' are equal and $\parallel$ to CO, $\because$ $ABA'B'$ is a ▱. Also, $BCB'C'$, $ACA'C'$ are ▱s. $\because$ AA', BB', CC' are diagonals in these ▱s taken in pairs, each pair bisects one another and consequently they are concurrent at their middle point Q. Now, $AA'^2 + BB'^2 = 2\,(AB^2 + A'B^2) = 2\,(AB^2 + CO^2)$. Similarly, $AA'^2 +$

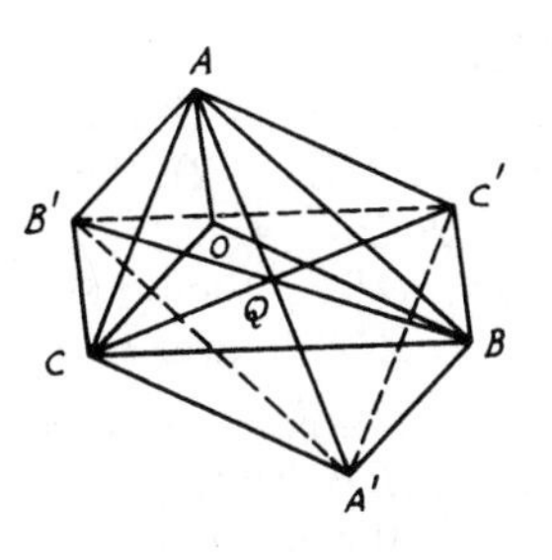

FIGURE 63

$CC'^2 = 2\,(AC^2 + A'C^2) = 2\,(AC^2 + BO^2)$ and $BB'^2 + CC'^2 = 2(BC^2 + B'C^2) = 2\,(BC^2 + AO^2)$. Adding yields $AA'^2 + BB'^2 + CC'^2 = (AB^2 + AC^2 + BC^2) + (AO^2 + BO^2 + CO^2)$.

Note: This theorem is interesting to consider in three dimensions by taking point O outside the plane of the $\triangle ABC$. Then the diagram can be interpreted as a plane diagram of a three-dimensional figure.

2.29. *ABCD is a square. Perpendiculars AA', BB', CC', DD' are drawn to any straight line outside the square. Prove that $A'A^2 + C'C^2 - 2\,BB' \cdot DD' = B'B^2 + D'D^2 - 2\,AA' \cdot CC'$ = area of ABCD.*

CONSTRUCTION: Join the diagonals AC, BD, to intersect in O. Draw $OO' \perp$ the straight line and the perpendiculars AX, BY, DZ on OO' or produced; then join AO', CO' (*Fig.* 64).

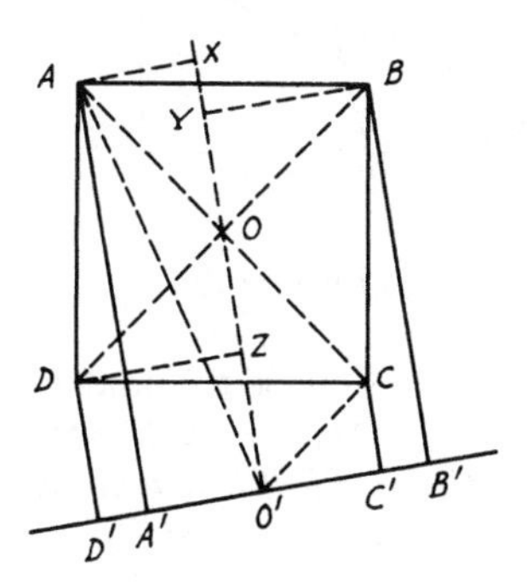

FIGURE 64

Proof: $\because$ O is the mid-point of AC, BD, $\therefore$ according to 1.17, $AA' + CC' = BB' + DD' = 2\,OO'$. Hence $(AA' + CC')^2 = (BB' + DD')^2$, yielding $A'A^2 + C'C^2 - 2\,B'B \cdot D'D = B'B^2 + D'D^2 -$

$2\,AA'\cdot CC'$. Now, $\because$ $A'A^2 + C'C^2 = AO'^2 + CO'^2 - 2\,A'O'^2 = 2\,OO'^2 + 2\,AO^2 - 2\,AX^2 = 2\,OO'^2 + 2\,OX^2$ (1), and $2\,BB'\cdot DD' = 2\,YO'\cdot ZO' = 2\,ZO'^2 + 2\,ZO'\cdot ZY = 2\,ZO'^2 + 4\,ZO\cdot ZO'$ ($\because$ $ZY = 2\,ZO$) $= 2\,OO'^2 - 2\,OZ^2$ (2), then from (1) and (2), $A'A^2 + C'C^2 - 2\,BB'\cdot DD' = 2\,(OX^2 + OZ^2)$ (3). $\because$ $OX = DZ$ from congruence of $\triangle$s AOX, ODZ, then (3) becomes $A'A^2 + C'C^2 - 2\,BB'\cdot DD' = 2\,DO^2 = AD^2 =$ area of $ABCD$.

2.30. *Describe a square equal to a given rectilinear figure.*

CONSTRUCTION: Let $ABCDE$ be the given figure. Convert this figure into an equal triangle through one of its vertices D, by drawing CF, $EG \parallel BD$, AD respectively, where F, G are on AB produced. $\therefore$ $\triangle DFG =$ fig. $ABCDE$ (2.4). Bisect FG in K and draw the rectangle $FIJK$ on FK to meet the $\parallel$ from D to AB in IJ. If $FK = FI$, then FJ would be the required square. If not, produce KF to N, so that $FN = FI$. Bisect KN in M, and with M as center and radius MK, describe a semi-circle KLN. Produce IF to L. Then the square on FL will be equal to fig. $ABCDE$. Join LM (*Fig.* 65).

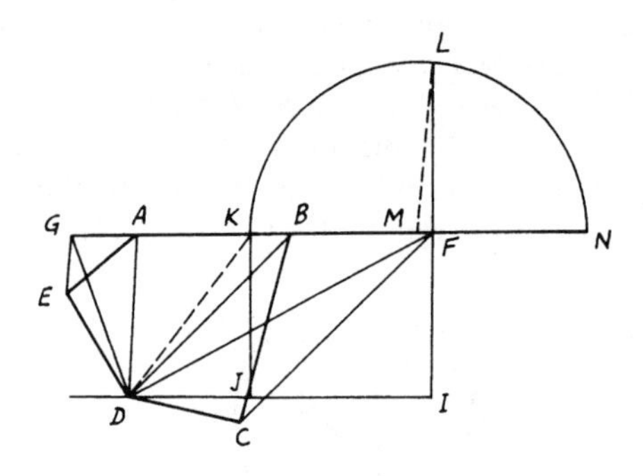

FIGURE 65

Proof: $\because$ M is the mid-point of KN and F is another point on KN, $\therefore$ $KF\cdot FN + FM^2 = MN^2 = ML^2$. $\because$ $MN = ML = FM^2 + FL^2$, $\therefore$ $FL^2 = FK\cdot FN = FK\cdot FI =$ rectangle FJ. But rectangle $FJ = 2\,\triangle DFK = \triangle DFG =$ fig. $ABCDE$. Hence $FL^2 =$ fig. $ABCDE$.

2.31. *ABCD is any quadrilateral. Bisect the sides AB, BC, CD and DA in E, F, G and H respectively. Join EG and FH, which intersect in O. If the diagonals AC, BD are bisected in L, M respectively show that (i) LOM is one straight line and O bisects LM; (ii)* $OA^2 + OB^2 + OC^2 + OD^2 = EG^2 + FH^2 + LM^2$.

CONSTRUCTION: Join EH, EF, LB, LD, LE, GL, EM, and GM (*Fig.* 66).

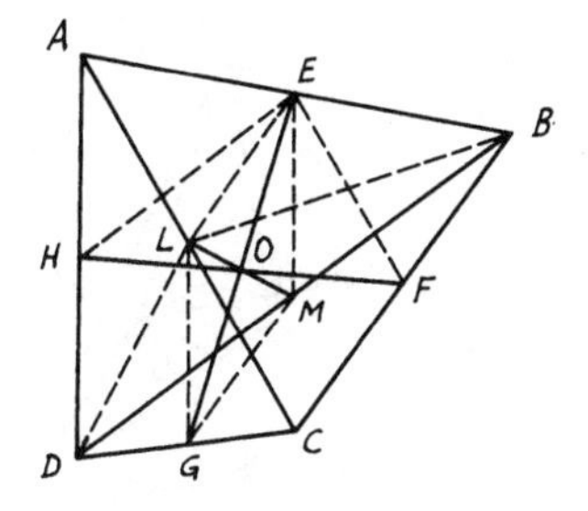

FIGURE 66

Proof: (i) Since L, E, G, and M are the mid-points of AC, AB and DC, DB in $\triangle$s ABC, DBC, $\therefore LE = GM = \frac{1}{2} BC$ and $\parallel$ to it. Therefore, $LEMG$ is a $\square$. $\therefore$ LOM is one straight line and O bisects LM, GE, FH (see Problem 1.14).

(ii) In $\triangle$s ABC, ADC, $AB^2 + BC^2 + CD^2 + DA^2 = 2\,BL^2 + 2\,DL^2 + 4\,AL^2$. $\because$ In $\triangle BLD$, $2\,BL^2 + 2\,DL^2 = 4\,LM^2 + 4\,BM^2$, hence $AB^2 + BC^2 + CD^2 + DA^2 = AC^2 + BD^2 + 4\,LM^2$. Now,

$$OA^2 + OB^2 = 2\,AE^2 + 2\,OE^2,$$
$$OB^2 + OC^2 = 2\,BF^2 + 2\,OF^2,$$
$$OC^2 + OD^2 = 2\,CG^2 + 2\,OG^2,$$
$$OD^2 + OA^2 = 2\,DH^2 + 2\,OH^2.$$

$$\begin{aligned}\therefore 4\,(OA^2 + OB^2 + OC^2 + OD^2) &= 4\,(OB^2 + OF^2 + OG^2 + OH^2) + 4\,(AE^2 + BF^2 + CG^2 + DH^2)\\ &= 2\,(EG^2 + HF^2) + (AB^2 + BC^2 + CD^2 + DA^2)\\ &= 2\,(EG^2 + HF^2) + (AC^2 + BD^2 + 4\,LM^2)\\ &= 2\,(EG^2 + HF^2) + 4\,(EF^2 + EH^2 + LM^2)\\ &= 2\,(EG^2 + HF^2) + 8\,(OE^2 + OH^2) + 4\,LM^2\\ &= 4\,(EG^2 + FH^2 + LM^2).\end{aligned}$$

Hence

$$OA^2 + OB^2 + OC^2 + OD^2 = EG^2 + FH^2 + LM^2.$$

2.32. *ABC, DEF are two triangles so located that the perpendiculars from A, B, C on EF, DF, DE respectively are concurrent. Show that the perpendiculars from D, E, F on BC, CA, AB respectively are also concurrent.*

CONSTRUCTION: Let AA', BB', CC' be the three concurrent $\perp$s from A, B, C on EF, DF, DE respectively and DD', EE', FF' the other three $\perp$s from D, E, F on BC, CA, AB respectively. Join AE, AF, BD, BF, CD, CE (*Fig.* 67).

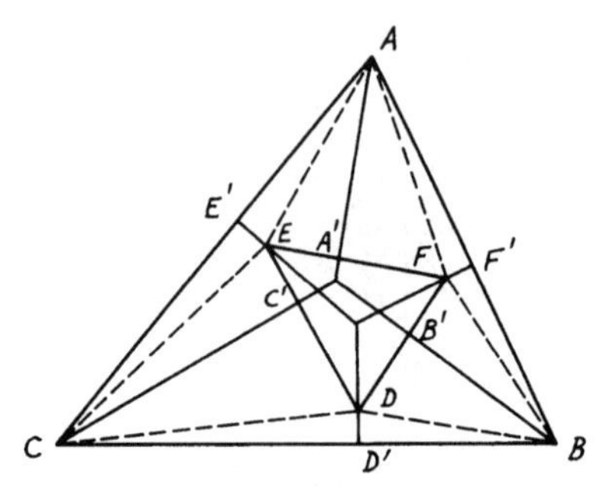

FIGURE 67

Proof: According to Problem 2.20, since AA', BB' and CC' are concurrent, $\therefore$ $(B'D^2 - B'F^2) + (A'F^2 - A'E^2) + (C'E^2 - C'D^2) = 0$. Now, in the $\triangle BDF$, $(B'D^2 - B'F^2) = (BD^2 - BF^2)$. Similarly in $\triangle$s AEF, CDE, $(A'F^2 - A'E^2) = (AF^2 - AE^2)$ and $(C'E^2 - C'D^2) = (CE^2 - CD^2)$. Adding yields $(BD^2 - BF^2) + (AF^2 - AE^2) + (CE^2 - CD^2) = 0$. By rearranging, $(BD^2 - CD^2) + (AF^2 - BF^2) + (CE - AE^2) = 0$, which is equal to $(D'B^2 - D'C^2) + (F'A^2 - F'B^2) + (E'C^2 - E'A^2) = 0$. Therefore, according to the converse of Problem 2.20, the perpendiculars DD', EE', FF' are concurrent.

Miscellaneous Exercises

1. If a square is described on a side of a rhombus, show that the area of this square is greater than that of the rhombus.
2. Of all parallelograms on the same base and of the same area, that which is rectangular has the smallest perimeter.
3. $ABCD$ is a square. E, F, G are the middle points of the sides AB, BC, CD respectively. Prove that $\triangle AFG = \triangle DEF$. (Both $= \frac{3}{8}$ $ABCD$.)
4. APB, ADQ are two straight lines such that the triangles PAQ, BAD are equal. If the parallelogram $ABCD$ be completed, and BQ joined cutting CD in R, show that $CR = AP$.
5. ABC is a triangle. DE is drawn parallel to BC meeting AB, AC or these produced in D, E respectively. If BE, CD intersect in G, show by means of Problem 2.2 that $\triangle AGD = \triangle AEG$ and that AG, produced if necessary, bisects BC.

6. Construct a parallelogram equal and equiangular to a given parallelogram and having one of its sides equal to a given straight line.
7. In the right-angled triangle ABC, $ABDE$, $ACFG$ are the squares on the sides AB, AC containing the right angle. DH, FK are the perpendiculars from D and F on BC produced. Prove that $\triangle$s DHB, CFK are together equal to $\triangle ABC$. (Draw $AN \perp BC$.)
8. ABC is a triangle and D is the middle of the base BC. If E is any point inside the triangle, then $\triangle ABE - \triangle ACE = 2\ \triangle ADE$.
9. $ABCD$, $AECF$ are parallelograms between the same parallels EAD, BCF. FG is drawn parallel to AC, meeting BA on G. Prove that the $\triangle$s ABE, ADG are equal.
10. F, D are two points in the side AC of a triangle ABC such that FC is equal to AD. FG, DE are drawn parallel to AB meeting BC in G, E. Show that the $\triangle$s ADE, AGF are equal.
11. Convert a given quadrilateral $ABCD$ to another quadrilateral $ABCE$ of equal area so that the angle BAE will be equal to a given angle.
12. ABC is a triangle and P is any point on AC. Divide the triangle into two equal parts by a straight line through P parallel to BC.
13. If straight lines be drawn from any point P to the vertices of a parallelogram $ABCD$, prove that $\triangle PBD$ is equal to the sum or difference of $\triangle$s PAB, PCB according as P is (a) outside; (b) inside the angle ABC or its vertically opposite angle.
14. Construct a parallelogram equal to a given parallelogram and having its sides equal to two given straight lines. Show when the problem is impossible.
15. Find a point O inside a triangle ABC such that the triangles OAB, OBC, OCA are equal.
16. $ABCD$ is a parallelogram. From A is drawn a straight line AEF cutting BC in E and DC produced in F. Prove that $\triangle BEF = \triangle DCE$.
17. $ABCD$ is a quadrilateral and E, F are the middle points of AC, BD respectively. Prove that if the diagonals meet in O, then $(\triangle AOB + \triangle COD) - (\triangle AOD + \triangle BOC) = 4\ \triangle EOF$.
18. Prove that the parallelogram formed by drawing straight lines through the vertices of a quadrilateral parallel to its diagonals is double the quadrilateral. Hence prove that two quadrilaterals are equal if their diagonals are equal and contain equal angles.
19. $ABCD$ is a rectangle and E, F are any two points on BC, CD respectively. Show that $\triangle AEF = ABCD - BE \cdot DF$. (Through E, F draw $\parallel$s to CD, BC to intersect in G. Join AG.)
20. $ABCD$ is a parallelogram. If AC is bisected in O and a straight line MON is drawn to meet AB, CD in M, N respectively, and OR parallel to AB meets AN in R, then the $\triangle$s ARM, CRN are equal.

21. On the sides BC, AC of a right-angled triangle ABC at A, squares $BCDE$, $ACFG$ are described. If L is any point taken on AC, show that $\triangle DCL = \triangle FCL$ and hence deduce that $\triangle ADC + \triangle AEB = \frac{1}{2} BC^2$.

22. Find the area of an isosceles trapezoid having the base angle equal to 60°, one of its non-parallel sides 12 inches, and its altitude equal to half the median. (*Answer*: 216 square inches.)

23. Construct a triangle equal in area to a given triangle ABC, so that (a) it will have a given altitude; (b) it will have a base on a part of BC or on BC produced.

24. Show that any line drawn from the mid-point of the median of any trapezoid and terminated by the parallel sides bisects the trapezoid into two equal parts.

25. $ABCD$ is a quadrilateral and E, F are the middle points of AB, CD. Show that if AF, DE intersect in G and CE, BF intersect in H, then the ▱ $EGFH = \triangle AGD + \triangle BHC$.

26. Inscribe a triangle CDE inside a given triangle ABC so that CD will lie on the base BC, its vertex E on AB, and be equal to half the triangle ABC.

27. P is a point inside a parallelogram $ABCD$ such that the area of the quadrilateral $PBCD$ is twice that of the figure $PBAD$. Find the locus of P. (Join AC, BD cutting in O. Trisect AO in P, so that $AP = 2\ PO$.)

28. If through the vertices of a triangle ABC there be drawn three parallel straight lines AD, BE, CF to meet the opposite sides or sides produced in D, E, F, then the area of the triangle DEF is double that of ABC.

29. ABC is a triangle. D, E, F are points on BC, CA, AB respectively such that BD is twice DC, CE twice EA, and AF twice FB. Prove that the triangle DEF is $\frac{1}{3}$ of the triangle ABC.

30. The areas of all quadrilateral figures, the sides of which have the same points of bisection, are equal.

31. Show by a figure how to divide a triangle by a line so that its two parts may be made to coincide with a parallelogram on the same base as the triangle and of half its altitude.

32. The diagonals of a trapezoid intersect in the straight line joining the middle points of its parallel sides.

33. In a given triangle inscribe a parallelogram equal to half the triangle so that one side is in the same straight line with one side of the triangle and has one end at a given point in that side.

34. ABC is a right-angled triangle at B. On BC describe an equilateral triangle BCD outside the triangle ABC and join AD. Show that $\triangle BCD = \triangle ACD - \triangle ABD$. (Bisect BC in F. Join AF, DF.)

35. Two parallelograms $ACBD$, $A'CB'D'$ have a common angle C. Prove that DD' passes through the intersection of $A'B$ and AB'.

36. Construct a triangle equal in area to a given triangle so that its vertex will be equidistant from two given intersecting straight lines.

37. Prove that the area of a trapezoid is equal to the rectangle contained by either of the non-parallel sides and the distance between that side and the middle point of the other side.

38. Describe a triangle equal in area to the sum or difference of two given triangles.

39. If O be any point in the plane of a parallelogram $ABCD$ and the parallelograms $OAEB$, $OBFC$, $OCGD$, $ODHA$ be completed, then $EFGH$ is a parallelogram whose area is double that of $ABCD$.

40. Through any point in the base of a triangle two straight lines are drawn in given directions terminated by the sides of the triangle. Prove that the part of the triangle cut off by them will be a maximum when the straight line joining their extremities is parallel to the base.

41. Through the middle point of the side AB of the triangle ABC, a line is drawn cutting CA, CB in D, E respectively. A parallel line through C meets AB or AB produced in F. Prove that the triangles ADF, BEF are equal.

42. D, E, F are the middle points of the sides BC, CA, AB of a triangle ABC. FG is drawn parallel to BE meeting DE produced in G. Show that the sides of the triangle CFG are equal to the medians of the triangle ABC. Hence show also that the area of the triangle CFG which has its sides respectively equal to the medians of the triangle ABC is $\frac{3}{4}$ that of ABC.

43. D, E, F are the middle points of the sides BC, CA, AB of a triangle. Any line through A meets DE, DF produced if necessary in G, H respectively. Show that CG is parallel to BH.

44. Two equal and equiangular parallelograms are placed so as to have a common angle BAC. P, Q are the intersections of their diagonals. If PQ produced cut AB, AC in M, N respectively, show that $PM = QN$. (Let $ABDC$, $AGEF$ be the ▱s; let G lie between A, B and $\therefore$ C between A, F. Join DE and produce it to meet AB, AC produced in R, S. Join GC, GD, CE.)

45. A', B', C' are the middle points of the sides BC, CA, AB of the triangle ABC. Through A, B, C are drawn three parallel straight lines meeting $B'C'$, $C'A'$, $A'B'$ respectively in a, b, c. Prove that bc passes through A, ca through B, and ab through C and that the triangle abc is half the triangle ABC. (Join Ba, ac.)

46. On the smaller base DC produced of a trapezoid $ABCD$ find a point P so that if PA is joined, it will divide the trapezoid into two equal parts. (Convert $ABCD$ into an equal triangle ADG where G is on DC produced using Problem 2.4. Bisect DG in Q and draw $QM \parallel BG$ and cut BC in M. Hence AM produced meets DG in P.)

47. Squares are described on the sides of a quadrilateral and the adjacent corners of the squares joined so as to form four triangles. Prove that two of these triangles are together equal to the other two.

48. Construct a square that will be $\frac{1}{3}$ of a given square.

49. If $ABCD$ is a straight line, prove that $AC \cdot BD = AB \cdot CD + BC \cdot AD$.

50. If two equal straight lines intersect each other anywhere at right angles, the quadrilateral formed by joining their extremities is equal to half the square on either line.

51. R is the middle point of the straight line PQ. PL, QM, RN are drawn perpendicular to another straight line meeting it in L, M, N respectively. Show that the figure $PLMQ = LM \cdot RN$.

52. If in an isosceles triangle a perpendicular be let fall from one of the equal angles on the opposite side, the square on this perpendicular is greater or less than the square on the line intercepted between the other equal angle and the perpendicular, by twice the rectangle contained by the segments of that side, according as the vertical angle of the triangle is acute or obtuse.

53. BA is divided in C so that the square on AC is equal to the rectangle AB, BC and produced to D so that AD is twice AC. Show that the square on BD is five times the square on AB.

54. An equilateral triangle is described, one of whose vertices is at the angle B of another equilateral triangle ABC, and whose opposite side PQ passes through C. Prove that $BP^2 = AB^2 + PC \cdot QC$.

55. On a given straight line describe a rectangle which will be equal to the difference of the squares on two given straight lines.

56. Any rectangle is half the rectangle contained by the diagonals of the squares on its adjacent sides.

57. A and B are fixed points, CD a fixed straight line of indefinite length. Find a point P in CD such that $(PA^2 + PB^2)$ is a minimum.

58. The squares on the diagonals of any quadrilateral are double the squares on the lines joining the middle points of the opposite sides. (See Problem 2.31.)

59. Show that the sum of the squares on the distances of any point from the angular points of a parallelogram is greater than the sum of the squares on two adjacent sides by four times the square on the line joining the point to the point of intersection of the diagonals.

60. ABC, DEF are two triangles having the two sides AB, AC equal to the two sides DE, DF, each to each, and having the angles BAC, EDF supplementary. Show that $BC^2 + EF^2 = 2\,(AB^2 + AC^2)$.

61. Two squares $ABCD$, $A'B'C'D'$ are placed with their sides parallel, AB parallel to $A'B'$, and so on. Prove that $AA'^2 + CC'^2 = BB'^2 + DD'^2$.

62. On the sides of a quadrilateral, squares are described outward, forming an eight-sided figure by joining the adjacent corners of consecutive squares. Prove that the sum of the squares on the eight sides with twice the squares on the diagonals of the quadrilateral is equal to five times the sum of the squares on its sides.

63. The diagonal AC of a square $ABCD$ is produced to E so that CE is equal to BC. Show that $BE^2 = AC \cdot AE$.

64. $ABCD$ is a parallelogram whose diagonals intersect in O. OL, OM are drawn perpendicular to AB, AD meeting them in L, M respectively. Show that $AB \cdot AL + AD \cdot AM = 2\,AO$.

65. AB, AC are the equal sides of an isosceles triangle ABC. BD is drawn perpendicular to AB meeting AC produced in D and the bisector of the angle A meets BD in E. Prove that the square on AB is equal to the difference of the rectangles DA, AC and DB, BE.

66. From the vertices of a triangle ABC, three perpendiculars AD, BE, CF are drawn on any straight line outside ABC. Show that the perpendiculars from D, E, F on BC, CA, AB are concurrent. (Join AE, AF, BD, BF, CD, CE and apply Problem 2.20, making use of Problem 2.32.)

67. $ABCD$ is a trapezoid having sides AB, CD parallel and the sides AD, BC perpendicular to each other. E, F, G, H are the middle points of AB, CD, AD, BC respectively. Prove that the difference of the squares on CD, AB is equal to four times the rectangle FE, GH.

68. When the perimeter of a parallelogram is given, its area is a maximum when it is a square. (If the two sides are given, the area is a maximum when the parallelogram is a rectangle. Hence the question resolves itself to: Of all rectangles with the same perimeter, the square has the greatest area.)

69. If P be the orthocenter of the triangle ABC, show that (a) $AP^2 + BC^2 = BP^2 + AC^2 = CP^2 + AB^2$; (b) $AP \cdot BC + BP \cdot AC + CP \cdot AB = 4\,\triangle ABC$.

70. Prove that three times the sum of the squares on the sides of a triangle are equal to four times the sum of the squares on its medians.

71. G is the centroid of the triangle ABC and P any point inside or outside the triangle. Show that (a) $AB^2 + BC^2 + AC^2 = 3\,(AG^2 + BG^2 + CG^2)$; (b) $PA^2 + PB^2 + PC^2 = AG^2 + BG^2 + CG^2 + 3\,PG^2$.

72. Divide the straight line AB into two parts at C so that the square on AC may be equal to twice the square on BC and prove that the squares on AB, BC are together equal to twice the rectangle AB, AC.

73. ABC is any triangle. Find a point P on the base BC so that the difference of the squares on AB, AC will be equal to $BC \cdot BP$. (Bisect BC in D and draw $AE \perp BC$. Take BP on $BC = 2\,DE$.)

74. $ABCD$ is a quadrilateral in which $AC = CD$, $AD = BC$, and $\angle ACB$ is the supplement of $\angle ADC$. Show that $AB^2 = BC^2 + CD^2 + DA^2$.

75. D is the foot of the perpendicular from A on the side BC of the triangle ABC; E is the middle point of AC. Prove that the square on BE is equal to the sum or difference of the square on half AC and the rectangle BC, BD, according as the angle B is acute or obtuse.

76. Find the locus of a point such that the sum of the squares on its distances from (a) two given points; (b) three given points, may be constant.

77. Of all parallelograms inscribed in a given rectangle, that whose vertices bisect the sides of the rectangle has the least sum of squares of sides.

78. *AB*, *CD* are two straight lines, and *F*, *G* the middle points of *AC*, *BD* respectively. *H*, *K* are the middle points of *AD*, *BC* respectively. Prove that $AB^2 + CD^2 = 2\,(FG^2 + HK^2)$. (Join *FH*, *HG*, *GK*, *KF*.)

79. Prove that the locus of a point such that the sum of the squares on its distances from the vertices of a quadrilateral is constant is a circle the center of which coincides with the intersection of the lines joining the middle points of opposite sides of the quadrilateral.

80. On *AB* describe the square *ABCD*, bisect *AD* in *E*, and join *EB*. From *EB* cut off *EF* equal to *EA* and from *BA* cut off *BG* equal to *BF*. Show that the rectangle *AB*, *AG* is equal to the square on *BG*.

81. Divide a straight line into two parts so that the square on the whole line with the square on one part shall be equal to three times the square on the other part.

82. Construct a right-angled triangle with given hypotenuse such that the difference of the squares on the sides containing the right angle may be equal to the square on the perpendicular from the right angle on the hypotenuse.

83. In Fig. 59(i), *GD* is joined cutting *AB*, *HL* in *g*, *d* respectively. Show that $Gg = Dd$. If *CK*, *DB* produced meet in *N*, prove also that $CK^2 + CN^2 = 5\,AB^2$.

84. *ABC* is a right-angled triangle, *BAC* being the right angle. Any straight line *AO* drawn from *A* meets *BC* in *O*. From *B* and *C*, *BM*, *CN* are drawn perpendicular to *AO* or *AO* produced. Prove that the squares on *AM*, *AN* are equal to the squares on *OM*, *ON* with twice the rectangle *BO*, *OC*.

85. Produce *AB* to *C* so that the rectangle *AB*, *AC* may be equal to the square on *BC*.

86. *AB*, one of the sides of an equilateral triangle *ABC*, is produced to *D* so that *BD* is equal to twice *AB*. Prove that the square on *CD* is equal to seven times the square on *AB*.

87. Two right-angled triangles *ACB*, *ADB* have a common hypotenuse *AB*. *AC*, *BD* meet in *E*, *AD*, *BC* in *F*. Prove that the rectangles *AE*, *EC* and *BE*, *ED* are equal, also the rectangles *AF*, *FD* and *BF*, *FC*.

88. *ABC* is a triangle having the angle *A* equal to half a right angle. *M*, *N* are the feet of the perpendiculars from *B*, *C* on the opposite sides. Prove that $BC^2 = 2\,MN^2$.

89. Given the sum of two lines and also the sum of the squares described on them. Obtain by a geometric construction the lines themselves.

90. If a straight line be bisected and produced to any point, the square on the whole line is equal to the square on the produced part with four times the rectangle contained by half the line bisected and the line made up of the half and the produced part.

91. Of all parallelograms of equal perimeter, the sum of the squares on the diagonals is least in those whose sides are equal.

92. *ABCD* is a square and *E* is any point in *DC*. On *AE*, *BE* outside the triangle *AEB* are described squares *APQE*, *BRSE*. Prove that the square on *QS* is less than five times the square on *AB* by four times the rectangle *DE*, *DC*.

93. If in a quadrilateral the sum of the squares on one pair of opposite sides is equal to the sum of the squares on the other pair, the lines joining the middle points of opposite sides are equal.

94. On the same base *AB* and on opposite sides of it are described a right-angled triangle *AQB*, *Q* being the right angle, and an equilateral triangle *APB*. Show that the square on *PQ* exceeds the square on *AB* by twice the rectangle contained by the perpendiculars from *P*, *Q* on *AB*. (Draw $PM, QN \perp AB$, then $AM = MB = MQ$. Draw $QR \parallel AB$ meeting *PM* produced in *R*.)

95. Show how to inscribe in a given right-angled isosceles triangle a rectangle equal to a given rectilinear figure. When is it impossible to do this?

CHAPTER 3

CIRCLES AND TANGENCY

Theorems and Corollaries

DIAMETERS, CHORDS, AND ARCS IN CIRCLES

3.56. (*i*) *The diameter of a circle which bisects a chord is perpendicular to the chord.* (*ii*) *The diameter which is perpendicular to a chord bisects it.* (*iii*) *The perpendicular bisector of any chord contains the center.*

COROLLARY. *A circle is symmetric with regard to any diameter.*

3.57. *One circle and only one, can be drawn through three given points not in the same straight line.*

COROLLARY 1. *Circles which have three common points coincide.*

COROLLARY 2. *If O is a point within a circle from which three equal straight lines OA, OB, OC can be drawn to the circumference, then O is the center of the circle.*

3.58. *In the same circle or in equal circles:* (*i*) *If two arcs subtend equal angles at the center, they are equal.* (*ii*) *Conversely, if two arcs are equal, they subtend equal angles at the center.*

3.59. *In the same circle or in equal circles:* (*i*) *If two arcs are equal, then the chords of the arcs are equal.* (*ii*) *Conversely, if two chords are equal, then the minor arcs which they cut off are equal and so are the major arcs.* (*iii*) *Equal arcs, or equal chords, determine equal sectors and equal segments of the circle.*

3.60. *In the same circle or in equal circles:* (*i*) *If two chords are equal, they are equidistant from the center.* (*ii*) *Conversely, if two chords are equidistant from the center they are equal.*

3.61. *Two chords of a circle, which do not both pass through the center, cannot bisect each other. Either chord may be bisected by the other, but they cannot both be bisected at their point of intersection.*

COROLLARY. *The diagonals of any parallelogram inscribed in a circle intersect in the center of the circle.*

3.62. *If two circles intersect, they cannot have the same center.*

3.63. *One circle cannot cut another in more than two points.*

3.64. *The diameter is the greatest chord of a circle and of all others the chord which is nearer to the center is greater than one more remote, and the greater is nearer the center than the less.*

Angles Subtended by Arcs in Circles

3.65. *The angle which an arc of a circle subtends at the center is double that which it subtends at any point on the remaining part of the circumference.*

3.66. (*i*) *Angles in the same segment of a circle are equal.* (*ii*) *The angle in a segment which is greater than a semicircle is less than a right angle.* (*iii*) *The angle in a semicircle is a right angle.* (*iv*) *The angle in a segment which is less than a semicircle is greater than a right angle.*

3.67. *In the same circle or in equal circles, equal arcs subtend equal angles at the circumference and equal angles at the center.*

Tangents and Touching Circles

3.68. *One tangent, and only one, can be drawn to a circle at any point on the circumference, and this tangent is perpendicular to the radius through the point of contact.*

Corollary 1. *The straight line joining the center of a circle to the point of contact of a tangent is perpendicular to the tangent.*

Corollary 2. *The perpendicular to a tangent to a circle at the point of contact passes through the center.*

3.69. *If two tangents are drawn to a circle from an external point:* (*i*) *The lengths of the tangents from the external point to the points of contact are equal.* (*ii*) *They subtend equal angles at the center of the circle.* (*iii*) *They make equal angles with the straight line joining the given point to the center.*

3.70. *If A is any point inside or outside a circle with center O, and OA or produced cuts the circumference in B, then AB is the shortest distance from the point A to the circumference of the circle.*

3.71. *Two circles cannot touch at more than one point, and if they touch, the point of contact lies in the straight line joining the centers, or in that line produced.*

3.72. *If through an extremity of a chord in a circle a straight line is drawn:* (*i*) *If the straight line touches the circle, the angles which it makes with the chord are equal to the angles in the alternate segments.* (*ii*) *Conversely, if either of the angles which the straight line makes with the chord is equal to the angle in the alternate segment, the straight line touches the circle.*

Cyclic Polygons

3.73. *If a quadrilateral is such that one of its sides subtends equal angles at the extremities of the opposite side, the quadrilateral is cyclic.*

Corollary 1. *The locus of points at which a finite straight line AB subtends a given angle consists of the arcs of two segments of circles on AB as base, each containing the given angle.*

COROLLARY 2. *The locus of points at which a given straight line AB subtends a right angle is the circle on AB as diameter.*

COROLLARY 3. *The straight line joining the middle point of the hypotenuse of a right-angled triangle to the vertex is equal to half the hypotenuse.*

3.74. *The opposite angles of a quadrilateral inscribed in a circle are supplementary.*

COROLLARY. *If a side of a quadrilateral inscribed in a circle is produced, the exterior angle is equal to the interior opposite angle.*

3.75. *If two opposite angles of a quadrilateral are supplementary, the quadrilateral is cyclic.*

COROLLARY. *If one side of a quadrilateral is produced and the exterior angle so formed is equal to the interior opposite angle, then the quadrilateral is cyclic.*

NINE-POINT CIRCLE

3.76. *The circle through the middle points of the sides of a triangle passes through* (*i*) *the feet of the perpendiculars from the vertices of the triangle on the opposite sides;* (*ii*) *the middle points of the lines joining the orthocenter to the vertices. This is called the nine-point circle.*

COROLLARY. *The radius of the nine-point circle of a triangle is equal to half of the circumcircle.*

3.77. (*i*) *The center of the nine-point circle of any triangle is the middle point of the line joining the circumcenter and orthocenter of the triangle.* (*ii*) *The centroid is a point of trisection of this line.*

SQUARES AND RECTANGLES RELATED TO CIRCLES

3.78. (*i*) *If two chords of a circle intersect at a point, either inside or outside the circle, the rectangle contained by the segments of the one is equal to that contained by the segments of the other.* (*ii*) *When this point is outside the circle, each of these rectangles is equal to the square on the tangent from the point of intersection of the chords to the circle.*

3.79. *If two straight lines intersect, or both being produced intersect, so that the rectangle contained by the segments of the one is equal to that contained by the segments of the other, the extremities of the lines are cyclic. This is the converse of Th.* 3.78(*i*).

3.80. *If a chord AB of a circle is produced to any point P and from this point a straight line PQ is drawn to meet the circle, such that the square on PQ is equal to the rectangle $AP \cdot PB$, then the straight line PQ touches the circle at Q.*

3.81. *In a right-angled triangle ABC in which A is the right angle and BC is the hypotenuse, if AP is drawn perpendicular to the hypotenuse BC then:* (*i*) $AB^2 = BP \cdot BC$; (*ii*) $AC^2 = CP \cdot BC$; (*iii*) $AP^2 = PB \cdot PC$.

COROLLARY. *If a line BC is divided internally in P, then* $BC^2 = BP \cdot BC + CP \cdot BC$. *If it is divided externally in P such that BP is greater than PC, then* $BC^2 = BP \cdot BC - CP \cdot BC$.

Solved Problems

3.1. *O is the center of a circle whose circumference passes through the vertices of an equilateral triangle ABC. If P is any point taken on the circumference and AD, BE, CF are three perpendiculars from A, B, C on the tangent from P to the circle, show that the sum of these perpendiculars is equal to twice the altitude of the* $\triangle ABC$.

CONSTRUCTION: Draw $AM \perp BC$; then produce it to meet the circle in N. Draw the perpendiculars MK, NL on the tangent from P and join OP (*Fig.* 68).

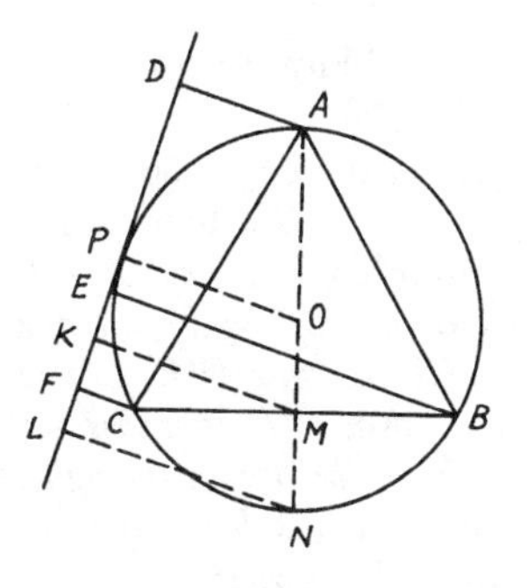

FIGURE 68

Proof: Since AM is an altitude in the equilateral $\triangle ABC$, $\therefore$ it bisects the base BC and hence passes through the center of the $\odot$. Also, $OP \perp$ tangent $DPEF$ (Th. 3.68, Cor.1). Since $\triangle ABC$ is equilateral, $\therefore$ $OM = MN = \frac{1}{2} OP$. $\therefore$ M is the mid-point of ON. $\because$ $BCFE$ is a trapezoid, $\therefore$ $BE + CF = 2\ MK$ (Th. 1.29). Similarly, in the trapezoid $OPLN$, $OP + NL = 2\ MK$. Hence $BE + CF = OP + NL$. Now, in a similar way, the trapezoid $ADLN$ gives $AD + NL = 2\ OP$. Adding OP to each side, $\therefore$ $AD + NL + OP = 3\ OP$. $\therefore$ $AD + BE + CF = 3\ OP = 2\ AM$.

3.2. *Through one of the points of intersection of two given circles, draw a straight line terminated by the two circles which will be equal to a given straight line.*

CONSTRUCTION: Let A be one of the points of intersection of $\odot$s

ABC, *ABD*; *O*, *O′* their centers. On *OO′* as diameter describe semi-circle *OEO′*. With center *O′* and distance equal to half the given straight line, describe an arc of ⊙ cutting the semi-circle in *E*. Join *O′E* and through *A* draw *CAD* ∥ *O′E*. Join *OE* and produce it to meet *CD* in *F*. Draw *O′G* ⊥ *CD* (*Fig.* 69).

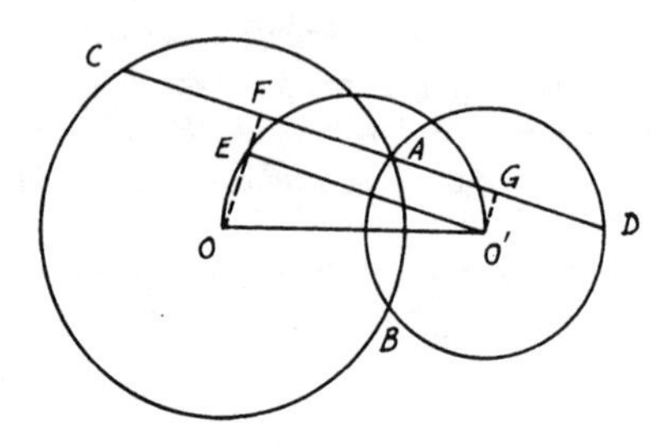

FIGURE 69

Proof: ∵ ∠*OEO′* is right [Th. 3.66 (iii)] and *CD* is ∥ *O′E*, ∴ ∠*OFA* is right. ∴ *OF* is ∥ *O′G*. Hence *EFGO′* is a rectangle. ∴ *FG* = *EO′*. But *AC* is double of *AF* and *AD* double of *AG*. ∴ *CD* is double of *FG*, i.e., of *EO′*. ∴ *CD* = given straight line.

COROLLARY. *Because CD is twice O′E, CD will be a maximum when O′E is a maximum, i.e., when O′E coincides with OO′. Therefore, the greatest straight line which can be drawn through A is parallel to the line joining the centers of the circles and is double that line. We shall leave for the student the question which is the shortest given straight line that can be used subject to the given conditions.*

3.3. *Construct a triangle given a vertex and (i) the circumscribed circle and center of its incircle; (ii) the circumscribed circle and the orthocenter.*

CONSTRUCTION: (i) Consider the vertex *A* on the given circumference. Join *AP*, *P* being the given center of the incircle of △. Produce *AP* to meet the circumference of the given circumscribed ⊙*O* in *Q*. From *Q* draw the two chords *QB*, *QC* equal to *QP*; then *ABC* is the required △ (*Fig.* 70).

Proof: ∵ *QB* = *QP* (construct), ∴ ∠*QBP* = ∠*QPB*. ∴ ∠*QBC* + ∠*CBP* = ∠*PBA* + ∠*PAB*. But, since *PA* bisects ∠*A*, ∴ ∠*PAB* = ∠*PAC* = ∠*QBC* (Th. 3.67). ∴ ∠*CBP* = ∠*PBA*; i.e., *PB* bisects ∠*B*, and similarly *PC* bisects ∠*C*. Hence *P* is the incenter of △*ABC* and therefore *ABC* is the required △.

CONSTRUCTION: (ii) Circumcircle *O* and orthocenter *D* are given. From the given vertex *A* on the given circumference, join *AD* and

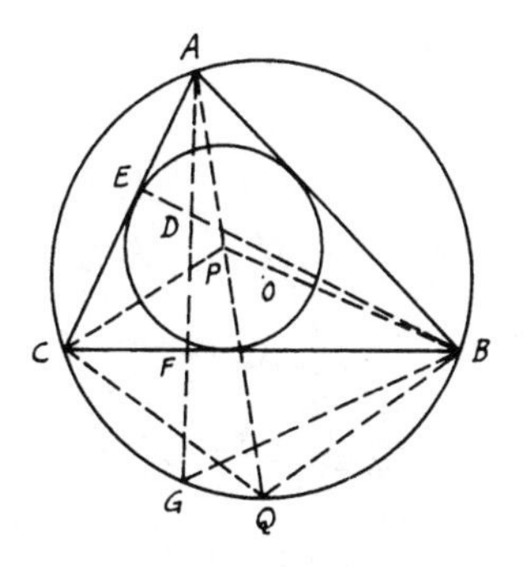

FIGURE 70

produce it to meet the circumference in G. Bisect DG in F and draw $BFC \perp ADG$. Then ABC is the required $\triangle$.

Proof: $\triangle$s BDF, BGF are congruent. $\therefore$ $\angle DBF = \angle GBF$. $\because$ $\angle GBF = \angle GAC$, $\therefore$ $\angle DBF = \angle GAC$ or $\angle FAE$. $\therefore$ quadrilateral $ABFE$ is cyclic (Th. 3.73). $\therefore$ $\angle BFA = \angle BEA$ = right angle. $\therefore$ $BE \perp AC$. $\because$ $AF \perp BC$, $\therefore$ D is the orthocenter of $\triangle ABC$.

3.4. *I is the incenter of a triangle ABC and the circumcircle of triangle BIC cuts AB, AC in D, E respectively. Show that (i) AB = AE and AC = AD. (ii) I is the orthocenter of the triangle formed by joining the circumcenters of $\triangle$s BIC, CIA, AIB. (iii) The circumcircle of $\triangle$ABC passes through the circumcenters of $\triangle$s BIC, CIA, AIB.*

CONSTRUCTION: Let P, Q, R be the circumcenters of $\triangle$s BIC, CIA, AIB respectively. Join IP, IQ, IR and produce them to meet the $\odot$s on BIC, CIA, AIB in F, G, H. Join FG, GH, HF, ID, IE, PB, PC (*Fig.* 71).

Proof: (i) $\because$ BI, AI bisects $\angle$s B, A respectively, $\therefore$ $\angle DBI = \angle IBC = \angle IEC$. Therefore, $\triangle$s AIB, AIE are congruent. $\therefore$ $AB = AE$. Similarly, $AD = AC$.

(ii) Since P is the center and IF is the diameter of $\odot BIC$, $\therefore$ $\angle ICF$ = right angle [Th. 3.66 (iii)]. Also, in $\odot CIA$, $\angle ICG$ = right angle. Hence $\angle ICF + \angle ICG$ = 2 right angles. $\therefore$ FCG is a straight line (Th. 1.2). Similarly, GAH, HBF are straight lines. $\because$ P, Q are the mid-points of IF, IG (since they are centers), $\therefore$ PQ is $\parallel FG$. $\because$ $IC \perp FCG$, $\therefore$ $PQ \perp IC$. Similarly, $QR \perp IA$ and $RP \perp IB$. Hence I is the orthocenter of $\triangle PQR$, and $AIPF$, $BIQG$, $CIRH$ are straight lines.

(iii) $\because$ $PB = PI = PC$ and PR, PQ $\perp$s IB, IC, $\therefore$ RP, QP bisect $\angle$s IPB, IPC. $\therefore$ $\angle QPC = \frac{1}{2} \angle CPI$; i.e., $\angle CPA = \frac{1}{2} \angle CBA$

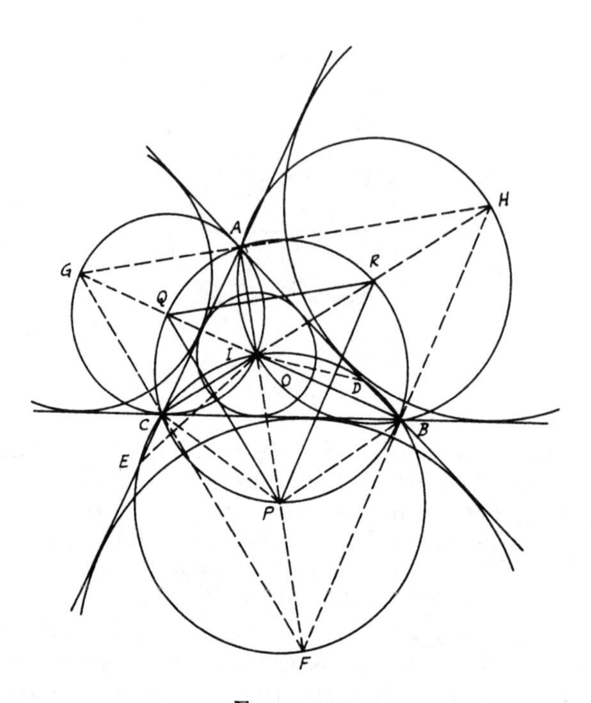

FIGURE 71

$= \angle QBC$. $\therefore$ fig. $PBQC$ is cyclic (Th. 3.73). Similarly, fig. $BCQR$ is cyclic. Therefore, $BPCQAR$ is cyclic or the circumcircle of $\triangle ABC$ passes through P, Q, R.

Note: The points F, G, H are the centers of the external $\odot$s of $\triangle ABC$. Also the circumcircle of $\triangle ABC$, which passes through P, Q, R, is the nine-point $\odot$ of $\triangle FGH$ (Th. 3.76).

3.5. *A, B are the two points of intersection of three circles. From A a straight line is drawn to meet the circles in D, E, F. If the tangents from D, E meet in P, those from E, F meet in Q and those from F, D meet in R, show that $PBQR$ is cyclic.*

CONSTRUCTION: Join BD, BA, BE, BF, BR (*Fig.* 72).

Proof: $\because$ PE, PD are tangents to the two $\odot$s, $\therefore$ $\angle PED = \angle EBA$. $\angle PDE = \angle DBA$ [Th. 3.72 (i)]. Hence, $\angle PED + \angle PDE = EBD$. Since, in the $\triangle PDE$, $\angle PED + \angle PDE + \angle DPE = 2$ right angles, $\therefore$ $\angle EBD + \angle DPE = 2$ right angles. $\therefore$ $DBEP$ is cyclic. Similarly, $DBFR$ is cyclic. Therefore, $\angle BDE = \angle BPE$ and $\angle BDF$ or $\angle BDE$

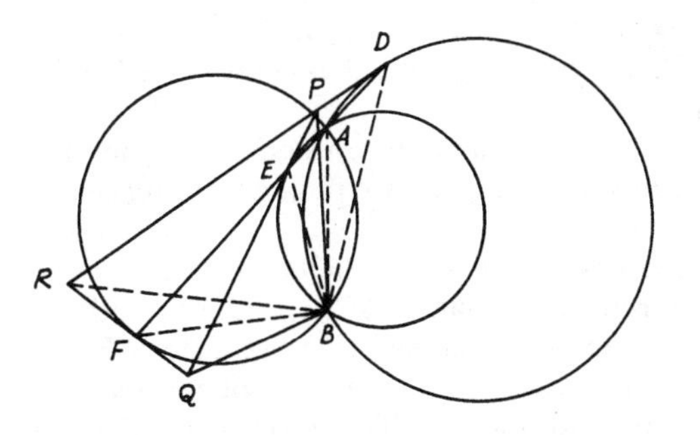

FIGURE 72

$= \angle BRF$ [Th. 3.66(i)]. $\therefore$ $\angle BPE$ or $BPQ = \angle BRF$ or BRQ. $\therefore$ fig. $PBQR$ is cyclic (Th. 3.73).

3.6. *Describe a circle of given radius to touch a given straight line so that the tangents drawn to it from two given points on the straight line will be parallel. How many solutions are there to this problem and when is it impossible to describe this circle?*

CONSTRUCTION: Let PQ be the given straight line and B, C the two given points on it. On BC as diameter describe a $\odot$. At distance from BC = given radius, draw $OO' \parallel BC$ and cutting the $\odot$ in O, O'. Draw $OA \perp BC$; then OA = given radius. Then, $\odot$ with O as center and OA as radius will be the required $\odot$ (*Fig.* 73).

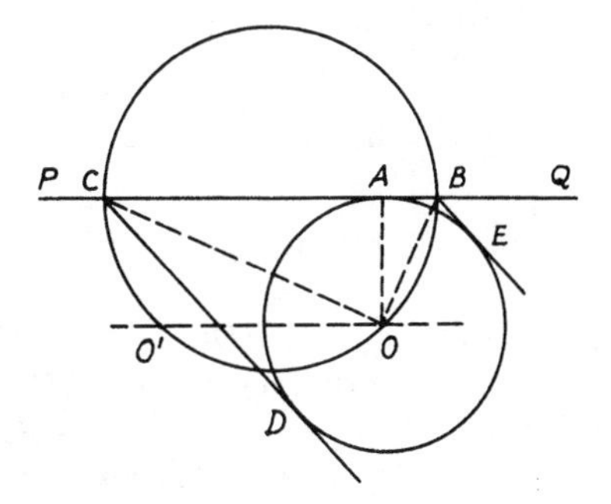

FIGURE 73

Proof: Draw BE, CD tangents to $\odot O$. $\because$ BA, BE are tangents to $\odot O$, $\therefore$ OB bisects $\angle ABE$. Similarly, OC bisects $\angle ACD$. Since

$\angle BOC$ = right angle [Th. 3.66(iii)], $\therefore$ in $\triangle BOC$, $\angle OBC + \angle OCB$ = right angle. Hence $\angle ABE + \angle ACD$ = 2 right angles. $\therefore BE \parallel CD$. Therefore $\odot$ with O as center and OA the given radius is the required $\odot$.

(i) In this case there are four solutions, since the $\parallel$ line to BC at the required radius will cut $\odot$ in two points O, O' on each side of the line PQ.

(ii) In the case when OO' touches the $\odot$ at one point, say O, its distance from BC = half of BC; i.e., the $\odot$ with O as center will be equal to $\odot$ on BC as diameter. The tangents BE, CD will be $\perp BC$ and there will be two $\odot$s, one at each side of PQ.

(iii) The solution is impossible, however, if the given radius of the required $\odot$ is greater than half BC. In this case, the $\parallel$ line to BC will never cut $\odot$ on BC.

3.7. *If circles are inscribed inside the six triangles into which a triangle ABC is divided by its altitudes AD, BE, CF, then the sum of the diameters of these six circles together with the perimeter of the triangle ABC equals twice the sum of the altitudes of triangle ABC.*

Construction: Let O_1, O_2 be the inscribed $\odot$s of $\triangle$s ABD, ACD, touching BC, AD in G, P, H, Q. Join O_1G, O_1H, O_2P, O_2Q (*Fig.* 74).

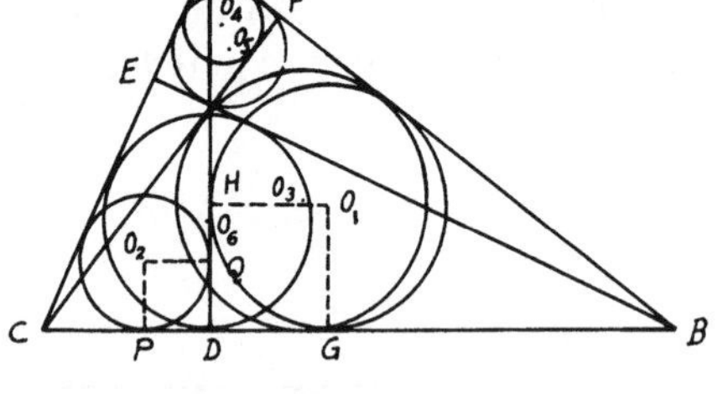

Figure 74

Proof: $\because$ O_1G, O_1H are $\perp BD$, AD, $\therefore$ O_1GDH is a square. $\therefore$ $O_1G + O_1H = GD + DH$ = diameter of $\odot O_1$. $\because$ $AB = BG + AH$, $\therefore$ diameter of $\odot O_1 + AB = BD + AD$. Similarly, in $\triangle ADC$, diameter of $\odot O_2 + AC = CD + AD$. Adding yields diameter of $\odot O_1$ + diameter of $\odot O_2 + AB + AC = BC + 2\,AD$ (1). Similarly, diameter of $\odot O_3$ + diameter of $\odot O_4 + BC + AC = AB + 2\,CF$ (2), diameter of $\odot O_5$ + diameter of $\odot O_6 + AB + BC = AC + 2\,BE$ (3). Adding (1) to (3) yields diameters of $\odot$s $O_1, O_2, O_3, \ldots, O_6 + (AB + BC + AC) = 2\,(AD + BE + CF)$.

3.8. *ABC is a triangle and AD, BE, CF its altitudes. If P, Q, R are the middle points of DE, EF, FD respectively, show that the perpendiculars from P, Q, R on the opposite sides of the triangle ABC, AB, BC, CA respectively, are concurrent.*

Construction: Let PX, QY, RZ be the three $\perp$s on AB, BC, CA respectively. Join PQ, QR, RP. Let O be the orthocenter of $\triangle ABC$ (*Fig.* 75).

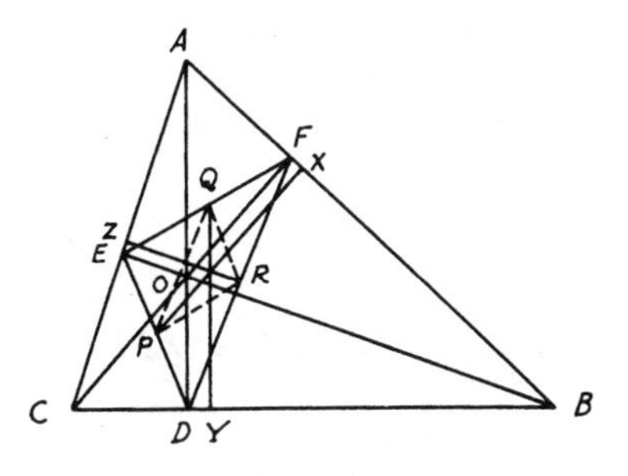

Figure 75

Proof: O is the point of concurrence of AD, BE, CF, and DEF is the pedal $\triangle$. $\because$ $\angle BDO = \angle BFO =$ right angle, $\therefore$ ⏢ $BDOF$ is cyclic (Th. 3.75). $\therefore$ $\angle FDO = \angle FBO$. Similarly, ⏢ $CDOE$ is cyclic. $\therefore$ $\angle EDO = \angle ECO$. But since ⏢ $FBCE$ is cyclic also, $\therefore$ $\angle FBO = \angle ECO$. Hence $\angle FDO = \angle EDO$; i.e., AD bisects $\angle EDF$ of the pedal $\triangle DEF$. Similarly, BE, CF bisect $\angle$s DEF, DFE. Now, since $PQ \parallel DF$ and $QY \parallel AD$ (both $\perp$s BC), $\therefore$ $\angle PQY = \angle ADF$. Since also $QR \parallel DE$ and $QY \parallel AD$, $\therefore$ $\angle RQY = \angle ADE$. $\because$ $\angle ADF = \angle ADE$, $\therefore$ $\angle PQY = \angle RQY$. Hence QY bisects $\angle PQR$. Similarly, PX, RZ bisect $\angle$s QPR, PRQ. Therefore, PX, QY, RZ are the angle bisectors of $\triangle PQR$ and consequently they are concurrent (Th. 1.32).

3.9. *AB is a fixed diameter in a circle with center O. C, D are two given points on its circumference and on one side of AB. Find a point P on the circumference so that (i) PC, PD will cut equal distances of AB from O whether P is on the same or opposite side of AB as C, D; (ii) PC, PD will cut a given length m of AB.*

Analysis: (i) Suppose P is the required point. PC, PD or produced will cut AB in Q, R, where $OQ = OR$. Join CO and produce it to meet the circumference in E, then join ER, ED (*Fig.* 76). Now, from the congruence of $\triangle$s OQC, ORE, $\angle QCO = \angle OER$. $\therefore$ $PQ \parallel ER$. $\because$ $\angle QCE = \angle PDE$ (Th. 3.66), $\therefore$ in (i) $\angle OER = \angle PDE$. $\therefore$

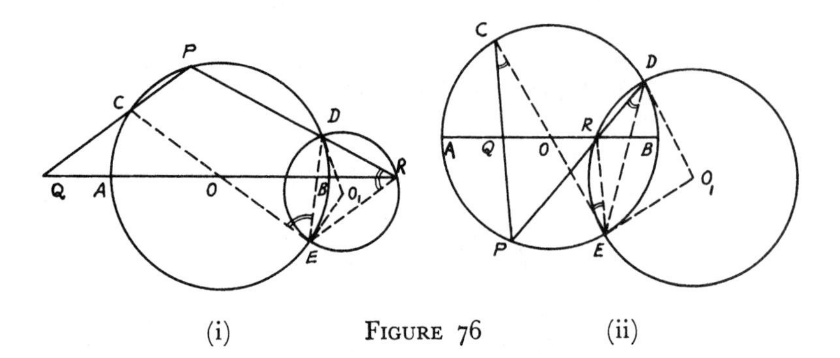

(i) FIGURE 76 (ii)

$\angle OED = \angle DRE$ and in (ii) $\angle OER = \angle PDE$. $\therefore$ OE touches $\odot$ circumscribed on $\triangle RDE$ with center O_1 [Th. 3.72(ii)]. Hence $O_1E \perp OE$ and $O_1E = O_1D$. Therefore,

SYNTHESIS: Join CO and produce it to E. Draw $EO_1 \perp EOC$ and make an $\angle EDO_1 = DEO_1$. Then O_1 is the center of $\odot$ touching OE in E. Draw $\odot$ with center O_1 and radius O_1E to cut AB or produced in R. Join DR and produce it to meet $\odot O$ in the required point P. Join CP to cut AB in Q. Now, since OE touches $\odot O_1$, $\therefore$ in (ia) $\angle OED = \angle DRE$. $\therefore$ $\angle OER = \angle PDE = \angle QCO$. $\therefore$ $ER \parallel PQ$ and in (ib) $\angle OER = \angle EDR$. $\because$ $\angle EDP = \angle ECP$, $\therefore$ $\angle OER = \angle ECP$. $\therefore$ $ER \parallel PQC$. Since COE is a diameter in $\odot O$, $\therefore$ $OQ = OR$.

ANALYSIS: (ii) Let P be the required point. Join PC, PD cutting AB in Q, R, where QR is the given length m. Draw CS equal and $\parallel QR$ (*Fig.* 77). $\therefore$ $CQRS$ is a $\square$. $\therefore$ $SR \parallel CQP$. $\therefore$ $\angle DRS = \angle DPC$,

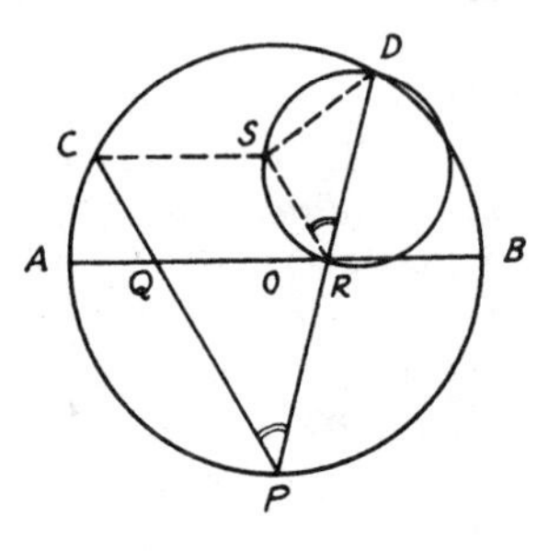

FIGURE 77

since D, S are two fixed points (because C and length CS are given).

SYNTHESIS: From C draw $CS \parallel AB$ and equal to the given length m on AB. On SD, draw a $\odot$ subtending $\angle SRD = \angle DPC$ to cut AB in R. Join DR and produce it to meet $\odot O$ in P. Therefore, $SR \parallel CP$ and $CQRS$ is a $\square$. $\therefore$ $QR = CS =$ given length m.

3.10. *If the vertex angle A of a triangle ABC is $60°$, prove that if O is the circumcenter, D is the orthocenter, E and F are the centers of the inscribed and escribed circles touching the base BC of the triangle, then $OEDF$ is cyclic. Show also that O, D are equidistant from E, F.*

CONSTRUCTION: Let AEF cut $\odot O$ in P. Join OP cutting BC in Q. Join also AO, OB, BE, BP, PD, BF (*Fig.* 78).

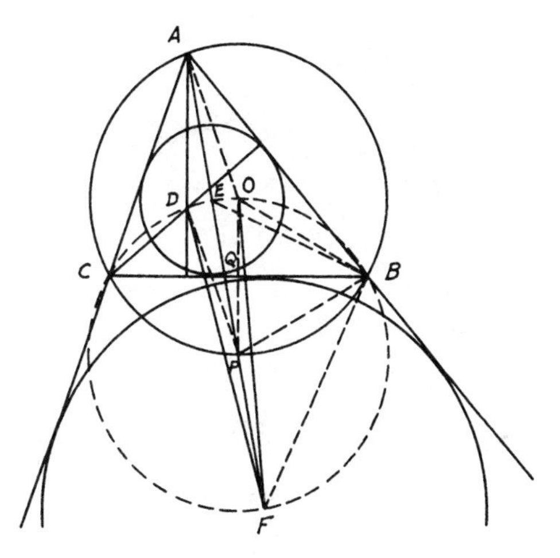

FIGURE 78

Proof: $\angle BOP = 2\angle BAP$ (Th. 3.65). $\because$ $AEPF$ bisects $\angle A$, which $= 60°$, $\therefore$ $\angle BOP = 60°$. Since $OB = OP$, $\therefore$ $\triangle OBP$ is equilateral.

Hence $PB = PO$. $\because$ $\angle PEB = \angle EAB + \angle EBA = \angle EAC + \angle EBC = \angle PBC + \angle EBC = \angle EBP$, $\therefore$ $PB = PE = PO$. $\because$ BE, BF are the internal and external bisectors of $\angle B$, they are $\perp$ to one another. Hence EBF is a right-angled triangle at B and $PB = PE$. $\therefore$ $PB = PE = PF$. But P is the mid-point of the arc BC. $\therefore$ $OP \perp BC$. $\angle PBC = PAC = 30°$. Since $\angle PBO = 60°$, $\therefore$ $\angle PBC = \frac{1}{2}\angle PBO$. Therefore, $PQ = QO$. But $AD = 2\,QO$ (Problem 1.32). $\therefore$ AD $=$ and $\parallel OP$. $\because$ $AO = OP$, $\therefore$ fig. $AOPD$ is a rhombus $\therefore$ $PO = PD$.

Since $PO = PE = PB = PF$, $\therefore$ P is the center of $\odot$ passing through $OEDF$, and also through B, C, with EF as diameter. Again, since $AOPD$ is a rhombus, ∴ diagonal PA bisects $\angle OPD$. ∵ $PO = PD$, ∴ △s OPE, DPE are congruent. ∴ $EO = ED$. Similarly, $FO = FD$.

3.11. *ABCD is a cyclic quadrilateral inscribed in a circle. E, F, G, H are the middle points of the arcs AB, BC, CD, DA. P, Q, R, S are the centers of the circles inscribed in the triangles ABC, BCD, DCA, DBA respectively. Show that the figure PQRS is a rectangle whose sides are parallel to EG, FH.*

CONSTRUCTION: Join AS, BP, AP, PF, BS, SH, CP, PE, DS, SE (*Fig.* 79).

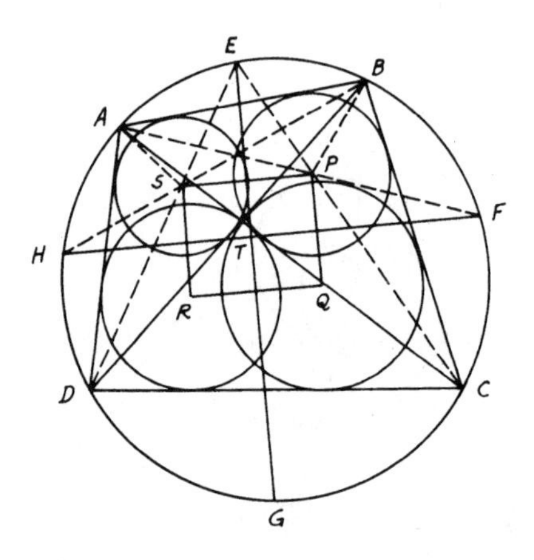

FIGURE 79

Proof: Let EG, FH meet in T. $\angle FTG$ is measured by half the sum of the arcs FG, $EH = \frac{1}{2}$ the sum of the arcs BC, CD, DA, AB = right angle. ∴ $EG \perp FH$. Again, APF, BSH, CPE, DSE are straight lines since they bisect ∠s BAC, ABD, BCA, BDA. Also, PB, SA bisect ∠s ABC, BAD. Since $\angle CBD = \angle CAD$ [Th. 3.66(i)], ∴ $\angle CBA - \triangle DBA = \angle BAD - \angle BAC$. ∴ $\frac{1}{2}(\angle CBA - \angle DBA) = \frac{1}{2}(\angle BAD - \angle BAC)$ or $\angle PBA - \angle SBA = \angle BAS - \angle BAP$. Hence $\angle PBS = \angle PAS$. ∴ fig. $BPSA$ is cyclic (Th. 3.73). ∴ $\angle BAP = \angle BSP$. ∵ $\angle BAP$ or $BAF = \angle BHF$, ∴ $\angle BSP = \angle BHF$. ∴ $PS \parallel FH$. Similarly, $QR \parallel FH$ and PQ, $RS \parallel GE$. ∴ fig. $PQRS$ is a rectangle whose sides $\parallel$ EG, FH, which are perpendicular to one another.

3.12. *Construct a triangle ABC having given the vertices of the three equilateral triangles described on its sides outside the triangle.*

CONSTRUCTION: Let D, E, F be the given vertices. On DE, EF, FD describe equilateral $\triangle$s DEG, EFH, FDJ. Then FG, DH, EJ will intersect in one point L and subtend $120°$ $\angle$s to each other. Bisect FE in K. Draw KM to make a $60°$ angle with FD cutting ELJ in A. Join AF and make $\angle AFC = 60°$ meeting DLH in C. Similarly, join CE and make $\angle ECB = 60°$ meeting GLF in B. Therefore, ABC is the required $\triangle$. Draw $FN \parallel ELJ$ to meet MAK produced in N (*Fig.* 80).

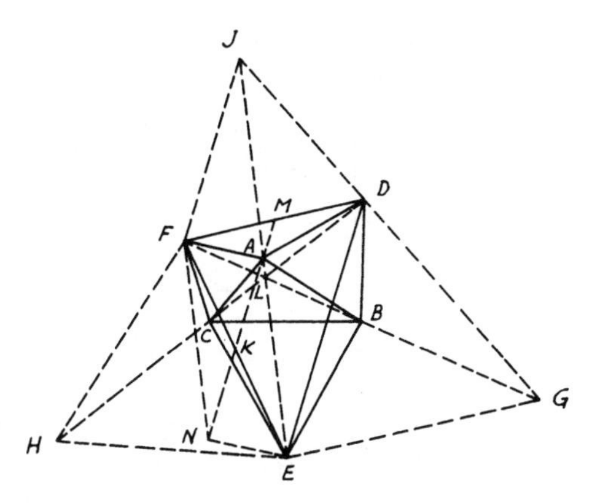

FIGURE 80

Proof: $\because$ $\angle FDJ = \angle EDG = 60°$, adding $\angle FDE$ gives $\angle EDJ = \angle GDF$. $\therefore$ $\triangle$s EDJ, GDF are congruent. $\therefore$ $\angle DJE = \angle DFG$. $\therefore$ quadrilateral $DJFL$ is cyclic. $\therefore$ $\angle DLF = 120°$. Hence EJ, FG and similarly DH meet in one point at $120°$ $\angle$s. But $\angle KAF = \angle KMF + \angle MFA = \angle MFC$ ($\because$ $\angle KMF = \angle AFC = 60°$). Since, also, quadrilateral $ALCF$ is cyclic, $\therefore$ $\angle LCF + \angle LAF = 2$ right angles. But $\angle LAF + \angle AFN = 2$ right angles ($FN \parallel AE$). $\therefore$ $\angle LCF = \angle AFN$. $\because \angle CAF = \angle CLF = 60°$, $\therefore$ $\triangle AFC$ is equilateral. $\therefore$ $\triangle$s DCF, NFA are congruent. $\therefore$ $DC = FN = AE$ ($AFNE$ is a ▱). Again, quadrilateral $LBEC$ is cyclic. $\therefore$ $\angle LCB = \angle LEB$ and $\angle CLE = \angle CBE = 60°$. $\therefore$ $\triangle CBE$ is equilateral. Now, $\triangle$s ABE, DBC are congruent ($\because$ $DC = AE$). $\therefore$ $\angle ABE = \angle DBC$. $\therefore$ $\angle CBE = \angle DBA = 60°$. $\because$ Quadrilateral $DBLA$ is cyclic, $\therefore$ $\angle BDA + \angle BLA = 2$ right angles since $\angle BLA = 120°$. $\therefore$ $\angle BDA = 60°$ also. Hence $\triangle DBA$ is equilateral.

3.13. *Construct a triangle having given its nine-point circle, orthocenter, and the difference between the base angles. How many solutions are possible?*

ANALYSIS: Assume ABC is the required $\triangle$ and that O_1 is the center of its nine-point circle, O being the orthocenter. Let G be the center of the circumcircle. Then O_1 is the middle point of OG [Th. 3.77(i)]. F, H, J are the mid-points of CB, BA, CA and GF, GH, GJ are $\perp$ to these sides of the triangle (*Fig.* 81). Since $ABDE$ is cyclic, $\therefore$ $\angle DAE$

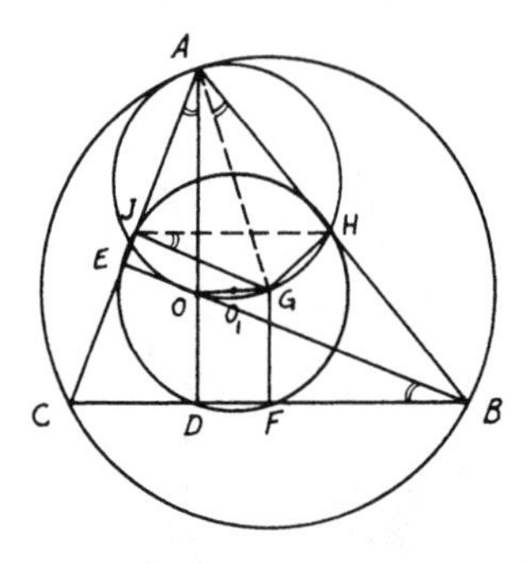

FIGURE 81

$= \angle DBE$. Also, $AHGJ$ is cyclic. $\therefore$ $\angle GAH = \angle GJH$. Since H, J are the mid-points of AB, AC, $\therefore$ $HJ \parallel BC$. But $GJ \parallel BE$, $\therefore$ $\angle GJH = \angle DBE$ (between $\parallel$s). $\therefore$ $\angle DAE = \angle GAH$. Since $\angle C - \angle B = \angle BAD - \angle CAD =$ given angle, $\therefore$ $\angle GAO =$ given angle. Also the radius of the nine-point $\odot O_1 =$ half that of circumcircle O (Th. 3.76, Cor.). Therefore,

SYNTHESIS: Join OO_1 and produce it to G, so that $OO_1 = O_1G$. On OG as a chord, draw a $\odot$ to subtend at the arc OG an $\angle GAO =$ given difference between base angles of $\triangle$. With G as center and radius $=$ given diameter of $\odot O_1$, draw a $\odot$ cutting $\odot GAO$ in A. Join AO and produce it to meet the nine-point $\odot$ in D. Draw BDC $\perp$ AD to meet $\odot G$ in B, C; then ABC is the required $\triangle$. Proof is obvious. If $\odot G$ cuts $\odot$ on GAO in two points the problem will have two solutions. In case $\odot G$ touches $\odot$ on GAO, the problem will then have only one solution. What would be the condition, if any, for which there will be no solution?

3.14. *O is the orthocenter of a triangle ABC, O_1 is the center of its circumscribed circle, and D, E, F are the centers of the circles drawn to circumscribe triangles BOC, COA, AOB. Show that (i) O_1 is the orthocenter of triangle DEF; (ii) O is the circumcenter of triangle DEF; (iii) A, B, C are the circumcenters of triangles EO_1F, FO_1D, DO_1E; (iv) all mentioned eight triangles have the same nine-point circle.*

CONSTRUCTION: Let AP, BQ, CR be the perpendiculars from A, B, C to corresponding sides of $\triangle ABC$. Let also FE, FD, DE cut AO, BO, CO at right angles in X, Y, Z. Join DO_1, EO_1, FO_1 to meet the opposite sides of $\triangle DEF$ in L, M, N respectively; then join BF, BO, and BD (*Fig.* 82).

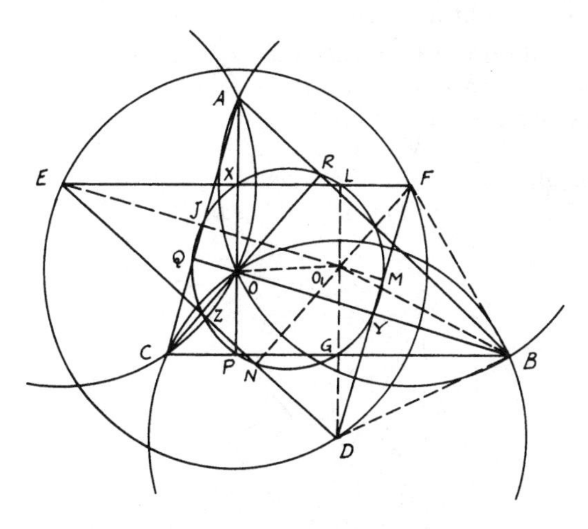

FIGURE 82

Proof: (i) $\because$ D, E are two circumcenters of $\triangle$s BOC, COA, $\therefore$ DE bisects the common chord CO and $\perp$ to it. $\because$ $AB \perp COR$, $\therefore$ $AB \parallel DE$. Similarly, $EF \parallel BC$ and $FD \parallel AC$. $\because$ AB is the common chord in the $\odot$s FBA, ABC, $\therefore$ likewise, $FO_1 \perp AB$. $\therefore$ $FO_1 \perp DE$. Similarly, DO_1, EO_1 when produced are $\perp EF$, FD. Hence O_1 is the orthocenter of $\triangle DEF$.

(ii) $\because$ $\angle BOC$ is supplementary to $\angle BAC$ and BC is a common base, $\therefore$ $\angle BAC =$ any $\angle$ on $\odot BOC$ on other side of BC. $\therefore$ $\angle BDG = \angle BO_1G$. Hence $\odot$s BOC, ABC are equal. Similarly, $\odot$s COA, BOA are each $= \odot ABC$. $\therefore$ $BD = BO_1 = BF = CE$. $\because$ $DF \perp BO$, $\therefore$ $OD = OF$ (from congruence of $\triangle$s BDO, BFO). Similarly, $OD = OE$. Hence O is the circumcenter of $\triangle DEF$.

(iii) Since $\odot$s ABC, AOB, BOC, COA, DEF are equal, $\therefore$ their radii are equal. $\therefore$ $BD = BO_1 = BF$. $\therefore$ B is the circumcenter of $\triangle FO_1D$. Similarly, A, C are circumcenters of $\triangle$s EO_1F, DO_1E, and these eight circles are equal.

(iv) $\because$ Nine-point $\odot$ of $\triangle ABC$ passes through mid-points of AO, BO, CO, i.e., X, Y, Z, which are also the mid-points of EF, FD, DE,

$\therefore$ ⊙PQR, the nine-point ⊙ of △ABC, is also the nine-point ⊙ of △DEF. Suppose DO_1 cuts BC in G. $\therefore$ G is the mid-point of BC. Since ⊙PQR passes through Y, G, Z, which are the mid-points of the sides of △BOC, $\therefore$ ⊙PQR is the nine-point ⊙ of △BOC and similarly of △s COA, AOB. Also, ⊙PQR is the nine-point ⊙ of △s EO_1F, FO_1D, DO_1E since it passes through the mid-points of their sides. Therefore, these eight equal circles have the same nine-point ⊙PQR.

3.15. *Describe a circle which will pass through a given point P and touch a given straight line MN and also a given circle ABC.*

Analysis: Suppose $PP'D$ is the required ⊙. Through centers O, O' draw $AOBE$, $O'D \perp MN$. $\because$ $O'D$ is $\parallel AB$, $\therefore$ straight line joining AD passes through F the point of contact of ⊙s. Join BF, AP. Let AP meet ⊙PFD again in P' (*Fig.* 83). $\because$ $\angle$s BFD, BED are right, hence

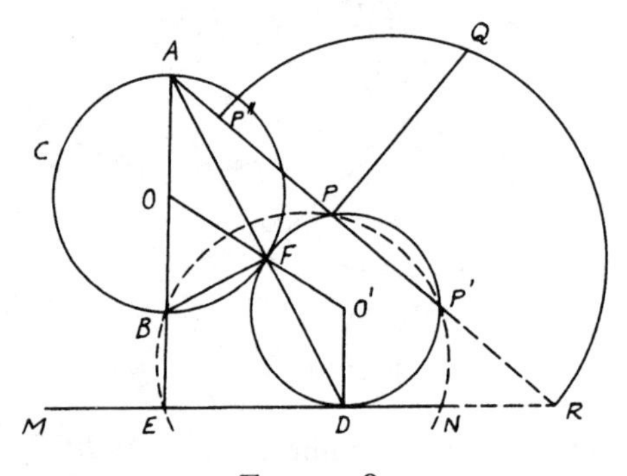

Figure 83

B, E, D, F are concyclic. $\therefore$ $AB \cdot AE = AF \cdot AD = AP \cdot AP'$. $\therefore$ P' is a known point. Therefore, the problem is reduced to a simpler one—to describe a ⊙ passing through P and P' and touching MN. Hence,

Synthesis: Find O the center of given ⊙ABC. Draw $AOBE \perp MN$. In AP, produced if necessary, find P' such that the rectangle $AP \cdot AP'$ = rectangle $AB \cdot AE$. This is done by describing ⊙EBP and producing AP to meet its circumference in P'. Produce APP' to meet MN in R. Take on AP the distance $PP'' = RP'$. On RP'' as diameter describe a semi-circle, then draw $PQ \perp APP'$. Take $RD = PQ$; then ⊙$PP'D$ touches MN. Since $PQ^2 = PP'' \cdot PR = RP' \cdot RP = RD^2$, $\therefore$ MN touches ⊙$PP'D$. Join AD and let AD cut the ⊙$PP'D$ in F. Now, $AD \cdot AF = AP \cdot AP' = AB \cdot AE$. $\therefore$ $BEDF$ is cyclic (Th. 3.79). $\therefore$ $\angle AFB = \angle BED$ = right angle. $\therefore$ F is on ⊙ABC. $\because$ $\angle O'FD = \angle O'DF$ = alternate $\angle OAF = \angle OFA$, $\therefore$ OFO' is a straight line. $\therefore$ ⊙$PP'D$ touches ⊙ABC at F.

Note: 1. If line AP intersects MN, then two circles can be described through P, P' touching MN. Also, if the points A, B are interchanged, P' will occupy a different position and two more circles will be obtained. In this case the contact will be internal. Hence the problem has four solutions.

2. This is one of a group of related problems known collectively as the *problems of Apollonius*. The problem is to construct a ⊙ either passing through one or more given points, tangent to one or more given lines or tangent to one or more given ⊙s in various combinations. This problem would be PLC, meaning that the required ⊙ must pass through a given point P, be tangent to a given line L, and be tangent to a given ⊙C. We have previously considered in different terminology the more elementary cases PPP and LLL.

3.16. *ABCD is a parallelogram. A straight line is drawn through A and meets CB, CD produced in E, F. Show that* $CB \cdot CE + CD \cdot CF = AC^2 + AE \cdot AF$.

CONSTRUCTION: Describe circumscribing ⊙CFE. Produce CA to meet ⊙ in G. Join GE, GF. Draw DP, BQ to make with AC ∠s CDP, CBQ = ∠s FGC, EGC respectively (*Fig.* 84).

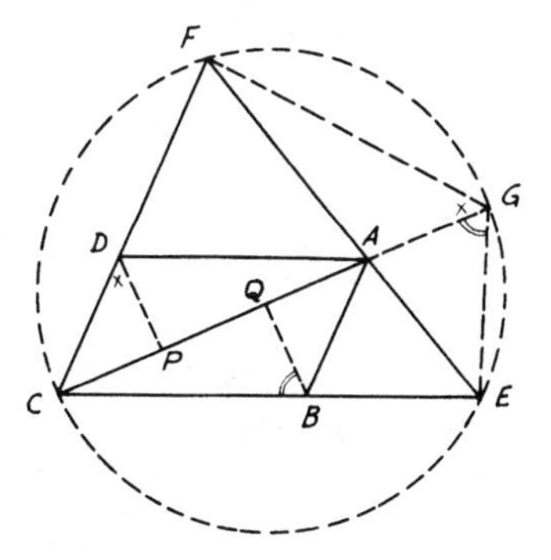

FIGURE 84

Proof: ∵ $\angle CBQ = \angle EGC$, ∴ quadrilateral $BECQ$ is cyclic. ∴ $\angle BEG + \angle BQG = 2$ right angles. But $\angle BEG = \angle FCG + \angle FGC = \angle FCG + \angle CDP$. ∴ Supplementary $\angle CPD$ = supplementary $\angle BQC$. ∴ BQ is ∥ DP. ∵ $ABCD$ is a ▱, ∴ $AP = CQ$. Since $BEGQ$ is cyclic, ∴ $CB \cdot CE = CQ \cdot CG$. Also, $DFGP$ is cyclic. ∴ $CD \cdot CF = CP \cdot CG$ [Th. 3.78(i)]. Adding yields $CB \cdot CE + CD \cdot CF = CG\,(CQ + CP) = CG \cdot AC = AC^2 + AC \cdot AG$. ∴ $CB \cdot CE + CD \cdot CF = AC^2 + AE \cdot AF$.

3.17. *ABC is any triangle. If AD, AE, AF are the median, bisector of the vertex angle, and altitude respectively and AG is the external bisector of the vertex angle, show that (i) $4\,DE \cdot DF = (AB - AC)^2$; (ii) $4\,DG \cdot DF = (AB + AC)^2$.*

CONSTRUCTION: Draw BR, CP $\perp$s AE, AG or produced respectively. BR, CP produced meet AC, BA in S, Q. Then join RD, RF, PD, PF. (*Fig.* 85).

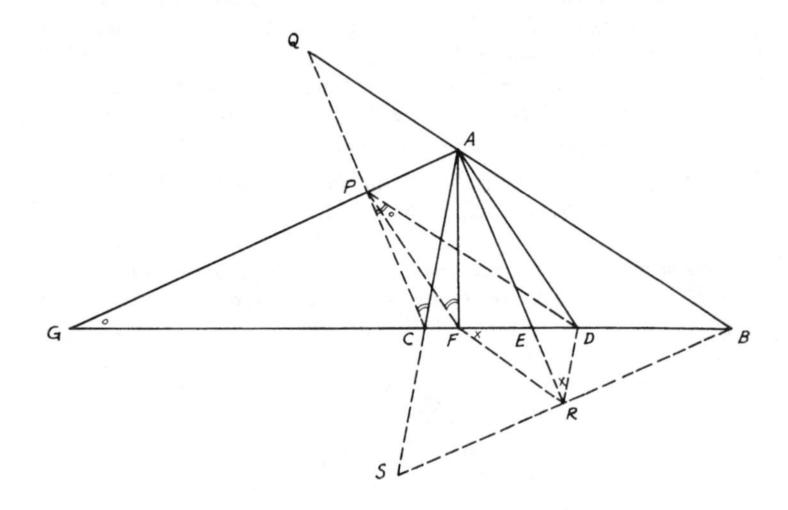

FIGURE 85

Proof: (i) From congruence of $\triangle$s ABR, ASR, $BR + RS$, and $AB = AS$. Hence DR is $\parallel$ and $= \frac{1}{2}\,CS$ or $DR = \frac{1}{2}\,(AB - AC)$. Since DR is $\parallel ACS$, $\therefore$ $\angle DRA = \angle CAE = \angle BAE$. $\because$ Quadrilateral $ABRF$ is cyclic, $\therefore$ $\angle BAE = \angle BFR$. $\therefore$ $\angle DRA = \angle BFR$. $\therefore$ DR touches $\odot ERF$ (Th. 3.72). $\therefore$ $DR^2 = DE \cdot DF$. $\therefore$ $4\,DE \cdot DF = (AB - AC)^2$.

(ii) Similarly, from congruence of $\triangle$s APC, APQ, $PC = PQ$, $AC = AQ$, and $\angle ACP = \angle AQP$. Hence DP is $\parallel$ and $= \frac{1}{2}\,BQ = \frac{1}{2}\,(AB + AC)$. $\therefore$ $\angle AQP = \angle CPD$. $\because$ Quadrilateral $AFCP$ is cyclic, $\therefore$ $\angle ACP = \angle AFP$. $\therefore$ $\angle CPD = \angle AFP$. Adding a right angle gives $\angle GPD = \angle DFP = \angle FGP + \angle FPG$, eliminating common $\angle FPG$. $\therefore$ $\angle DPF = \angle FGP$. $\therefore$ DP touches $\odot FGP$. $\therefore$ $4\,DP^2 = 4\,DG \cdot DF = (AB + AC)^2$.

3.18. *AB is a chord at right angles to a diameter CD of a given circle, and AB is nearer C than D. Draw through C a chord CQ cutting AB in P so that PQ is a given length.*

ANALYSIS: Suppose CPQ is the required chord, PQ being a given length. Join DQ, and bisect PQ in R (*Fig.* 86). Now, $\because$ $\angle CQD$

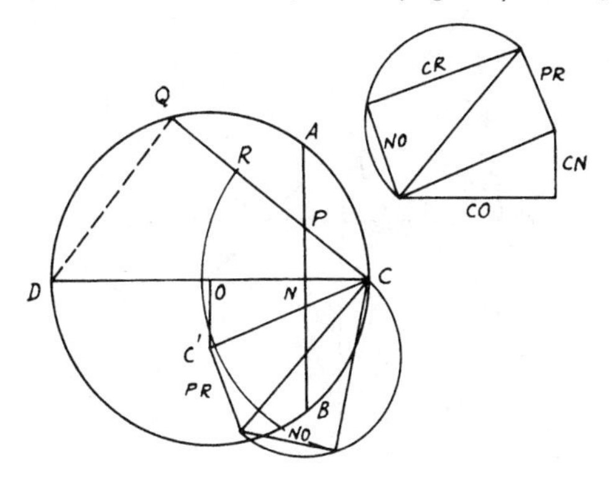

FIGURE 86

= right angle, $\therefore$ quadrilateral $PNDQ$ is cyclic. $\therefore$ $CP \cdot CQ = CN \cdot CD$. Analyzing, $(CR - PR)(CR + PR) = CN^2 + CN \cdot ND$. $\therefore$ $CR^2 - PR^2 = CN^2 + CO^2 - NO^2$. Hence $CR^2 = CN^2 + CO^2 + PR^2 - NO^2$, which is known.

Also, $CP = (CR - PR)$ is known. Therefore,

SYNTHESIS: Draw from C the line

$$CP = \sqrt{CN^2 + CO^2 + PR^2 - NO^2} - PR.$$

This is done geometrically outside the figure and can be easily followed. Then produce CP until it cuts the circumference in Q. Hence CPQ is the required chord. The construction of CR can also be shown on the figure.

3.19. *D, E are the points of contact of the inscribed and escribed circles of any triangle ABC with the side AB. Show that $AD \cdot DB = AE \cdot EB =$ the rectangle of the radii of the two circles.*

CONSTRUCTION: Let O, O_1 be the centers of the inscribed and escribed ⊙s touching side AB of the $\triangle ABC$. Join AO, OB, AO_1, O_1B and draw a ⊙ through A, O, B, O_1. Produce OD to meet this ⊙ in F and join FO_1 (*Fig.* 87).

Proof: $\because$ AO, AO_1 are the bisectors of $\angle A$ and BO, BO_1 are the bisectors of $\angle B$, $\therefore$ $\angle OAO_1 = \angle OBO_1 =$ right angle. Hence quadrilateral $AOBO_1$ is cyclic, and OO_1 is a diameter of this ⊙. $\therefore$ $\angle OFO_1 =$ right angle and $EDFO_1$ is ▭. $\therefore$ $EO_1 = DF$. $\therefore$ $AD \cdot DB = OD \cdot DF = OD \cdot EO_1$. Similarly, by producing O_1E to meet ⊙O_1AO at F_1 etc., we have $AE \cdot EB = OD \cdot EO_1$.

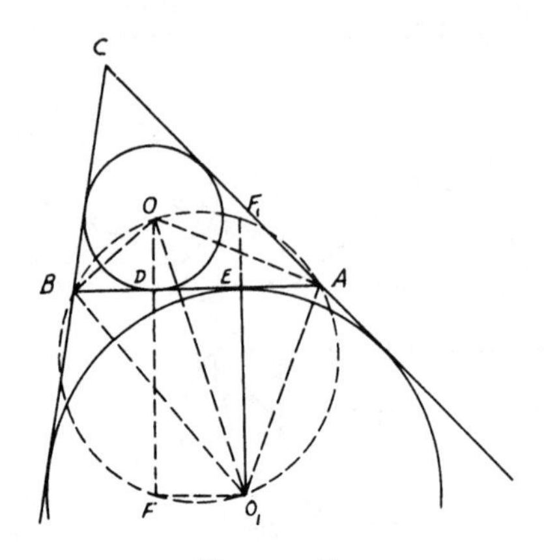

FIGURE 87

3.20. *AB is a diameter of a circle and CDC′ a chord perpendicular to it. A circle is inscribed in the figure bounded by AD, DC and the arc AC and it touches AB at E. Show that BE = BC, and hence give a construction for inscribing this circle.*

CONSTRUCTION: Draw diameter $POQ \perp AB$. Let M be the center of inscribed ⊙ touching ⊙O in K. Draw $NME \perp AB$ and $QR \perp NME$ produced. Let also ⊙M touch CDC' in F. Join MF (*Fig.* 88). The analysis is carried out for the smaller arc, but could just as well be extended to the larger arc BC.

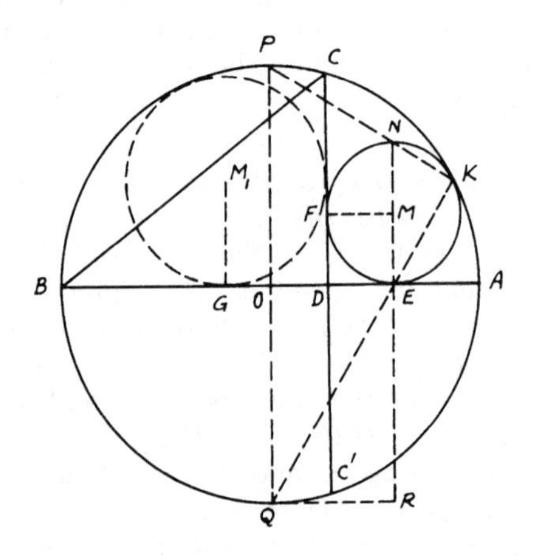

FIGURE 88

Proof: $\because$ POQ, NME are diameters in the $\odot$s O, M which touch one another at K, $\therefore$ PNK, QEK are straight lines ($\because$ $\odot$s M, O touch at K). $\angle NKE$ = right angle = $\angle QRE$. $\therefore$ Quadrilateral $KNRQ$ is cyclic. $\therefore NE \cdot ER = KE \cdot EQ$. Hence $NE \cdot OQ = AE \cdot EB$. $\because$ $MEDF$ is a square, $\therefore$ $NE = 2\,DE$. $\therefore$ $DE \cdot PQ = AE \cdot EB$. $\therefore$ $DE \cdot AB = AE \cdot EB$. Hence $AE \cdot DB = AE \cdot EB - AE \cdot ED = DE \cdot AB - AE \cdot DE$. $\therefore$ $AE \cdot DB = DE \cdot EB = DE^2 + DE \cdot DB$. Adding $(DE \cdot DB + DB^2)$ to both sides gives $AB \cdot DB = BE^2$. Since $BC^2 = AB \cdot DB$, $\therefore$ $BE = BC$. To construct the inscribed $\odot M$, take on AB the distance $BE = BC$. Draw $EM \perp AB$ and $= ED$; then with M as center and ME as radius draw the inscribed $\odot$ required. An illustration of the inscribed $\odot M_1$ on opposite side of CDC' is also given.

3.21. *Construct a triangle having given the base, the median which bisects the base, and the difference of the base angles.*

ANALYSIS: Suppose ABC is the required $\triangle$. AD is the given median bisecting the given base BC. Produce AD to meet the circumscribing $\odot O$ in E. Draw $AF \parallel BC$ and join DO and produce it to meet AF in G. Join OA, OF, EF, EB, EC (*Fig.* 89). Now, since D is the mid-point

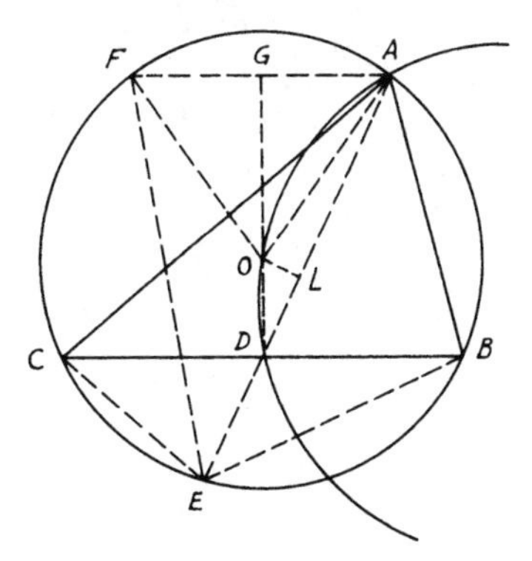

FIGURE 89

of BC and AF is $\parallel BC$, $\therefore$ DOG is $\perp AF$. $\therefore \angle AOG = \angle GOF$. But $\angle AOF = 2\,\angle AEF$ (Th. 3.65). $\therefore$ $\angle AOG = \angle AEF$, since $\angle AEF = \angle AEC - \angle FEC = \angle B - \angle AEB = \angle B - \angle C$ = given. Hence $\angle AOG$ = given $(\angle B - \angle C)$. $\therefore$ $\angle AOD$ = 2 right angles $- (\angle B - \angle C)$. Also, $4\,AD \cdot DE = 4\,BD^2 = BC^2$. $\therefore$ $DE = BC^2/4\,AD$ = given. Hence,

SYNTHESIS: Draw the median AD to the given length. Produce it to E such that $DE = BC^2/4\,AD$. On AD as a chord, describe an arc of a ⊙ subtending an angle = 2 right angles − ($\angle B - \angle C$). Bisect AE in L and draw $LO \perp AD$ to meet this arc in O. Join OD and draw $BDC \perp$ to it, cutting the ⊙ with center O and radius OA in B, C. ABC is then the required △.

3.22. *AB, AC are drawn tangents to a circle with center O from any point A. AB, AC are bisected in D, E. If any point F is taken on DE or produced, show that AF is equal to the tangent from F to the circle. Prove also that if a straight line is drawn from F to cut the circle in G, H, then AF will be tangent to the circle described on the triangle AGH.*

CONSTRUCTION: Let FK be the tangent from F to the circle. Join OB, OF, OD, OA, OK, CB (*Fig.* 90).

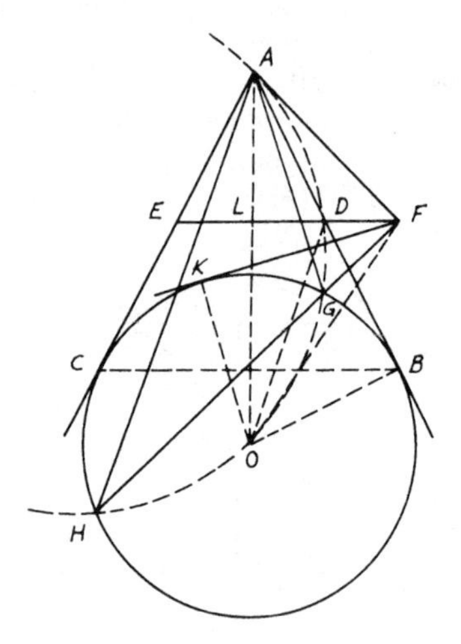

FIGURE 90

Proof: Let OA cut ED in L. Since $AB = AC$, it is easily shown that OA is the $\perp$ bisector of BC. $\because$ D, E are the mid-points of AB, AC, $\therefore$ DE is $\parallel BC$ and hence $\perp AO$, since $\angle OBA$ = right angle (Th. 3.68). Therefore, $DO^2 = DB^2 + BO^2$. But $DO^2 = DL^2 + LO^2$. Also, $DB^2 = AD^2 = AL^2 + DL^2$. $\therefore$ $LO^2 = AL^2 + OK^2$ ($BO = OK$) or $LO^2 - AL^2 = OK^2$. Hence $FO^2 - AF^2 = OK^2$. $\therefore$ $FO^2 - OK^2 = AF^2$. Since $FK^2 = FO^2 - OK^2$ (in right-angled triangle FKO), $\therefore$ $AF = FK$. $\because$ $FK^2 = FG \cdot FH$ [Th. 3.78(ii)], $\therefore$ $AF^2 = FG \cdot FH$. Therefore, AF is tangent to the circle described on $\triangle AGH$ (Th. 3.80).

3.23. *$ABCD$ is a quadrilateral drawn inside a circle. If BA, CD are* *produced to meet in E and AD, BC in F, prove that the circle with EF as diameter cuts the first circle orthogonally.*

CONSTRUCTION: Let O be the center of the $\odot ABCD$ and M be the center of the $\odot$ with diameter EF which cuts the first one in N. Make AG meet EF in G, so that $\angle AGE = \angle ABC$. Join OB, ON, OE, OM, OF, MN (*Fig.* 91).

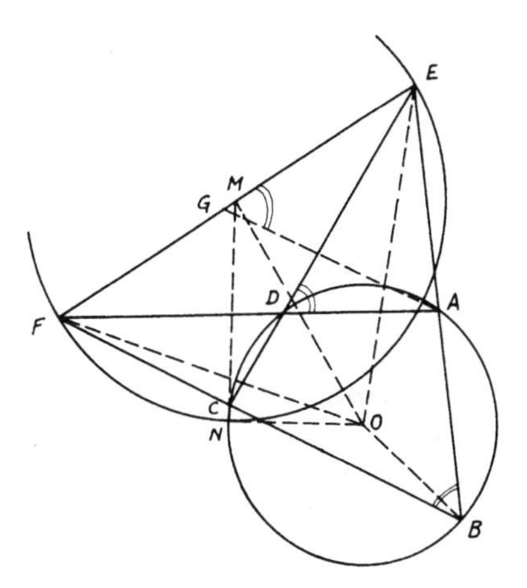

FIGURE 91

Proof: Quadrilateral $ABFG$ is cyclic (Th. 3.75). $\therefore$ $EA \cdot EB = EG \cdot EF$ [Th. 3.78(i)]. Since $\angle ABC = \angle ADE$, $\therefore$ $\angle AGE = \angle ADE$. Hence quadrilateral $ADGE$ is cyclic also. $\therefore$ $FD \cdot FA = FG \cdot FE$. Adding gives $EA \cdot EB + FD \cdot FA = EG \cdot EF + FG \cdot EF = EF^2$. $\because$ $OE^2 = \text{rad.}^2 \odot O + EA \cdot EB$ and $OF^2 = \text{rad.}^2 \odot O + FD \cdot FA$, $\therefore$ $OE^2 + OF^2 = 2\,OB^2 + EF^2 = 2\,OB^2 + 4\,ME^2 = 2\,OM^2 + 2\,ME^2$. Therefore, $OM^2 = OB^2 + ME^2 = ON^2 + MN^2$. $\therefore$ $\angle ONM$ is right (Th. 2.52). $\because$ ON, MN are radii of both $\odot$s, $\therefore$ each is tangent to the other $\odot$.

Hence $\odot$ on EF as diameter cuts $\odot ABCD$ orthogonally.

3.24. *From the middle point C of an arc AB of a circle, a diameter CD is drawn and also a chord CE which meets the straight line AB in F. If a circle, drawn with center C to bisect FE, meets BD in G, prove that* $EF = 2\,BG$.

CONSTRUCTION: Suppose $\odot$ with center C bisects FE in K. Join BE, BC, GC (*Fig.* 92).

Proof: Since CD is a diameter in the $\odot CBD$, $\therefore$ $\angle DBC$ = right angle. Hence in $\triangle GBC$, $CG^2 = BG^2 + BC^2$. But $CG = CK$ = radii in $\odot C$. $\therefore$ $CK^2 = BG^2 + BC^2$ (1). $\because$ $CK^2 = CE^2 + EK^2 +$

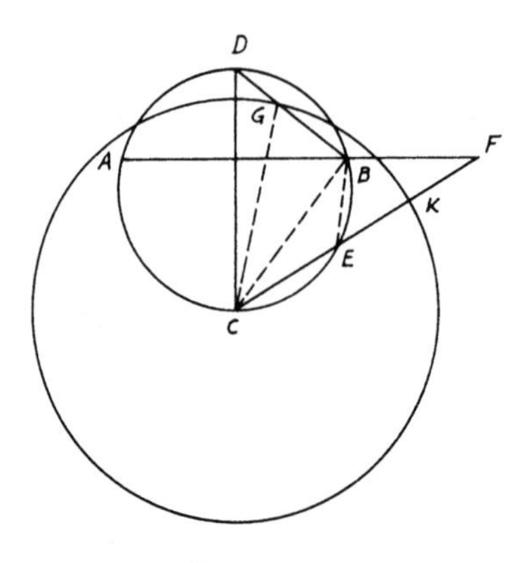

FIGURE 92

$2\,CE \cdot EK$ and $CE^2 + 2\,CE \cdot EK = CE\,(CE + 2\,EK) = CE \cdot CF$, $\therefore$ $CK^2 = EK^2 + CE \cdot CF$ (2).

Since $\angle BFE$ is measured by half the difference of the arcs AC, BE or BC, BE, i.e., CE, $\therefore$ $\angle BFE = \angle CBE$. $\therefore$ CB touches ⊙ on $\triangle BFE$ [Th. 3.72(ii)]. $\therefore$ $CB^2 = CE \cdot CF$.

Hence (2) becomes $CK^2 = EK^2 + CB^2$ (3). From (1) and (3), $\therefore$ $BG = EK$ or $EF = 2\,BG$.

3.25. *ABC is a triangle drawn in a circle, having AC greater than AB. DE is a diameter in the circle drawn at a right angle to the base BC, so that A, D are at the same side of BC. From A a perpendicular AF is drawn to DE. If G is the point of intersection of DE and BC, show that (i)* $4\,DG \cdot EF = (AB + AC)^2$*; (ii)* $4\,DF \cdot GE = (AC - AB)^2$.

CONSTRUCTION: Draw EP, EQ ⊥s AC, AB or produced is necessary. Join AE, BE, CE, FP, PG, CD (*Fig.* 93).

Proof: (i) $\because$ DE is a diameter ⊥ BC, $\therefore$ D, E are the mid-points of the larger and smaller arcs BC. $\therefore$ AE bisects $\angle A$. Hence △s AEP, AEQ are congruent $\therefore$ $AP = AQ$ and $EP = EQ$. Similarly, △s ECP, EBQ are congruent. $\therefore$ $CP = BQ$. $\therefore$ $2\,AP = (AB + AC)$ and $2\,CP = (AC - AB)$. $\because$ AF, EP are ⊥s DE, AC respectively, $\therefore$ quadrilateral $AFPE$ is cyclic. $\therefore$ $\angle EAP = \angle EFP$. But $\angle EAP$ or $EAC = \angle EDC = \angle ECG$. Since $ECPG$ is cyclic ⏢, $\therefore$ $\angle ECG = \angle EPG$. Hence $\angle EFP = \angle EPG$. $\therefore$ EP touches ⊙PGF. $\therefore$ $EP^2 = EG \cdot EF$. Since $AP^2 = AE^2 - EP^2 = ED \cdot EF - EG \cdot EF = DG \cdot EF$, $\therefore$ $4\,DG \cdot EF = (AB + AC)^2$.

(ii) Similarly, $CP^2 = CE^2 - EP^2 = EG \cdot ED - EG \cdot EF = EG \cdot DF$. $\therefore$ $4\,DF \cdot GE = (AC - AB)^2$.

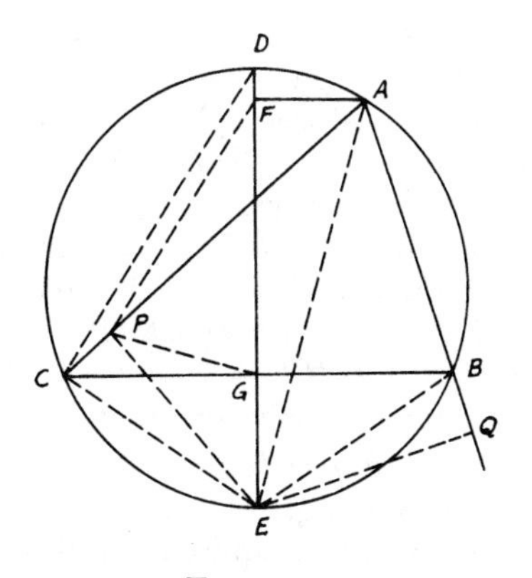

FIGURE 93

3.26. *Show that in any triangle, the perpendiculars from the vertices to the opposite sides of its pedal triangle are concurrent and that the area of the triangle is equal to the product of the radius of the circumscribed circle and half the perimeter of the pedal triangle.*

CONSTRUCTION: *ABC* is a triangle drawn inside a circle with center *O*. *AD*, *BE*, *CF* are the ⊥s to the opposite sides and *DEF* its pedal triangle. Join *AO*, *BO*, *CO* cutting *EF*, *FD*, *DE* in *G*, *K*, *L* respectively. Produce *AO* to meet ⊙ in *N*, then join *CN*, *OD*, *OE*, *OF* (*Fig.* 94).

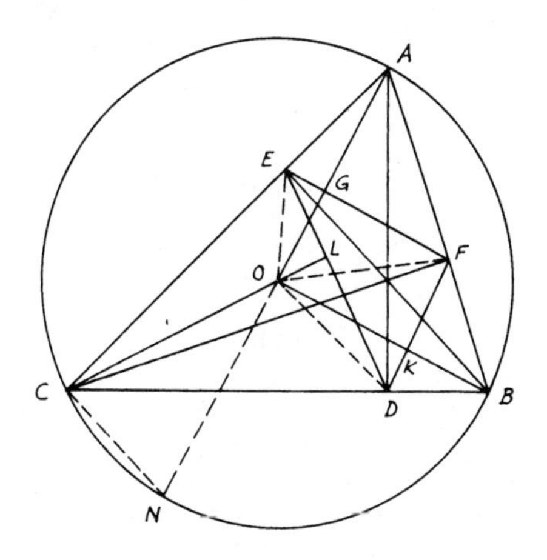

FIGURE 94

Proof: △s *ACN*, *CFB* have ∠*C* = ∠*F* = right and ∠*ACN* = ∠*CBF* or *CBA*. ∴ ∠*CAN* = ∠*BCF*. But, since *BFEC* is a cyclic ▱, ∴ ∠*BCF* = ∠*BEF* = ∠*CAN* or *EAG*. ∵ ∠*BEG* + ∠*GEA* = right, ∴ ∠*GEA* + ∠*EAG* = right. ∴ *AG* ⊥ *EF* and passes through *O*. Similarly, *BK*, *CL* are ⊥s *DF*, *DE* and also pass through *O*. Hence *AG*, *BK*, *CL* are concurrent at the center *O*. In the quadrilateral *AFOE*, the diagonals *AO*, *FE* are at right angles. ∴ quadrilateral *AFOE* = $\frac{1}{2}$ *AO*·*FE*. Similarly, quadrilateral *FBDO* = $\frac{1}{2}$ *BO*·*FD* and quadrilateral *EODC* = $\frac{1}{2}$ *CO*·*ED*. Hence △*ABC* = quadrilaterals *AFOE* + *FBDO* + *EODC* = $\frac{1}{2}$ radius *AO*(*FE* + *FD* + *ED*) = radius ⊙*O* × half the perimeter of pedal △*DEF*.

3.27. *Construct a right-angled triangle having given the hypotenuse and the bisector of either one of the base angles.*

Analysis: Suppose *ABC* is the required △ right-angled at *A*. Draw ⊙ with center *O* to circumscribe △*ABC* and bisect ∠*C* by *CD*, which when produced meets ⊙*O* in *E* and the tangent at *B* in *F* (*Fig.* 95). ∴ ∠*FBE* = ∠*BCE* = ∠*ECA* = ∠*EBA*. ∴ ∠*BEC* = right

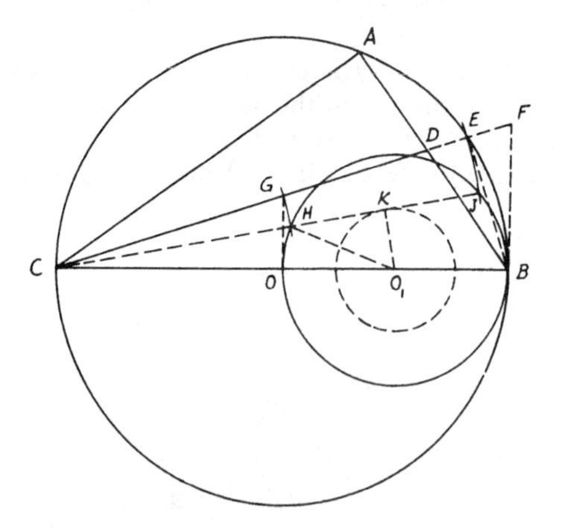

Figure 95

angle, i.e., *BE* ⊥ *DF*, ∴ △s *BED*, *BEF* are congruent. ∴ *DE* = *EF*. Draw *OG* ⊥ *CB* to meet *CD* in *G*. Since *OG* ∥ *BF* (both ⊥*CB*), ∴ *G* is the mid-point of *CF*. Hence *CD* = *CG* + *GD* = *GF* + *GD* = 2 *GE*, which is given. Therefore,

Synthesis: Bisect *BO* in O_1 and describe a circle center O_1 and radius equal to $\sqrt{OO_1^2 - (CD/4)^2}$, which is known, since *BC*, *CD* are given. From C draw a tangent CK to $\odot O_1$ to cut $\odot$ on OB as diameter in H, J. With C as center, and radii CH, CJ, draw the arcs HG, JE to meet $\perp$ from O to BC in G, and CG produced in E respectively (i.e., rotating HJ about C to positive GE). Therefore, $HK = CD/4$, i.e., $\frac{1}{4}$ of the given base angle bisector. $\because CH \cdot CJ = CG \cdot CE = CO \cdot CB$, $\therefore$ quadrilateral $EBOG$ is cyclic (Th. 3.79). $\therefore \angle BEG = \angle GOC =$ right. Hence E lies on the $\odot O$ with BC as diameter. Draw $\angle BCE = \angle ECA$. Therefore, ABC is the required $\triangle$.

3.28. *D is any point on the diameter AB of a circle, and DC is drawn perpendicular to AB to meet the circle in C. A semi-circle is described on BD as diameter on the same side as ABC. A circle with diameter ME is described to touch DC in E and the two circles in F, G. Show that* $AB \cdot ME = CD^2$ *and hence derive a method of describing the circle MEF.*

CONSTRUCTION: Let O, O_1 be the centers of $\odot$s with BD, ME as diameters. Draw their common tangent at G to meet DC in K. Join OO_1, OK, O_1K, FM, MB, FE, EA. Draw $AH \perp ME$ produced and produce CD to meet $\odot$ in L (*Fig.* 96).

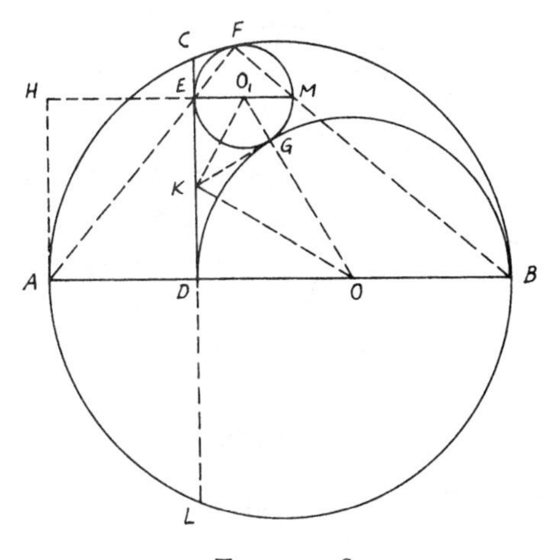

FIGURE 96

Proof: KG is $\perp$ line of centers OO_1 at G. Also, KO, KO_1 bisect $\angle$s DKG, EKG (since KE, KG, KD are tangents to $\odot$s O_1, O). $\therefore OK$ is $\perp KO_1$. $\therefore KG^2 = OG \cdot O_1G$. $\because EK = KG = KD$, $\therefore DE^2 = 4\,KG^2 = 4\,OG \cdot O_1G = DB \cdot ME$ (1).

Now, since $\odot$s ME, AB touch at F and ME, AB are diameters. $\therefore FMB$, FEA are straight lines. $\therefore CE \cdot EL = FE \cdot EA$. $\because \angle MFE$

$= \angle AHE =$ right, $\therefore$ quadrilateral $FMAH$ is cyclic. $\therefore$ $FE \cdot EA = ME \cdot EH$. But $ADEH$ is a rectangle. $\therefore$ $AD = HE$. $\therefore$ $CE \cdot EL = ME \cdot AD$. $\therefore$ $ME \cdot AD = CE^2 + 2\,CE \cdot ED$ (2). Adding (1) and (2) yields $ME \cdot DB + ME \cdot AD = DE^2 + CE^2 + 2\,CE \cdot DE$ or $AB \cdot ME = CD^2$. Therefore, $ME = CD^2/AB$ and $DE^2 = DB \cdot ME = (DB \cdot CD^2/AB)$. Accordingly, in order to describe $\odot O_1$, take a distance DE on DC equal to $\sqrt{(DB \cdot CD^2/AB)}$, which is known. Draw $EM \perp CD$ and equal to CD^2/AB, which is also known. Hence $\odot$ on EM as diameter is the required $\odot$.

3.29. *ABC, $A'B'C'$ are equilateral triangles inscribed in two concentric circles $ABCP$, $A'B'C'P'$. Prove that the sum of the squares on PA', PB', PC' is equal to the sum of the squares on $P'A$, $P'B$, $P'C$.*

CONSTRUCTION: Join $C'O$ and produce it to meet $A'B'$ in E; then join OB', PE, PO, $P'O$ (*Fig.* 97).

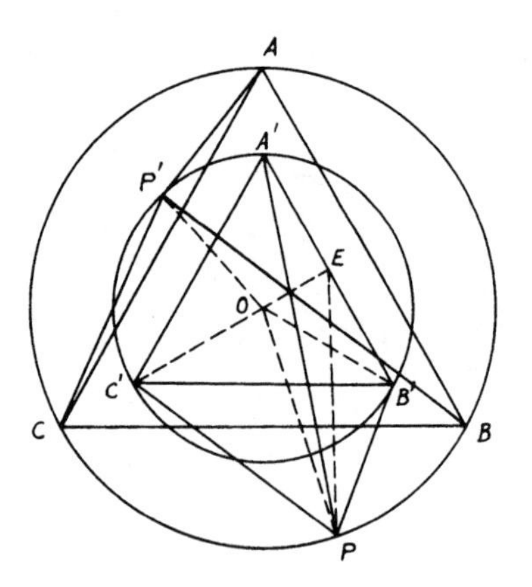

FIGURE 97

Proof: $\because$ $A'B'C'$ is an equilateral $\triangle$, $\therefore$ $C'OE$ is $\perp$ bisector of $A'B'$. In $\triangle PA'B'$, $PA'^2 + PB'^2 = 2\,PE^2 = 2\,EB'^2$. Adding PC'^2 to both sides, $\therefore$ $PA'^2 + PB'^2 + PC'^2 = 2\,PE^2 + 2\,EB'^2 + PC'^2$. $\because$ $OE = \frac{1}{2}\,OC'$, since O is the centroid of $\triangle$s $A'B'C'$, ABC, it is easily proved that $PC'^2 + 2\,PE^2 = 3\,PO^2 + 6\,OE^2$. $\therefore$ $PA'^2 + PB'^2 + PC'^2 = 2\,EB'^2 + 3\,PO^2 + 6\,OE^2 = 2\,(EB'^2 + OE^2) + 3\,PO^2 + 4\,OE^2 = 2\,OB'^2 + OC'^2 + 3\,PO^2 = 3\,(P'O^2 + PO^2)$. Similarly, $P'A^2 + P'B^2 + P'C^2 = 3\,(P'O^2 + PO^2)$.

3.30. *Four circles are described to touch the sides of a triangle. Show that the square on the distance between the centers of any two circles with the square* *on the distance between the centers of the other two circles is equal to the square on the diameter of the circle through the centers of any three.*

CONSTRUCTION: Let M, D, E, F be the centers of the inscribed and escribed ⊙s opposite A, B, C respectively. Join AMD, CMF and the sides of $\triangle DEF$ and let EN be a diameter in ⊙DEF. Join also ND, NF (*Fig.* 98).

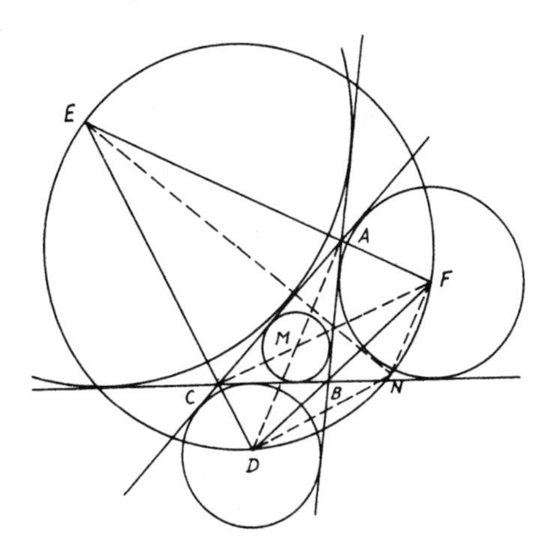

FIGURE 98

Proof: AMD, CMF are straight lines bisecting $\angle$s A, C. Also, EAF, FBD, DCE are straight lines bisecting $\angle$s A, B, C externally. Hence AMD, CMF are $\perp$s EAF, DCE. $\therefore$ M is the orthocenter of $\triangle DEF$. $\because$ EN is a diameter in ⊙DEF, $\therefore$ $\angle NDE$ = right angle. $\therefore$ ND is $\parallel$ CF. Hence $FNDM$ is a ▱. $\therefore$ $DN = FM$. Hence $EN^2 = ED^2 + DN^2 = ED^2 + FM^2$ and, similarly, $EN^2 = EF^2 + MD^2$ (see Problem 3.14).

3.31. *AD, BE, CF are perpendiculars drawn from the vertices of a triangle ABC on any diameter in its inscribed circle. Show that the perpendiculars DP, EQ, FR on BC, CA, AB respectively are concurrent.*

CONSTRUCTION: Join AE, AF, BD, BF, CD, CE (*Fig.* 99).

Proof: In $\triangle DBC$, $DP \perp BC$. $\therefore$ $BP^2 - PC^2 = DB^2 - DC^2 = BE^2 + ED^2 - DF^2 - FC^2$. Similarly, $CQ^2 - QA^2 = EC^2 - EA^2 = EF^2 + FC^2 - ED^2 - AD^2$ and $AR^2 - RB^2 = AF^2 - FB^2 = AD^2 + DF^2 - EF^2 - BE^2$.

Adding yields $(BP^2 - PC^2) + (CQ^2 - QA^2) + (AR^2 - RB^2) = 0$. Therefore, DP, EQ, FR are concurrent (see Problem 2.20).

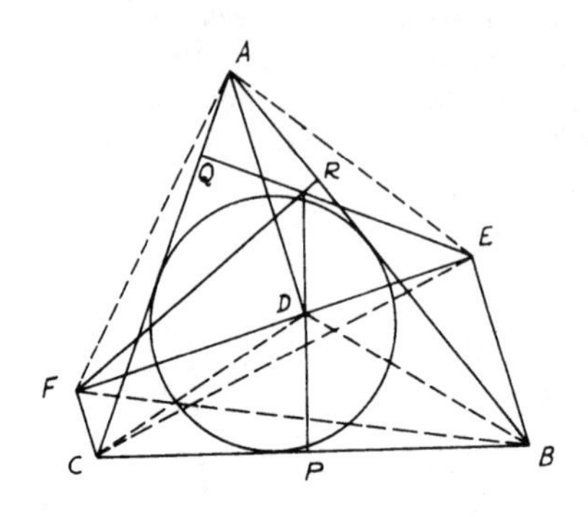

FIGURE 99

3.32. *D is a fixed point inside a given circle with center O. AB is a chord in the circle which always subtends a right angle at D. Show that the rectangle contained by the two perpendiculars OC, DE on AB is constant and that the sum of the squares of the perpendiculars on AB from two other fixed points, which can be determined, is also constant.*

CONSTRUCTION: Join OD and bisect it in K. Draw $KL \perp AB$ and $PKQ \parallel AB$ meeting OC, DE or produced in P, Q. Join CD, CK, EK, OA. (*Fig.* 100).

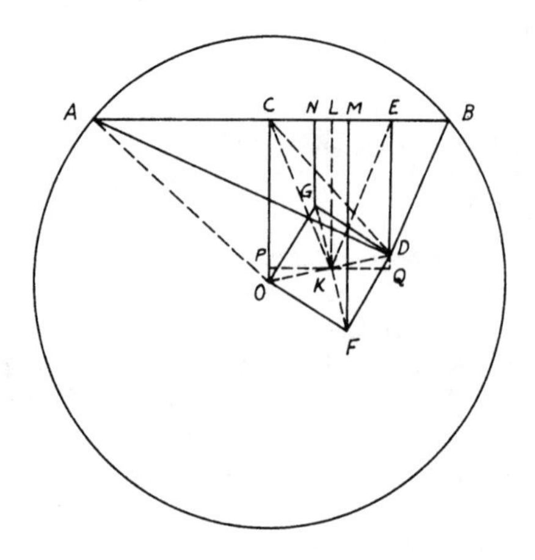

FIGURE 100

Proof: (i) $OC \cdot DE = DE\,(EQ + DQ) = DE^2 + 2\,DE \cdot DQ = EK^2 - DK^2$. Since $\angle ADB$ = right angle, $\therefore\ CD = \frac{1}{2}\,AB = AC$ (Th. 1.27). $\because\ CD^2 + CO^2 = 2\,CK^2 + 2\,DK^2 = AC^2 + CO^2 = AO^2$ = constant, $\because\ DK$ is fixed (since OD is of a fixed length), $\therefore\ CK = EK$ is fixed. Therefore, $(EK^2 - DK^2)$ or $OC \cdot DE$ is constant.

(ii) On OD as diagonal, construct a square $OGDF$; then F, G are also fixed points. Draw FM, $GN \perp$s AB and join FKG. $\therefore FM^2 + GN^2 = 2\,((FM + GN)/2)^2 + 2\,((FM - GN)/2)^2 = 2\,KL^2 + 2\,KQ^2 = 2\,EQ^2 + 2\,KQ^2 = 2\,EK^2$. Since EK is fixed, $\therefore\ (FM^2 + GN^2)$ is constant.

Miscellaneous Exercises

1. If the perpendiculars from A, B, C on the opposite sides of the triangle ABC meet the circumscribed circle in G, H, K, show that the area of the hexagon $AHCGBK$ is twice that of the triangle ABC.
2. P is the orthocenter of the triangle ABC, D any point in BC. If a circle be described with center D and radius DP meeting AP produced in E, then E lies on the circumscribing circle of the triangle.
3. AD the bisector of the angle BAC cuts the base BC in D, and BH a parallel to AD meets CA produced in H. Prove that the circles circumscribing the triangles BAC, HDC cut AD produced in points equidistant from A.
4. AB is a fixed chord of a circle, AP, BQ any two chords parallel to each other. Prove that PQ touches a fixed circle.
5. A straight line LPM meets the lines CX, CY in L, M. At M make the angle YMD equal to the angle CPM. At C make the angle YCD equal to the angle PCL and let MD, CD meet in D. Prove that the angles DLP, PCL are equal.
6. The bisectors of the angles of the triangle ABC inscribed in a circle intersect in O and being produced meet the circle in D, E, F. Prove that O is the orthocenter of the triangle DEF.
7. P is any point on the circumference of a circle which passes through the center C of another circle. PQ, PR are tangents drawn from P to the other circle. Show that CP, QR meet on the common chord of the circles.
8. The angle A of the triangle ABC is a right angle. D is the foot of the perpendicular from A on BC and DM, DN are drawn perpendicular to AB, AC respectively. Show that the angles BMC, BNC are equal.
9. Through the point of contact of two given circles which touch each other, either externally or internally, draw a straight line terminated by the circles which shall be equal to a given straight line.
10. If two triangles equal in every respect be placed so as entirely to coincide, and one be turned in its plane about one angular point, show that

the line joining that angular point to the point of intersection of the opposite sides will bisect the angle between those sides.

11. From a point P inside a triangle ABC perpendiculars PD, PE, PF are drawn to BC, CA, AB respectively. If the angle EDF is equal to A, prove that the locus of P is an arc of the circle passing through B, C and the center of the circle circumscribing ABC.

12. With any point G on the circumference of a circle as center, a circle is described cutting the former in B, C. From a point H on the second circle as center, a circle is described touching BC. Prove that the other tangents from B, C to the third circle intersect on the circumference of the first circle.

13. A triangle ABC is described in a circle. From a point P on the circumference perpendiculars are drawn to the sides, meeting the circle again in A', B', C' respectively. Show that AA', BB', CC' are parallel.

14. From the vertices B, C of the triangle ABC perpendiculars BE, CF are drawn to the opposite sides meeting them in E, F. Show that the tangents at E, F to the circle through A, E, F intersect in BC.

15. ABC is a triangle, Q any point on the circumscribing circle, QM the perpendicular from Q on AB. If CQ meets AB in L, and if the diameter through C meet the pedal of Q in N, prove that C, L, M, N lie on a circle.

16. The lines joining the points where the bisectors of the angles between the opposite sides of a quadrilateral inscribed in a circle meet the sides form a rhombus.

17. A, B, C are any points on the circumference of a circle. D is the middle point of the arc AB, E the middle point of the arc AC. If the chord DE cuts the chords AB, AC in F, G respectively, prove that $AF = AG$.

18. ABC is an arc of a circle whose center is O. B is the middle point of the arc, and the whole arc is less than a semi-circumference. From P, any point in the arc, PM, PN, PQ are drawn perpendicular respectively to OA, OB, OC or produced if necessary and NR is drawn perpendicular to OA. Show that PM and PQ are together double of NR.

19. O is the center of the circle inscribed in the triangle ABC. Straight lines are drawn bisecting AO, BO, CO at right angles. Show that these straight lines intersect on the circle ABC.

20. A, B, C lie on a circle. Through the center, lines are drawn parallel to CA, CB meeting the tangents at A, B in D, E respectively. Prove that DE touches the circle.

21. ABC is a circle whose center is O. Any circle is described passing through O and cutting the circle ABC in A, B. From any point P on the circumference of this second circle, straight lines are drawn to A, B and, produced if necessary, meet the first circle in A', B' respectively. Prove that AB' is parallel to $A'B$.

22. If the chords which bisect two angles of a triangle inscribed in a circle be equal, prove that either the angles are equal, or the third angle is equal to the angle of an equilateral triangle.

23. ABC is a triangle inscribed in a circle. From D the middle point of one of the arcs subtended by BC, perpendiculars are drawn to AB, AC. Prove that the sum of the distances of the feet of these perpendiculars from A is equal to the sum or difference of the sides AB, AC according as A and D are on opposite sides or on the same side of BC.

24. AB is a diameter of a circle, CD a chord perpendicular to AB, DP any other chord meeting AB in Q. Prove that CA and also CB make equal angles with CP, CQ.

25. O is a point on the circumference of the circle circumscribing the triangle ABC. Prove that if the perpendiculars dropped from O on BC, CA, AB respectively meet the circle again in a, b, c, the triangle abc is equal in all respects to ABC.

26. Construct a square such that two of its sides shall pass through two points B, C respectively and the remaining two intersect in a given point A.

27. From any point P perpendiculars PA', PB', PC' are drawn to the sides BC, CA, AB of a triangle, and circles are described about the triangles $PA'B'$, $PB'C'$, $PC'A'$. Show that the area of the triangle formed by joining the centers of these circles is $\frac{1}{4}$ of the area of the triangle ABC.

28. Draw a straight line cutting two concentric circles such that the chord of the outer circle will be twice that of the inner.

29. Draw through B, one of two fixed points A, B, a line which will cut the circle on AB as diameter in C and the perpendicular from A to AB in D so that BC will be equal to AD.

30. ABC, $A'B'C'$ are two triangles equiangular to each other inscribed in two concentric circles. Show that the straight lines AA', BB', CC' form a triangle equiangular with the triangle formed by joining the feet of the perpendiculars from the vertices of ABC on the opposite sides.

31. Two circles touch each other externally at C and a straight line in A and B. AC, BC produced meet the circles again in E, F respectively. Show that the square on EF is less than the sum of the squares on the diameters of the circles by the rectangle contained by the diameters.

32. Two circles, lying wholly outside one another, are touched by four common tangents. Show that if AB be an outer common tangent, the two inner common tangents meet AB in points P, Q, such that AP is equal to BQ.

33. Given two parallel straight lines and a point between them, draw a straight line parallel to a given straight line, the part of which intercepted between the parallels will subtend a given angle at the given point.

34. In any triangle *ABC* if the internal and external bisectors of the angle *A* meet the opposite side in *I*, *K* respectively and if *M* be the middle point of *IK*, the triangles *ACM*, *BAM* are equiangular to each other and *MA* touches the circle *ABC*.

35. The circumference of one circle passes through the center of a second circle and the circles intersect in *A*, *B*. Prove that any two chords through *A*, *B* of the second circle which intersect on the circumference of the first circle are equal.

36. A circle revolves around a fixed point in its circumference. Show that the points of contact of tangents to the circle which are parallel to a fixed line lie on one or other of two fixed circles.

37. Describe a circle with given radius to pass through a given point and touch a given circle.

38. Of all triangles on the same base and having the same vertical angle, the isosceles has the greatest perimeter.

39. Two equal circles *EAB*, *FAB* intersect in *A*, *B*. *BE* is drawn touching the circle *BAF* at *B* and meeting the circle *BAE* in *E*. *EA* is joined and produced to meet the circle *BAF* in *F*. A line *BC′MC* is drawn through *B* at right angles to *EAF*, meeting it in *M*, the circle *BAE* in *C*, and the circle *BAF* in *C′*. *C′A* is joined and produced to meet *BE* in *K*. Prove that *KM* is parallel to *BF*.

40. *A* is a fixed point on a circle. From any point *B* on the circle, *BD* is drawn perpendicular to the diameter through *A*. Prove that the circle through *A* touching the chord *BD* at *B* is of constant magnitude.

41. *ABC* is a triangle and *AL*, *BM*, *CN*, its perpendiculars, meet the circumscribing circle in *A′*, *B′*, *C′*. *S* is any point on the circle. Show that *SA′*, *SB′*, *SC′* meet *BC*, *CA*, *AB* in points in a straight line which passes through the orthocenter of the triangle *ABC*.

42. From a point on a circle three chords are drawn. Prove that the circles described on these as diameters will intersect in three points in a straight line.

43. The alternate angles of any polygon of an even number of sides inscribed in a circle are together equal to a number of right angles less by two than the number of sides of the polygon.

44. Given a point *P* either inside or outside a given circle. Show how to draw through *P* straight lines *PA*, *PB* cutting the circle in *A*, *B* containing a given angle so that the circle circumscribing the triangle *PAB* will pass through the center of the given circle.

45. *A′*, *B′*, *C′* are the vertices of equilateral triangles described externally on the sides of a triangle *ABC*. Prove that *AA′*, *BB′*, *CC′* are equal, that they are equally inclined to each other, and that they are concurrent.

46. The sides of a triangle are cut by a circle concentric with the inscribed circle, and each vertex of the hexagon formed by the intersections is

joined to the opposite vertex. Show that the triangle so formed is equiangular to the triangle formed by the points of contact of the inscribed circle with the sides.

47. Two equal circles touch at A. A circle of twice the radius is described having internal contact with one of them at B and cutting the other in P, Q. Prove that the straight line AB will pass through P or Q.

48. From an external point P tangents are drawn to a circle whose center is C and CP is joined. If the points of contact be joined with the ends of the diameter perpendicular to CP, prove that the points of intersection of the joining lines and the points of contact are equally distant from P.

49. Describe a triangle equiangular to a given triangle whose sides pass through three given points and whose area shall be a maximum.

50. Show how to draw a pair of equal circles on two parallel sides of a parallelogram as chords, so as to touch each other, and show that the circles so drawn on the two pairs of parallel sides intersect at angles equal to those of the parallelogram.

51. E is the intersection of the diagonals of a quadrilateral inscribed in a circle. FEG is the chord which is bisected in E. Prove that the part of this chord intercepted between the opposite sides of the quadrilateral is also bisected in E.

52. (a) P is the orthocenter of the triangle ABC inscribed in a circle whose center is O. If the parallelogram $BACG$ be completed, show that G is a point on the circumference of the circle which passes through B, P, C, and hence prove that AP is twice the perpendicular drawn from O to BC.

    (b) $ABCD$ is a quadrilateral inscribed in a circle. Prove that the orthocenters of the triangles ABC, BCD, CDA, DAB lie on an equal circle.

53. Two segments of circles on the same straight line and on the same side of it, which contain supplementary angles, intercept equal lengths on perpendiculars to their common chord.

54. Prove that if two adjacent sides of a square pass through two fixed points, the diagonal also passes through a fixed point. Hence show how to describe a square about a given quadrilateral.

55. A', B', C' are the feet of the perpendiculars from the vertices of a triangle ABC on the opposite sides BC, CA, AB, D, E, F the middle points of those sides respectively. O is the orthocenter and G, H, K the middle points of AO, BO, CO. Show that GD, HE, FK are equal and concurrent.

56. Through a fixed point which is equidistant from two parallel straight lines, a straight line is drawn terminated by the two fixed straight lines and on it as base is described an equilateral triangle. Prove that the vertex of this triangle will lie on one of two straight lines.

57. Describe a circle to touch a given circle and a given straight line, and to have its center in another given straight line.

58. A straight line AB slides between two fixed parallel straight lines to which it is perpendicular. Find the position of AB when it subtends the greatest possible angle at a fixed point.

59. Construct a triangle given the base, the vertical angle, and the sum of the squares on the sides.

60. If a circle touch a given circle and also touch one of its diameters AB at C, prove that the square on the straight line drawn from C at right angles to AB to meet the circumference of the given circle is equal to half the rectangle contained by the diameters of the circles.

61. One diagonal of a quadrilateral inscribed in a circle is bisected by the other. Show that the squares on the lines joining their point of intersection with the middle points of the sides are together half the square on the latter diagonal.

62. A circle passing through the vertex A of an equilateral triangle ABC cuts AB, AC produced in D, E respectively and BC produced both ways in F, G. Show that the difference between AD and AE is equal to the difference between BF and CG.

63. A, B are points outside a given circle. C is a point in AB such that the rectangle AB, AC is equal to the square on the tangent from A to the circle. CD is drawn to touch the circle in D and AD is drawn cutting the circle again in E. If BE cut the circle again in F, show that DF is parallel to AB.

64. CD is a chord of a given circle parallel to a given straight line AB. G is a point in AB such that the rectangle AB, BG is equal to the square on the tangent from B to the circle. If DG cuts the circle in F, prove that AC, BF intersect on the circle.

65. AB is a diameter of a semi-circle on which P and Q are two points. From AQ, the distance QR is cut off equal to QB. Prove that if AR is equal to AP, the tangent from Q to the circle around ARP is the side of a square of which BP is a diameter.

66. A is the center of a circle. PN is a perpendicular let fall on the radius AB from a point P on the circle. Show that the tangent from P to the circle of which AB is a diameter is equal to the tangent from B to the circle of which AN is a diameter.

67. A, B, C, D are four fixed points in a straight line. A circle is described through A, B and another through C, D to touch the former. Prove that the point of contact lies on a fixed circle.

68. Two chords AB, CD of a circle intersect at right angles in a point O either inside or outside the circle. Prove that the squares on AB, CD are together less than twice the square on the diameter by four times the square on the line joining O to the center of the circle.

69. From a point P two tangents PT, PT' are drawn to a given circle whose center is O and a line PAB cutting the circle in A, B. If PO, TT' intersect in C, prove that TC bisects the angle ACB.

70. If perpendiculars are drawn from the orthocenter of a triangle ABC to the bisectors of the angle A, their feet are collinear with the middle point of BC and the nine-point center.

71. ABC is a right-angled triangle at A inscribed in a circle. D is any point on the smaller arc AC, and DE is drawn perpendicular to BC cutting AC in F. From F a perpendicular is drawn to AC meeting the circle on AC as diameter in G. Show that $DC = CG$.

72. $ABCD$ is a quadrilateral inscribed in a circle and P any point on the circumference. From P perpendiculars PM, PN, PQ, PR are drawn to AB, BC, CD, DA respectively. Show that $PM \cdot PQ = PN \cdot PR$.

73. ABC is a right-angled triangle at A. From any point D on the hypotenuse BC, a perpendicular DFG is drawn to BC to meet AB, AC or produced if necessary in F, G respectively. Show that (a) $DF^2 = BD \cdot DC - AF \cdot FB$; (b) $DG^2 = BD \cdot DC + GA \cdot GC$.

74. $ABCD$ is a concyclic quadrilateral. If BA, CD are produced to meet in K and also AD, BC in L, show that the square on KL is equal to the sum of the squares on the tangents from K and L to the circle.

75. C is the center and AB a diameter of the circle $ADEB$, and the chord DE is parallel to AB. Join AD and draw AP perpendicular to AD to meet ED produced in P and join PC and AE. Show that $PC^2 = AP^2 + AC^2 + AE^2$.

76. AB is a diameter of a circle with center O. CD is a chord parallel to AB. If P is any point on AB and Q is the mid-point of the smaller arc CD, prove that (a) $AP^2 = PB^2 = CP^2 + PD^2$; (b) $CP^2 + PD^2 = 2\, PQ^2$.

77. If two circles touch externally, show that the square on their common tangent is equal to the rectangle contained by the diameters.

78. A, B are two fixed points on a diameter of a circle with center C such that $CA = CB$. If any chord DAE is drawn through A in the circle, show that the sum of the squares on the sides of the triangle BDE is constant.

79. If two chords of a circle intersect at right angles, show that the sum of the squares on the four segments is equal to the square on the diameter.

80. A circle touches one side BC of a triangle and the other sides AB, AC produced, the points of contact being D, F, E. If I be the center of the inscribed circle, prove that the areas of the triangles IAE, IAF are together equal to that of the triangle ABC.

81. M, N are the centers of two intersecting circles in A, B. From A, CAB, DAE are drawn at right angles such that D, B lie on the circumference

of the circle M and C, E on that of circle N. Prove that $CB^2 + DE^2 = 4\,MN^2$.

82. If from a fixed point T, without a circle whose center is O, TA, TB are drawn equally inclined to TO to meet the concave and convex arcs respectively in A and B, show that AB, TO meet in a fixed point.

83. If the inscribed circle in a right-angled triangle at A touches the hypotenuse BC at D, then $DB \cdot DC =$ the area of triangle ABC.

84. Prove that if three circles intersect, their three common chords are concurrent.

85. OA, OB are tangents to a given circle whose center is C and CO cuts AB in D. Prove that any circle through O, D cuts the given circle orthogonally.

86. Through a point on the smaller of two concentric circles, draw a line bounded by the circumference of the larger circle and divided into three equal parts at the points of section of the smaller circle.

87. Describe a circle which will pass through two given points and cut a given circle orthogonally.

88. If the square on the line joining two points P, Q be equal to the sum of the squares on the tangents from P, Q to a circle, then Q is on the straight line joining the points of contact of tangents drawn to the circle from P.

89. OA, OB are straight lines touching a circle in A, B. OC is drawn perpendicular to AB and bisected in D. DF is drawn touching the circle in F. Prove that CFO is a right angle.

90. If P be any point in the circumference of a circle described about an equilateral triangle ABC, show that the sum of the squares on PA, PB, PC is constant.

91. If O be any point in the circumference of the circle inscribed in an equilateral triangle ABC, prove that the sum of the squares on OA, OB, OC is constant.

92. The tangents at A, B, the ends of a chord AB of a circle, whose center is C, intersect in E. Prove that the tangents at the ends of all chords of the circle which are bisected by AB intersect on the circle whose diameter is CE.

93. In a circle the arc AB is equal to the arc BC. P is any other point on the circle. From B let fall BQ, BR perpendiculars on AP, CP respectively. Show that $OQ^2 + OR^2 = 2\,OP^2$, where O is the center of the circle.

94. One circle cuts another at right angles. Show that, if tangents be drawn from any point in one circle to meet the other, then the chord of contact passes through the opposite extremity of the diameter of which the first point is one extremity.

95. The circle circumscribing the triangle ABC is touched internally at the point C by a circle, which also touches the side AB in F and cuts the

sides BC, AC in D, E. If the tangent at D to the inner circle cuts the outer circle in G, H, prove that BH, BG, BF are all equal, and that CF bisects the angle ACB.

96. Two circles, whose centers are A, B, intersect in C, D. E is the middle point of AB. If F be any point in CD, then the chords intercepted by the circles on a line through F perpendicular to EF are equal.

97. Describe through two given points a circle such that the chord intercepted by it on a given unlimited straight line may be of given length.

98. ABC is a triangle right-angled at A. From D, any point in the circumference of the circle described on BC as diameter, a perpendicular is drawn to BC meeting AB in E. From E is drawn a perpendicular to AB, meeting in F the circle described on AB as diameter. Prove that $BF = BD$.

99. Tangents are drawn to a circle at the ends of a chord PQ, and through O, a point in PQ, a straight line $COAB$ is drawn parallel to one of the tangents, meeting the other in B and the circle in A, C. Show that if A bisects OB, then O will bisect BC. Show also how to determine the point O that this may be the case.

100. Draw through a given point P a straight line PQR to meet two given lines in Q, R so that the rectangle PQ, PR will be equal to a given rectangle.

101. ABC is a triangle, D, E, F the middle points of its sides. With the orthocenter of the triangle as center any circle is described cutting EF, FD, DE in P, Q, R respectively. Prove that $AP = BQ = CR$.

102. AB is a chord of a circle whose center is O. P is a point in AB. If OP be produced to Q so that the rectangle $OP \cdot OQ$ is equal to the square on the radius of the circle, prove that QA, QB make equal angles with QO.

103. C is the middle point of AB, a chord of a circle whose center is O. A point P is taken in the circumference, whose distance PD from AB is equal to AC. M is the middle point of PD and CF is drawn parallel to OM to meet PD in F. Show that $CF = FP$.

104. A, B are two fixed points taken on a diameter of a semi-circle whose center is C such that they are equidistant from C. If AP, BQ are two parallel lines terminated by the circumference, show that $AP \cdot BQ$ is constant.

105. D is the middle point of a straight line AB. If AM, BN, DE are tangents to any circle, show that $AM^2 + BN^2 = 2\,(AD^2 + DE^2)$.

106. P is any point outside a circle whose center is O. From P, PA and $PBOC$ are drawn tangent and transversal through O respectively. Through P and A another circle is drawn tangent to PBC at P, and intersects the first circle in D. Show that AD produced bisects PE, where E is the projection of A on BC.

107. O, M are the centers of two intersecting circles in A, B. Draw a transversal from A or B to the two circles so that the rectangle of the two segments contained in the circles is equal to a given rectangle.

108. M, N are the centers of two intersecting circles in A, B. From A a straight line CAD is drawn and terminated by the circles M, N in C, D respectively. With C and D as centers, describe two circles to cut separately the circles N, M orthogonally. Show that these two circles together with the circle on CD as diameter are concurrent.

109. Describe a triangle ABC having given the rectangle of the sides AB, AC, the difference of the base angles and the median bisecting the base BC.

110. O is the center of a circle and BC is any straight line drawn outside it. OD is drawn perpendicular to BC; then the distance DE is taken on OD equal to the tangent from D to the circle. If F is any other point on BC, show that FE is equal to the tangent from F to the circle.

111. A, B, C, D are points lying in this order in a straight line, and $AC = a$ inches, $CD = b$ inches, $BC = 1$ inch. A perpendicular through C to the line cuts the circle on BD as diameter in E. EFG passes through the center of the circle on AC as diameter, cutting the circumference in F, G. Show that the area of the rectangle $EF \cdot EG$ is b square inches, and also that the number of inches in the length of EF is a root of the equation $x^2 + ax - b = 0$.

112. Two circles are drawn touching the sides AB, AC of a triangle ABC at the ends of the base BC and also passing through D the middle point of BC. If E is the other point of intersection of the circles, prove that the rectangle DA, DE is equal to the square on DC.

113. Construct a triangle given the altitude to the base, the median of the base, and the rectangle of the other two sides.

114. A, B are fixed points and AC, AD fixed straight lines such that BA bisects the angle CAD. If any circle through A, B cuts off chords AK, AL from AC, AD, prove that $(AK + AL)$ is constant.

115. Construct a triangle having given the bisector of the vertical angle, the rectangle of the sides containing this angle, and the difference of the base angles.

116. In a given circle, draw two parallel chords from two given points on its circumference such that their product may be of given value.

117. Draw two intersecting circles having their centers on the same side of the common chord AB, and draw a diameter of the smaller circle. Describe a circle within the area common to the two circles, which will touch them both and have its center in the given diameter. When will there be two circles fulfilling the given conditions?

118. O is any point in the plane of the triangle ABC. Perpendiculars to OA, OB, OC drawn through A, B, C form another triangle $A'B'C'$. Prove that the perpendiculars from A', B', C' on BC, CA, AB respectively meet in a point O', and that the center of the circle ABC is the middle point of OO'.

CHAPTER 4

RATIO AND PROPORTION

Theorems and Corollaries

RATIOS

4.82. *A.* (*i*) *If* $a/b > 1$ *and* $x > 0$, *then* $(a + x)/(b + x) < a/b$.
(*ii*) *If* $0 < a/b < 1$ *and* $x > 0$, *then* $(a + x)/(b + x) > a/b$.
B. *If* $a/b = c/d$, *then* $b/a = d/c$ *and* $a/c = b/d$.
C. (*i*) *If* $a/b = c/d$, *then* $ad = bc$.
(*ii*) *Conversely, if* $ad = bc$, *then* $a/b = c/d$ *or* $a/c = b/d$.
D. (*i*) *If* $a/b = b/c$, *then* $b^2 = ac$.
(*ii*) *Conversely, if* $b^2 = ac$, *then* $a/b = b/c$.
E. *If* $a/b = b/c$, *then* $a/c = a^2/b^2 = b^2/c^2$.
F. *If* $a/b = c/d$, *then* $(a + b)/b = (c + d)/d$, $(a - b)/b = (c - d)/d$ *and* $(a + b)/(a - b) = (c + d)/(c - d)$.
G. *If* $a/b = x/y$ *and* $b/c = y/z$, *then* $a/c = x/z$.
H. *If* $x/a = y/b = z/c$, *then each of these ratios* $= (lx + my + nz)/(la + mb + nc)$, *where* l, m, n *are any quantities whatever, positive or negative.*

PROPORTION AND SIMILAR POLYGONS

4.83. (*i*) *If a straight line is drawn parallel to one side of a triangle, it divides the other two sides proportionally.* (*ii*) *Conversely, if a straight line divides two sides of a triangle proportionally, it is parallel to the third side.*

COROLLARY. *If two straight lines* PT, $P'T'$ *cut three parallel straight lines* AB, CD, EF *at* Q, R, S; Q', R', S' *respectively, then* $QR/RS = Q'R'/R'S'$ *and* $QS/RS = Q'S'/R'S'$ *and, conversely, if* AB *is parallel to* EF *and* $QR/RS = Q'R'/R'S'$ *or* $QS/RS = Q'S'/R'S'$, *then* CD *is parallel to* AB.

4.84. *If two triangles are mutually equiangular, their corresponding sides are proportional, and the triangles are similar.*

4.85. *If the three sides of one triangle are proportional to the three sides of another triangle, the two triangles are mutually equiangular.*

4.86. *If two triangles have an angle of the one equal to an angle of the other, and the sides about these equal angles proportional, the triangles are mutually equiangular and similar.*

4.87. *If the straight lines joining a point to the vertices of a given polygon are divided (all internally or all externally) in the same ratio, the points of division are the vertices of a polygon which is mutually similar to the given polygon.*

4.88. *If O is any point and $ABC \cdots$ is any polygon, and straight lines AOA', BOB', COC', $\cdots$ are drawn such that $AO : A'O = BO : B'O = CO : C'O = \cdots = k$, then the resulting polygon $A'B'C' \cdots$ is said to be homothetic to the original polygon $ABC \cdots$; O is the center of homothecy and k the homothetic ratio. The polygons are similar, and when k equals 1, congruent.*

COROLLARY. *If two polygons are similar and similarly placed, the straight line joining any pair of opposite vertices of one figure is parallel to that joining the corresponding opposite vertices of the other.*

4.89. *If a polygon is divided into triangles by lines joining a given point to its vertices, any similar polygon can be divided into corresponding similar triangles.*

4.90. (*i*) *If the vertex angle of a triangle is bisected, internally or externally, by a straight line which cuts the base, or the base produced, it divides the base, internally or externally, in the ratio of the other sides of the triangle.* (*ii*) *Conversely, if a straight line through the vertex of a triangle divides the base, internally or externally, in the ratio of the other sides, it bisects the vertex angle, internally or externally.*

AREAS OF SIMILAR POLYGONS

4.91. *The ratio of the areas of two triangles or of two parallelograms of equal or the same altitude is equal to the ratio of their bases.*

4.92. *The ratio of the areas of two similar triangles, or of two similar polygons, is equal to the ratio of the squares on corresponding sides.*

COROLLARY. *The ratio of the areas of two similar triangles is equal to the ratio of* (*i*) *the squares of any two corresponding altitudes;* (*ii*) *the squares of any two corresponding medians;* (*iii*) *the squares of the bisectors of any two corresponding angles.*

4.93. *The areas of two triangles which have an angle of one equal to an angle of the other are to each other as the products of the sides including the equal angles.*

COROLLARY. *The areas of two triangles that have an angle of one supplementary to an angle of the other are to each other as the products of the sides including the supplementary angles.*

4.94. *If a triangle ABC equals another triangle $A'B'C'$ in area, and the angle A equals angle A', then $AB \cdot AC = A'B' \cdot A'C'$. Conversely, if in two triangles ABC, $A'B'C'$ angles A, A' are equal and $AB \cdot AC = A'B' \cdot A'C'$, then the two triangles are equal.*

4.95. *If two parallelograms are mutually equiangular, then the ratio of their areas is equal to the product of the ratios of two pairs of corresponding sides.*

4.96. *If $ABCD$ is a parallelogram and E is any point on the diagonal BD from which FEL and GEK are drawn parallel to AB and BC respectively and terminated by the sides, then the parallelograms $BGEF$ and $DKEL$ are similar to $ABCD$.*

4.97. *In any right-angled triangle, any rectilinear figure described on the hypotenuse is equal to the sum of two similar and similarly described figures on the sides containing the right angle.*

Ratio and Proportion in Circles

4.98. *In equal circles, angles either at the centers or at the circumferences have the same ratio to one another as the arcs on which they intercept; so also have the sectors.*

Corollary. *The sectors are to each other as their angles.*

4.99. *If any angle of a triangle is bisected by a straight line which cuts the base, the rectangle contained by the sides of the triangle is equal to the sum of the rectangle contained by the segments of the base and the square on the line which bisects the angle.*

4.100. *If from any vertex of a triangle a straight line is drawn perpendicular to the base, the rectangle contained by the sides of the triangle is equal to the rectangle contained by the perpendicular and the diameter of the circle described about the triangle.*

4.101. *Ptolemy General Theorem. The rectangle contained by the diagonals of a quadrilateral is less than the sum of the rectangles contained by opposite sides unless a circle can be circumscribed about the quadrilateral, in which case it is equal to their sum.*

The Measure of a Circle

4.102. *The circumferences of any two circles have the same ratio as their radii.*

Corollary 1. *The ratio of the circumference of any circle to its diameter is a constant denoted by π and equals approximately* 3.14.

Corollary 2. *If C is the circumference, D the diameter, and R the radius of a circle, then $C = \pi D = 2\pi R$.*

Corollary 3. *If A is the number of degrees in an arc of a circle, or in its subtended central angle, then the length of the arc may be expressed by* $(A/360)2\pi R$.

4.103. *The area of a circle is equal to half the product of its circumference and radius and expressed by πR^2.*

Corollary 1. *The area of a sector of a circle may be expressed in the form $(A/360)\pi R^2$, where A is the central angle of the sector.*

Corollary 2. *The area of a segment is (i) area of a sector − area of triangle (if central angle $<180°$); (ii) area of a sector + area of triangle (if central angle $>180°$).*

COROLLARY 3. *The area of any two circles are to each other as the squares of their radii or as the squares of their diameters.*

Solved Problems

4.1. *Two straight lines AD, AE are drawn from the vertex A of a triangle ABC to make equal angles with AB, AC and meet the base BC in D, E. If D is nearer to B and E nearer to C, show that $AB^2 : AC^2 = BD \cdot BE : CE \cdot CD$.*

CONSTRUCTION: On $\triangle ADE$ draw a circle cutting AB, AC in F, G. Join FG, DG (*Fig.* 101).

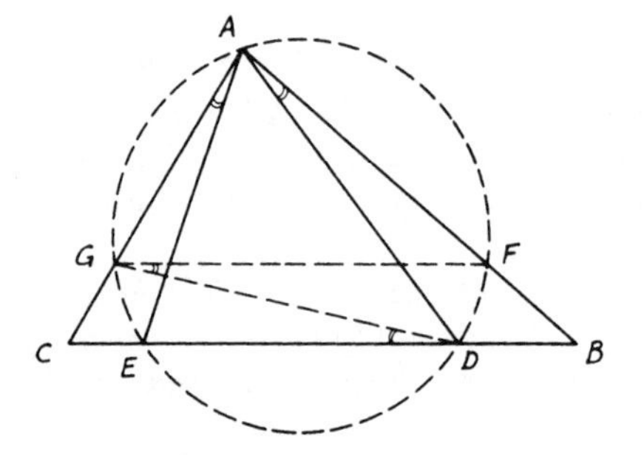

FIGURE 101

Proof: $\angle FAD = \angle FGD$ (in $\odot ADE$). Also, $\angle GAE = \angle GDE$. Since $\angle FAD = \angle GAE$ (hypothesis), $\therefore$ $\angle FGD = \angle GDE$. $\therefore$ FG is $\parallel DE \parallel BC$. Hence in $\triangle ABC$, $BF/AB = CG/AC$ (Th. 4.83). $\therefore$ $BF/CG = AB/AC$. Multiplying yields $(BF/CG) \cdot (AB/AC) = AB^2/AC^2$. But since $BF \cdot AB = BD \cdot BE$ and $CG \cdot AC = CE \cdot CD$, $\therefore$ $AB^2/AC^2 = (BD/CE) \cdot (BE/CD)$.

4.2. *ABC is any triangle. From the vertices three equal straight lines AD, BE, CF are drawn to meet BC, CA, AB in D, E, F respectively. If another three lines are drawn from any point M inside the triangle parallel to these equal lines and meeting BC, CA, AB in P, Q, R respectively, show that $AD = MP + MQ + MR$.*

CONSTRUCTION: Draw from M the lines GMH, MJ, $MK \parallel BC$, AC, AB (*Fig.* 102).

Proof: In similar $\triangle$s MKP, ABD, $MP/AD = MK/AB$. Also in similar $\triangle$s MKJ, ABC, $MK/AB = KJ/BC$ (Th. 4.84). $\therefore$ $KJ/BC = MP/AD$ (1). Similarly, in similar $\triangle$s MQH, BEC, $MQ/BE = MH/BC$. Since $MJCH$ is a $\square$, $\therefore$ $MH = CJ$. $\therefore$ $MQ/BE = CJ/BC$ (2). Again, in similar $\triangle$s MGR, CBF, $MR/CF = MG/BC$. Since $MGBK$ is another $\square$, $\therefore$ $MG = BK$. $\therefore$ $MR/CF = BK/BC$ (3).

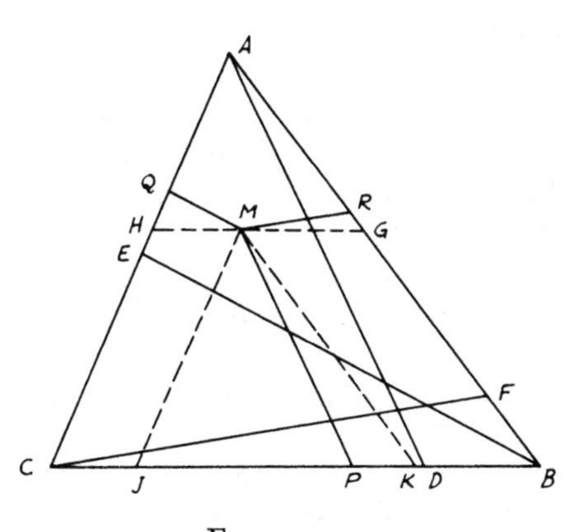

FIGURE 102

Since AD, BE, CF are equal, adding ratios (1), (2), (3) gives $(MP + MQ + MR)/AD = (KJ + CJ + BK)/BC = BC/BC$. $\therefore AD = MP + MQ + MR$.

COROLLARY. *If from M any point inside an equilateral △ perpendiculars are drawn to the three sides, then the sum of the three perpendiculars is always equal to any altitude in the triangle (see Exercise 1.46).*

4.3. *ABCD is a quadrilateral and a transversal line is drawn to cut AB, AD, CD, BC, AC, BD or produced in E, F, G, H, I, J. Show that* $EF/GH = (FI/GI)\cdot(EJ/HJ)$.

CONSTRUCTION: From D draw the line $LDMN \parallel HF$ to meet BC, BA, CA produced in L, M, N (*Fig.* 103).

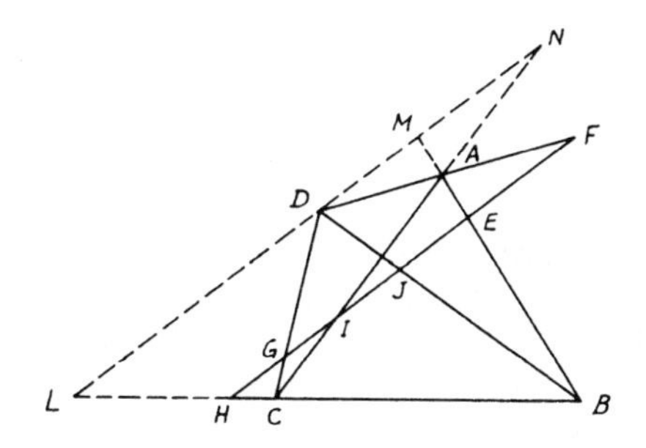

FIGURE 103

Proof: In similar $\triangle$s FAI, DAN, $FI/DN = AF/AD$. Also, in similar $\triangle$s AFE, ADM, $EF/DM = AF/AD$. $\therefore$ $EF/DM = FI/DN$ (1). In $\triangle BLM$, EH is $\parallel$ LM and BD is a transversal. Hence $DM/DL = EJ/HJ$ (2). Similarly, in $\triangle CLN$, $HI \parallel LN$ and CD is a transversal. $\therefore$ $DL/GH = DN/GI$ (3). Multiplying (1), (2), (3) gives $(EF/DM)\cdot(DM/DL)\cdot(DL/GH) = (FI/DN)\cdot(EJ/HJ)\cdot(DN/GI)$. $\therefore$ $EF/GH = (FI/HJ)\cdot(EJ/GI)$.

4.4. *Bisect a triangle ABC by (i) a straight line perpendicular to its base BC; (ii) a straight line parallel to one side; (iii) a straight line parallel to a given direction.*

CONSTRUCTION: (i) Bisect BC in M. Draw $MN \perp CB$ to meet AC in N. NP is drawn $\perp$ to CA cutting the circumference of $\odot$ on CA as diameter in P. Join CP and take CE on $CA = CP$. Then perpendicular ED to BC is the required bisector. Draw $AF \perp BC$ and join AM (*Fig.* 104(i)).

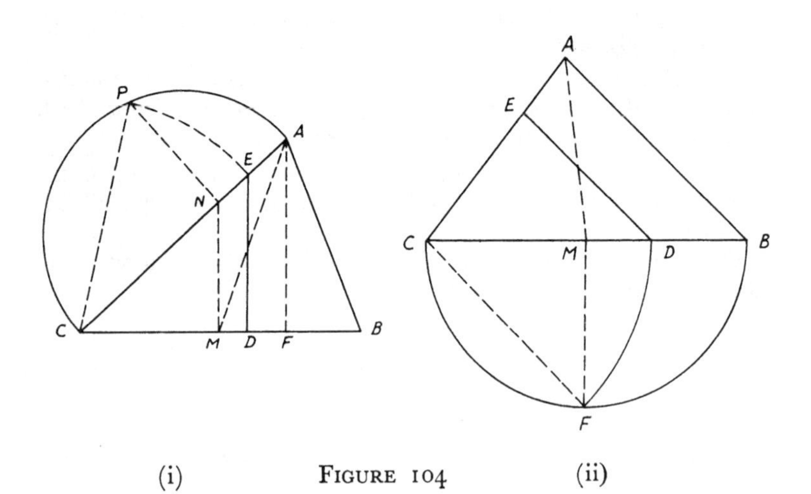

(i) FIGURE 104 (ii)

Proof: $\triangle ACM/\triangle ACF = CM/CF$ (Th. 4.91). Since $CM/CF = CN/CA = CN\cdot CA/CA^2 = CP^2/CA^2 = CE^2/CA^2$, $\therefore$ $\triangle ACM/\triangle ACF = CE^2/CA^2$. But $\triangle CED/\triangle ACF = CE^2/CA^2$, since $DE \parallel AF$ (Th. 4.92). $\therefore$ $\triangle CED = \triangle ACM = \frac{1}{2}\triangle ABC$.

CONSTRUCTION: (ii) Bisect the base BC in M. Then draw $MF \perp BC$ to meet the circle on BC as diameter in F. Join CF and take CD on BC equal to CF. Draw $DE \parallel AB$ meeting AC in E. DE is the required b sector. Join AM [Fig. 104(ii)].

Proof: Since M is the mid-point of BC, $\therefore\ \triangle ACM = \frac{1}{2}\triangle ABC$. Since DE is $\parallel AB$, $\therefore\ \triangle CDE/\triangle ABC = CD^2/CB^2$ (Th. 4.92) $= CF^2/CB^2 = CM \cdot CB/CB^2 = CM/CB = \frac{1}{2}$. $\therefore\ \triangle CDE = \frac{1}{2}\triangle ABC$.

(iii) This is a more general case of (i) and is always possible for some given direction. Since in *Fig.* 104(i), if from A a parallel is drawn to the given direction to cut BC in a point F, a similar construction of $DE \parallel$ to the line AF can be established as in case (i) by making $MN \parallel AF$ also. From some other given direction, for example, $\parallel BC$, it may be necessary to use a median other than AM.

4.5. *From a given point P in the base BC produced of a triangle ABC, draw a straight line to cut the sides of the triangle, so that if lines be drawn parallel to each side from the point where it intersects the other, they shall meet on the base BC.*

ANALYSIS: Suppose PEF be the required line, so that D the point of intersection of ED drawn $\parallel AB$ and FD drawn $\parallel AC$, is in BC (*Fig.* 105). $\because FB$ is $\parallel DE$, $\therefore PB/PD = PF/PE$. $\because FD$ is $\parallel EC$, $\therefore PD/PC$

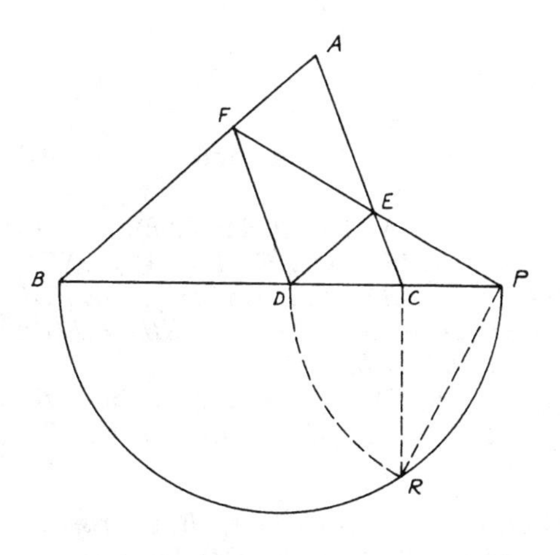

FIGURE 105

$= PF/PE$. $\therefore PB/PD = PD/PC$. Hence PD is the mean proportional between PB and PC; i.e., $PD^2 = PB \cdot PC$.

SYNTHESIS: On PB as diameter, describe a semi-circle. Draw $CR \perp BP$ to meet the circle in R. With P as center draw $PD = PR$; hence $PD^2 = PR^2 = PB \cdot PC$ or PD is the mean proportional between PB, PC. Then draw $DE \parallel AB$. Join PE and produce it to meet AB in F. Join DF. DF will be $\parallel AC$, $\because PB/PD = PD/PC$ and PB/PD

$= PF/PE$. $\because$ DE is $\parallel AB$, $\therefore PD/PC = PF/PE$. $\therefore DC/PC = FE/PE$. $\therefore FD$ is $\parallel AC$.

4.6. *A, B, C, D are any four points on a straight line. On AB, CD as diameters, two circles are described and a common tangent EF is drawn to touch them in E, F respectively and meet AD produced in G. Show that (i) $AC \cdot BG = BD \cdot AG$; (ii) $EF^2 = AC \cdot BD$.*

CONSTRUCTION: Let M, N be the centers of the two circles on AB, CD as diameters. Join AE, BE, ME, CF, DF, NF (*Fig.* 106).

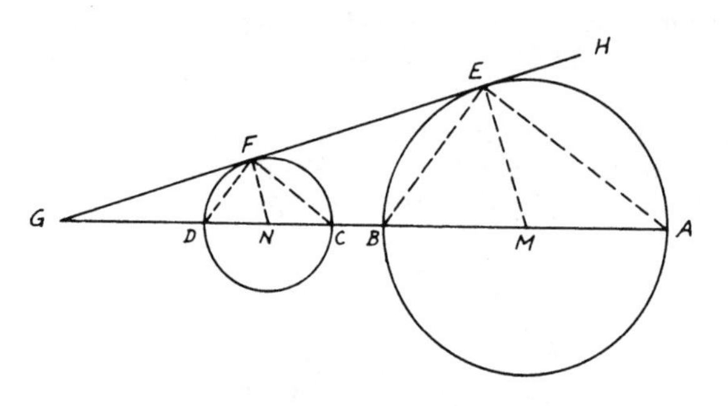

FIGURE .106

Proof: (i) Since $ME \parallel NF$ ($\perp EF$, Th. 3.68), $\therefore \angle AME = \angle CNF$. $\because \angle AEH = \frac{1}{2} \angle AME$ and $\angle CFH = \frac{1}{2} \angle CNF$ (Th. 3.72), $\therefore \angle AEH = \angle CFH$. $\therefore AE \parallel CF$. Similarly, $BE \parallel DF$. In $\triangle AEG$, $AG/AC = EG/EF$. Also, in $\triangle BEG$, $BG/BD = EG/EF$. $\therefore AG/AC = BG/BD$. $\therefore AC \cdot BG = BD \cdot AG$.

(ii) In $\triangle AEG$, $AC/EF = CG/FG$. Also, in $\triangle BEG$, $EF/BD = FG/GD$. Since $GF^2 = GD \cdot GC$ or $GC/FG = FG/GD$, $\therefore AC/EF = EF/BD$. $\therefore EF^2 = AC \cdot BD$.

4.7. *P and Q are two points in the sides AB, CD respectively of a quadrilateral $ABCD$ such that $AP : PB = CQ : QD$. Prove that if QA, QB, PC, PD be joined, the sum of the areas of the triangles QAB, PCD is equal to the area of the quadrilateral $ABCD$.*

CONSTRUCTION: Join AC, PQ (*Fig.* 107).

Proof: Assume $AP/PB = CQ/QD = a/b$. $\therefore \triangle APQ/\triangle QAB = a/(a+b)$ (Th. 4.91). $\therefore \triangle QAB = \triangle APQ\,((a+b)/a)$. Similarly, $\triangle CPQ/\triangle PCD = a/(a+b)$. $\therefore \triangle PCD = \triangle CPQ\,((a+b)/a)$. Hence, by adding, $\triangle QAB + \triangle PCD = [\triangle APQ + \triangle CPQ]((a+b)/a)$. But $\triangle ACQ/\triangle ACD = a/(a+b)$. $\therefore \triangle ACD = \triangle ACQ\,((a+b)/a)$. Also, $\triangle ABC = \triangle ACP\,((a+b)/a)$. $\therefore \triangle ACD + \triangle ABC = [\triangle ACQ$

$+ \triangle ACP]\ ((a + b)/a)$. $\therefore\ \triangle QAB + \triangle PCD = \triangle ACD + \triangle ABC$ $=$ quadrilateral $ABCD$.

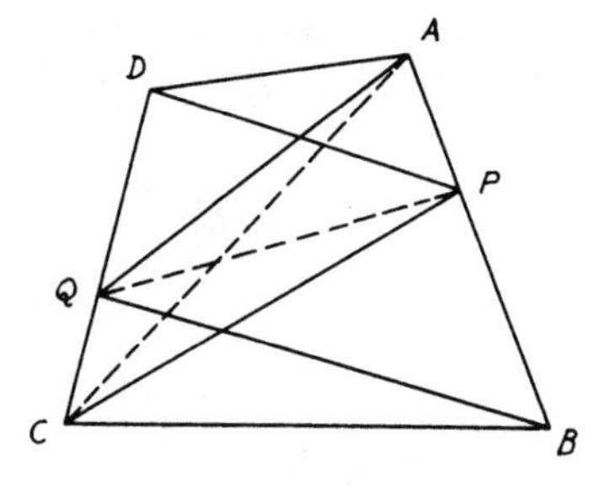

FIGURE 107

4.8. *A point P is given in the base of a triangle. Show how to draw a straight line cutting the sides of the triangle and parallel to the base, which will subtend a right angle at P.*

CONSTRUCTION: Let ABC be the given triangle and P is in BC. Bisect BC in D. With center D and radius DB describe a $\odot BEC$. Produce AP to meet the circle in E and draw $PF \parallel DE$ meeting AD in F. Then draw $GFH \parallel BC$ meeting AB, AC in G, H respectively. Join GP, HP (*Fig.* 108).

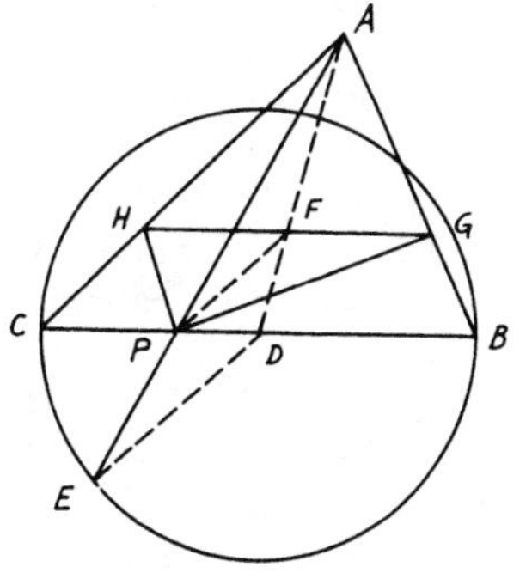

FIGURE 108

Proof: In $\triangle ABD$, $AF/FG = AD/DB$. Also, in $\triangle ADE$, $AF/FP = AD/DE$ (Th. 4.83). Since $DB = DE$ (radii in $\odot D$), $\therefore\ AF/FG = AE/FP$. $\therefore\ FG = FP$. Similarly, $FH = FP$. $\because\ FG = FH$ (since AD is a median and GFH is $\parallel BC$), $\therefore\ FG = FH = FP$. $\therefore\ F$ is the center of the circle GPH of which GH is a diameter. Hence $\angle GPH$ is a right angle.

4.9. *AB is a straight line and C is any point on it. On AB, BC, AC three equilateral triangles ABD, BCE, ACF are drawn such that the two small triangles are on the opposite side of the large one. If M, N, L are the centers of the inscribed circles in these triangles respectively, show that* $MN = ML$.

CONSTRUCTION: Join AM, BM, BN, CN, CD, MD (*Fig.* 109).

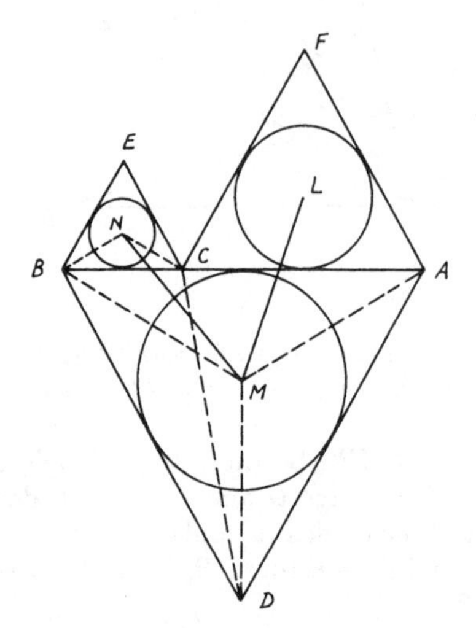

FIGURE 109

Proof: Since MB, NB bisect $\angle$s ABD, ABE respectively and $\because$ each $= 60°$, $\therefore$ $\angle MBD = \angle NBC$. Similarly, $\angle NCB = \angle MDB$. $\therefore$ $\triangle$s MBD, NBC are mutually equiangular and hence similar. $\therefore$ $MB/BD = NB/BC$. Again, $\angle CBD = \angle NBM = 60°$. Hence $\triangle$s CBD, NBM are similar (Th. 4.86). $\therefore$ $CD/BD = MN/BM$. Similarly, $ML/AM = CD/AD$. Since $BD = AD$, $\therefore$ $MN/BM = ML/AM$. $\because$ $BM = AM$ also, $\therefore$ $MN = ML$.

4.10. *Describe an equilateral triangle which will be equal in area to a given triangle ABC.*

CONSTRUCTION: Describe on BC an equilateral triangle BCG. Draw $AE \parallel BC$ meeting CG in E, and $EH \perp CG$ to meet the semi-circle described on CG as diameter in H. Take $CD = CH$ on CG and draw $DF \parallel BG$. $\triangle CDF$ is the required equilateral $\triangle$ (*Fig.* 110).

Proof: $CH^2 = CD^2 = CE \cdot CG$ (Th. 3.81). In $\triangle BCG$, $DF \parallel BG$, $\therefore$ $\triangle CDF/\triangle CBG = CD^2/CG^2 = CE \cdot CG/CG^2 = CE/CG$ (Th. 4.92).

But $\triangle CEB/\triangle CBG = CE/CG = \triangle ABC/\triangle CBG$ (Th. 4.91) (since

$\triangle CEB = \triangle ABC$ between the $\|$s AE, BC). Hence $\triangle CDF = \triangle ABC$ and it is an equilateral $\triangle$ (because $DF \parallel BG$). Therefore, $\triangle CDF$ is the required $\triangle$.

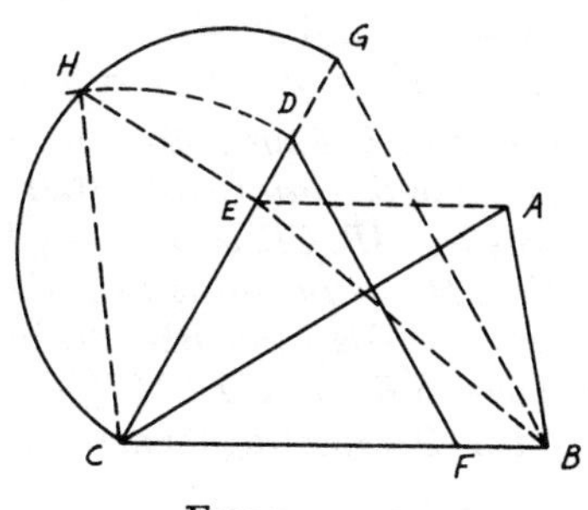

FIGURE 110

4.11. *AB, AC are two radii in a circle with center A making an angle* 120°. *Show how to inscribe a circle inside the sector ABC and, if D is the middle point of the arc BC of the sector, find the ratio of the area of this inscribed circle to that of the other circle inscribed in the segment BDC.*

CONSTRUCTION: Bisect $\angle BAC$ by AD. Draw $DE \perp AB$; then bisect $\angle ADE$ by DF meeting AB in F. FM is drawn $\perp AB$ meeting AD in M. Then M is the center of the inscribed $\odot$ in the sector ABC whose radius is MF or MD. Let AD, BC intersect in G and bisect DG in N. Describe another $\odot$ with center N and radius ND inscribed in the segment BDC (*Fig.* 111).

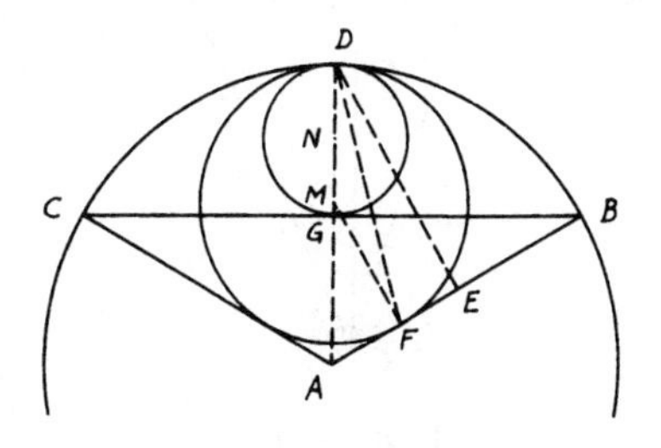

FIGURE 111

Proof: $\because MF \parallel DE$, $\therefore \angle MFD = \angle FDE$. But $\angle FDE = \angle FDA$ (since DF bisects $\angle ADE$). $\therefore \angle MFD = \angle FDA$. $\therefore MF = MD$. $\because MF \perp AB$ and D is the middle point of the arc BC, $\therefore \odot$ with M as center and $MF = MD$ as radii touches AB, arc BC, and similarly AC. In $\triangle AMF$, $\angle FAM = 60°$ and $\angle F =$ right angle. Hence MF

$= \sqrt{3}/2\, AM$. $\therefore AM = 2/\sqrt{3}\, MF$. $\therefore AD = AM + MD = (2/\sqrt{3})$ $MF + MF = ((2 + \sqrt{3})/\sqrt{3})\, MF$. $\therefore \odot M : \odot A = MF^2 : AD^2$ (Th. 4.103, Cor. 3) $= MF^2 : ((2 + \sqrt{3})/\sqrt{3})^2\, MF^2 = 1 : 4.64$ (approx.) (1). Again, in $\triangle GBA$, $\angle GAB = 60°$ and $AG \perp CB$. $\therefore AG = \frac{1}{2} AB$ $= GD$. $\therefore \odot N : \odot A = ND^2 : AD^2 = 1 : 16$ (2). Hence, dividing (1) by (2) yields $\odot M : \odot N = 16 : 4.64 = 3.45 : 1$ (approx.).

4.12. *ABCD is a quadrilateral in a circle whose diagonals intersect at right angles. Through O the center of the circle GOG′, HOH′ are drawn parallel to AC, BD respectively, meeting AB, CD in G, H and DC, AB, produced in G′, H′. Show that GH, G′H′ are parallel to BC, AD respectively.*

CONSTRUCTION: Let OG, OH meet BD, AC in E, F respectively. Produce BA, CD to meet in K. Draw $KLM \perp CA$, OG meeting them in L, M and $KPQ \perp BD$, OH meeting them in P, Q respectively (*Fig.* 112).

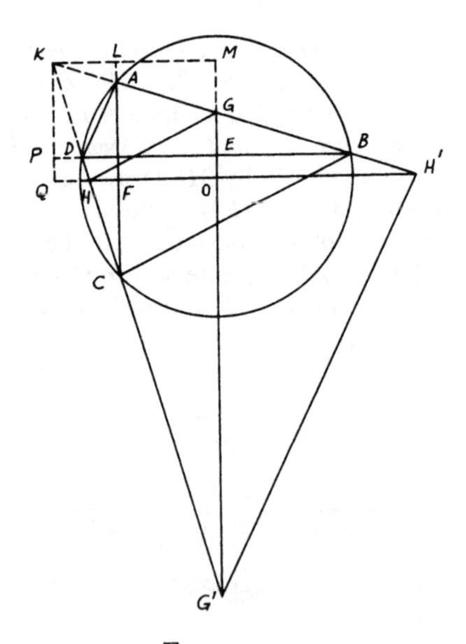

FIGURE 112

Proof: $\angle ALK = \angle DPK$ = right angle and $\angle LAK = \angle BAC$ $= \angle BDC = \angle KDP$. Hence $\triangle$s LAK, PDK are similar. $\therefore AL/DP$ $= AK/DK$. But $\triangle$s KAC, KDB are also similar. $\therefore AK/DK = AC/BD$ $= AF/DE$ (since F, E are the mid-points of AC, BD). $\therefore AL/DP$ $= AF/DE$ or $AL/AF = DP/DE$. $\therefore (AL + AF)/AF = (DP + DE)/DE$ or $KQ/AF = KM/DE$. $\therefore KQ/KM = AF/DE = FC/BE$. $\because$ $\triangle$s

BEG, CFH are similar, $\therefore FC/BE = CH/BG$. $\because$ $\triangle$s KHQ, KGM are similar, $\therefore KQ/KM = KH/KG = CH/BG$. $\therefore GH \parallel BC$ [Th. 4.83(ii)]. Again, $\triangle$s AFH', DEG' are similar. $\therefore AH'/DG' = AF/DE = CF/BE = CH/BG = KH/KG$. $\because$ $KA \cdot KG = KD \cdot KH$ ($\because$ $ADHG$ is cyclic), $\therefore KH/KG = KA/KD = AH'/DG'$. $\therefore G'H' \parallel AD$.

4.13. *Describe a right-angled triangle of given perimeter such that the rectangle contained by the hypotenuse and one side will be equal to the square on the other side.*

CONSTRUCTION: Let AB be the given perimeter. On AB describe a semi-circle AEB. Find E such that if ED be drawn $\perp AB$, $EB = AD$. This is similar to dividing AB in D such that $AD^2 = AB \cdot BD$ (see Problem 2.24) ($AJ = \frac{1}{2} AB$ and $JB = JK$). Bisect $\angle$s EAB, EBA by AF, BF. Draw $FG \parallel AE$, and $FH \parallel BE$ meeting AB in G, H (*Fig.* 113).

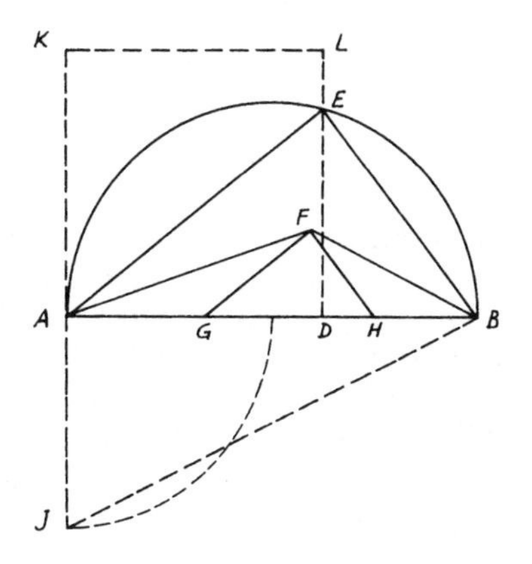

FIGURE 113

Hence FGH is the required $\triangle$. (Note that $\triangle$s GFH, AEB are similar.)

Proof: $\angle GFH$ is right. Also, $\angle GAF = \angle FAE =$ alternate $\angle AFG$. $\therefore GF = GA$. Similarly, $HF = HB$. $\therefore GH + GF + HF = AB =$ given perimeter. $\because$ $AB \cdot BE = AE^2$, $\therefore AB/AE = AE/BE$ and $AB/AE = GH/GF$. Also, $AE/BE = GF/FH$. $\therefore GH/GF = GF/FH$. Hence $GF^2 = GH \cdot FH$.

4.14. *In a triangle ABC, D is the middle point of BC. G, H are the points in which the inscribed and escribed circles touch BC. E is the foot of the perpendicular from A to BC, F the point in which the bisector of the angle A meets BC. Prove that* $EG \cdot FH = EH \cdot FG$.

CONSTRUCTION: Let M, N be the centers of the inscribed and escribed ⊙s. Join FN, NH, NB, BM, MG. Draw $DR \perp BC$ meeting FN in R (*Fig.* 114).

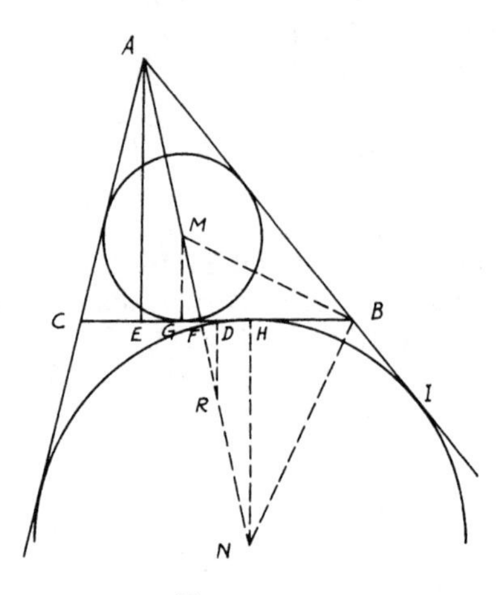

FIGURE 114

Proof: $\because$ BM bisects $\angle ABC$ and BN bisects $\angle CBI$, and since AMN is a straight line bisecting $\angle A$, $\therefore$ in $\triangle ABF$, $AM/MF = AN/NF = AB/BF$ (Th. 4.90). In $\triangle AEF$, $\because$ $MG \parallel AE$ (both $\perp BC$), $\therefore$ $AM/MF = EG/GF$. But, since $\triangle$s AFE, HFN are similar, $\therefore$ $AF/FN = FE/FH$. $\therefore$ $(AF + FN)/FN = (FE + FH)/FH$ or $AN/FN = EH/FH$. $\therefore$ $EH/FH = EG/GF$. $\therefore$ $EG \cdot FH = EH \cdot FG$.

4.15. *ABC is a right-angled triangle at A, and AD is perpendicular to BC. M, N are the centers of the circles inscribed in the triangles ABD, ACD. If ME, NF are drawn parallel to AD meeting AB, AC in E, F respectively, then AE = AF.*

CONSTRUCTION: Let the ⊙s M, N touch BC, AD in G, H and P, Q respectively. Join MG, MP, PB, MB, NQ, AN, AH, NH (*Fig.* 115).

Proof: $\angle MBG = \frac{1}{2} \angle B = \frac{1}{2} \angle CAD = \angle NAQ$. $\therefore$ right $\triangle$s MBG, NAQ are similar. $\therefore$ $MB/MG = AN/NQ$. $\therefore$ $MB/MP = AN/NH$. Since $\angle BMG = \angle ANQ$, $\therefore$ $\angle BMP = \angle ANH$. $\therefore$ $\triangle$s BMP, ANH are similar. $\therefore$ $\angle MBP = \angle NAH$. $\therefore$ $\angle PBD = \angle HAD$. Hence $\triangle$s PBD, HAD are similar. $\therefore$ $BD/DP = AD/HD$. $\therefore$ $BD \cdot DH = AD \cdot DP = AD \cdot DG$. $\therefore$ $BD \cdot DH/CD = AD \cdot DG/CD$. $\therefore$ $BD \cdot DH/CD \cdot DG = AD/CD = AB/AC$. $\because$ EMG is a straight line $\parallel AD$, $\therefore$ $BD/DG = AB/AE$. Similarly, $DH/CD = AF/AC$. Multiplying, $\therefore$ $BD \cdot DH/CD \cdot DG = AB \cdot AF/AE \cdot AC = AB/AC$. $\therefore$ $AE = AF$.

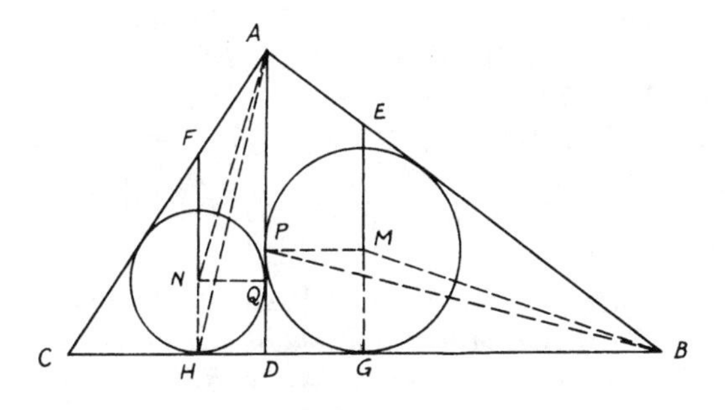

FIGURE 115

4.16. *ABC is a triangle and PQR is the triangle formed by the points of contact of the inscribed circle with AB, BC, CA respectively. Show that the product of the three perpendiculars drawn from any point in its circumference on the sides of triangle ABC is equal to the product of the other three perpendiculars drawn from that point on the sides of triangle PQR. If M is the center of this circle and BT, RS are perpendiculars, to AC, PQ respectively, then* $BT \cdot MR = MB \cdot RS$.

CONSTRUCTION: Let D be any point on the inscribed $\odot M$. Draw DE, DF, DG $\perp$s AB, BC, CA respectively and DH, DK, DL $\perp$s PQ, QR, RP respectively. Join DP, DR, LE, LG, MP. Draw also MN, MO $\perp$s RS, BT, and let BM cut PQ in J (*Fig.* 116).

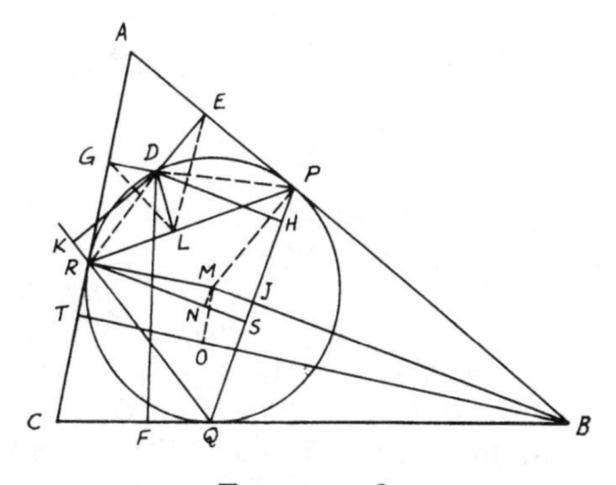

FIGURE 116

Proof: (i) Quadrilaterals $DEPL$, $DGRL$ are cyclic. $\therefore \angle DLE = \angle DPE$ and $\angle DGL = \angle DRL$. $\because \angle DPE = \angle DRL$ or DRP, $\therefore \angle DLE = \angle DGL$. Similarly, $\angle DEL = \angle DLG$. $\therefore \triangle$s DLE, DGL are similar. $\therefore DL/DE = DG/DL$. $\therefore DL^2 = DE \cdot DG$. Similarly,

$DH^2 = DE \cdot DF$ and $DK^2 = DF \cdot DG$. Multiplying and taking square roots, $\therefore DL \cdot DH \cdot DK = DE \cdot DF \cdot DG$.

(ii) $\triangle$s BMO, RMN are similar. $\therefore BM/RM = BO/RN$, since $\angle MPB$ is right and MB is $\perp PQ$ at J. $\therefore MP^2 = MJ \cdot MB = MR^2$. $\therefore MB/MR = MR/MJ = BO/RN$. $\therefore MB/MR = (MR + BO)/(MJ + RN) = (TO + BO)/(NS + RN) = BT/RS$. $\therefore BT \cdot MR = MB \cdot RS$.

4.17. *D, E are two points on the sides AB, AC of a triangle ABC such that $\angle AED = \angle B$. BE, CD intersect in F and BC, AF are bisected in M, N. If MN is produced to meet AB, AC or produced in R, Q respectively, show that AF is a common tangent to the circles circumscribing triangles ARQ, FRQ.*

CONSTRUCTION: Bisect AB, AC in G, H and join NG, NH. Through N draw LNP, $SNT \parallel$ to AB, AC meeting AC, GM, AB, MH in L, P, S, T respectively (*Fig.* 117).

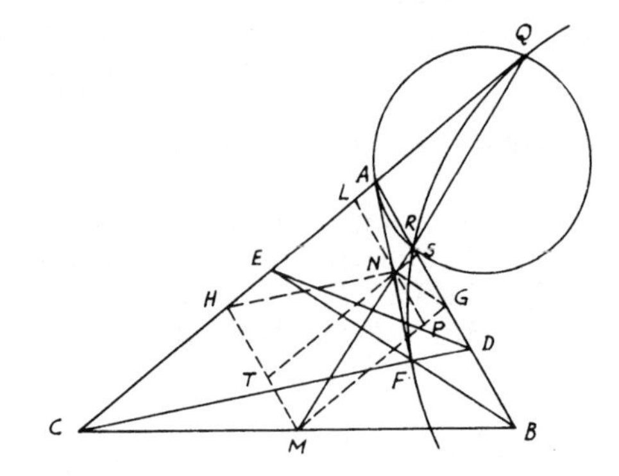

FIGURE 117

Proof: $\because GN \parallel BF$ and $HN \parallel CF$, $\therefore \angle AGN = \angle ABF$ and $\angle AHN = \angle ACF$. Since quadrilateral $DBCE$ is cyclic, $\therefore \angle ABF = \angle ACF$. $\therefore \angle AGN = \angle AHN$. $\because \angle GSN = \angle A = \angle HLN$, $\therefore \triangle$s GSN, HLN are similar. $\therefore NS/NL = GS/HL = NP/PM$. $\therefore NS/AS = NP/PM$, since $\angle ASN = \angle AGM = \angle NPM$. Hence $\triangle$s ASN, MPN are similar. $\therefore \angle PMN = \angle NAS = \angle NQA$ (since $GM \parallel CAQ$). $\therefore AN$ is tangent to $\odot ARQ$ [Th. 3.72(ii)]. $\therefore AN^2 = NR \cdot NQ = NF^2$. $\therefore NF$ is tangent to $\odot FRQ$ (Th. 3.80); i.e., AF is a common tangent to $\odot$s circumscribing $\triangle$s ARQ, FRQ.

4.18. *Points D, E, F are taken in the sides BC, CA, AB respectively of a triangle ABC so that BD, CE, AF may be equal. Through D, E, F lines are drawn parallel to CA, AB, BC so as to form a triangle GHK in which KH is parallel to BC (Fig.* 118). *Show that* (i)$2a - KH : 2b - GH : 2c - GK$

$= a : b : c$; *(ii) area* $GHK = \triangle ABC\{2 - (p/a + p/b + p/c)\}^2$, *where* a, b, c *denote the sides* BC, CA, AB *respectively and* p *stands for* BD.

Proof: (i) $\because$ $\triangle$s ABC, GKH are similar, $\therefore$ $BC/KH = AC/GH = AB/GK$. $\therefore$ $(2\,BC - KH)/BC = (2\,AC - GH)/AC = (2\,AB - GK)/AB$. Hence $2a - KH : 2b - GH : 2c - GK = a : b : c$.

$$\text{(ii)}\ \triangle ABC\left\{2 - \left(\frac{p}{a} + \frac{p}{b} + \frac{p}{c}\right)\right\}^2$$

$$= \triangle ABC\left\{2 - \left(\frac{BD}{BC} + \frac{CE}{CA} + \frac{AF}{AB}\right)\right\}^2$$

$$= \triangle ABC\left\{2 - \left(\frac{BD}{BC} + \frac{CL}{CB} + \frac{FM}{BC}\right)\right\}^2$$

$$= \triangle ABC\left\{2 - \left(\frac{BD}{BC} + \frac{CD + DL}{BC} + \frac{MH + FK + KH}{BC}\right)\right\}^2$$

$$= \triangle ABC\left\{2 - \left(\frac{BD + CD + DL + CD + BL}{BC} + \frac{KH}{BC}\right)\right\}^2$$

$$= \triangle ABC\frac{KH^2}{BC^2} = \triangle ABC\,\frac{\triangle GHK}{\triangle ABC} = \triangle GHK.$$

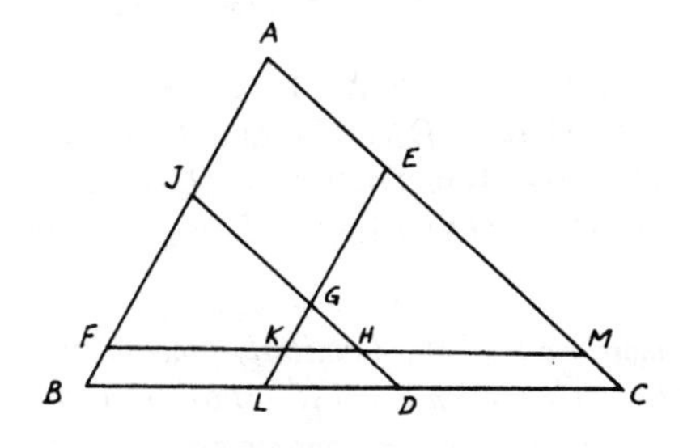

FIGURE 118

4.19. *Given two straight lines* OA, OB *and two points* A, B *in them, and a point* P *between them. It is required to draw through* P *a straight line,* XPY, *so that* XB *is parallel to* AY.

CONSTRUCTION: Join OP. Draw OD making an $\angle DOA = \angle POB$. Take OD a fourth proportional to OA, OB, OP; i.e., $OD/OA = OB/OP$. Then draw on PD a ⊙ subtending an $\angle = \angle DOB$. If this ⊙ does not cut OA, it is impossible to solve the problem. If on the other hand, the ⊙ touches OA, it has one solution, and if ⊙ cuts OA in two points, it has two solutions. Suppose ⊙ cuts OA in X, X'; join XP and produce it to meet OB in Y; then XB is $\parallel AY$ and similarly $X'B$ is $\parallel AY'$ (*Fig.* 119).

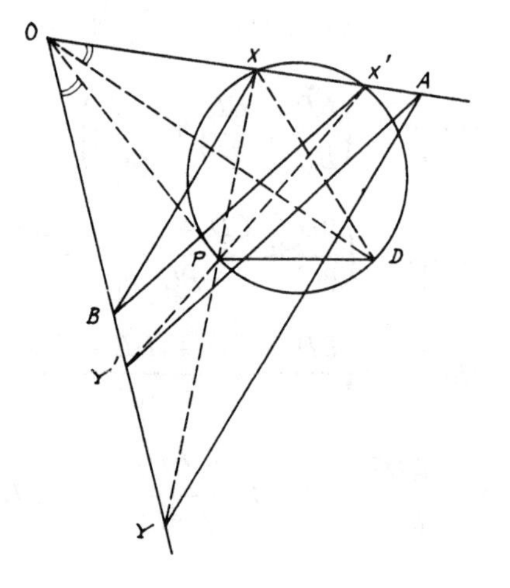

FIGURE 119

Proof: $\because$ $\angle OPY = \angle POX + \angle PXO = \angle DOB + PXO = \angle PXD + \angle PXO = \angle DXO$. Hence △s OPY, OXD are similar. $\therefore$ $OP/OY = OX/OD$. But $OB/OP = OD/OA$ (construct). $\therefore$ by multiplying, $OB/OY = OX/OA$. $\therefore$ XB is $\parallel AY$ [Th. 4.83(ii)]. Also, $X'B$ is $\parallel AY'$.

4.20. *If two homothetic triangles (similarly placed) are described such that one lies completely inside the other and if a third triangle can be constructed to circumscribe the smaller triangle and has its vertices on the sides of the larger triangle, then this third triangle will be the mean proportional of the first two.*

CONSTRUCTION: Let ABC, DEF be the two homothetic △s and GHK the third △. Join AE, AF and produce them to meet BC in M, N; then join AD, FM. Draw $DL \parallel FE$ and join FL (Fig. 120).

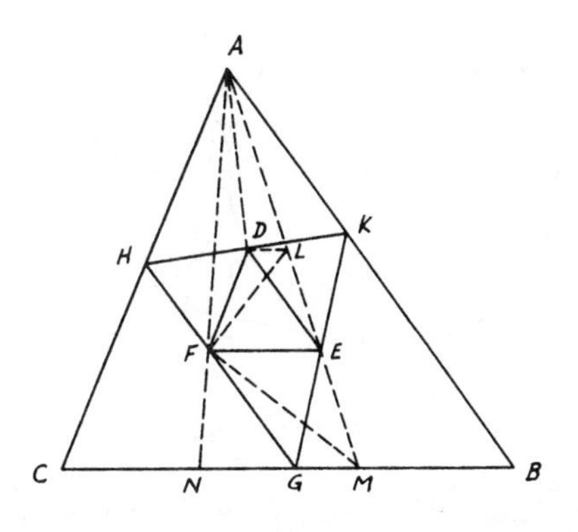

FIGURE 120

Proof: $\triangle DHF = \triangle DAF$ (Th. 2.39). Similarly, $\triangle DKE = \triangle DAE$ and $\triangle FGE = \triangle FME$. Hence $\triangle GHK = \triangle AFM$. Also, $\triangle DEF = \triangle LEF$.

$$\therefore \frac{\triangle GHK}{\triangle DEF} = \frac{\triangle AFM}{\triangle DEF} = \frac{\triangle AFM}{\triangle LEF} = \frac{AM}{LE}$$

$$= \frac{\text{Alt. } \triangle AFM \text{ from } A \text{ on } BC}{\text{Alt. } \triangle LEF \text{ from } L \text{ on } FE} = \frac{BC}{FE}$$

Since $\triangle ABC/\triangle GHK = \triangle ABC/\triangle AFM$, also $\triangle AFM/\triangle AMN = AF/AN = FE/MN$ and $\triangle AMN/\triangle ABC = MN/BC$. $\therefore$ $\triangle AFM/\triangle ABC = FE/BC = \triangle GHK/\triangle ABC = \triangle DEF/\triangle GHK$. Hence the result is obvious.

4.21. *A, B are the centers of two circles which touch externally. If the two common tangents CD, EF are drawn to the circles so that C, E lie on circle A, find the area of the trapezoid CEFD in terms of the radii.*

CONSTRUCTION: From G the contact point of $\odot$s, draw the third common tangent PGQ. Join AB, BF and draw AK, PR $\perp$s BF, FE respectively (*Fig.* 121).

Proof: Let r, r' be the radii of $\odot$s A, B respectively. Since PGQ bisects CD, EF, $\therefore$ $\triangle QCD + \triangle PEF =$ trapezoid $CEFD = 2\triangle PEF$. $\therefore$ Trapezoid $CEFD = PR \cdot EF$. $\because$ $AB = r + r'$ and $BK = r' - r$, $\therefore$ $AK^2 = AB^2 - BK^2 = 4rr'$. Hence $AK = 2\sqrt{rr'} = EF$. Since $\triangle$s PQR, ABK are similar ($GBFQ$ is cyclic), $\therefore$ $PR^2/PQ^2 = AK^2/AB^2$. $\because$ $EF = AK = PQ$, $\therefore$ $PR^2/EF^2 = EF^2/AB^2$. Therefore, $PR^2/4rr'$

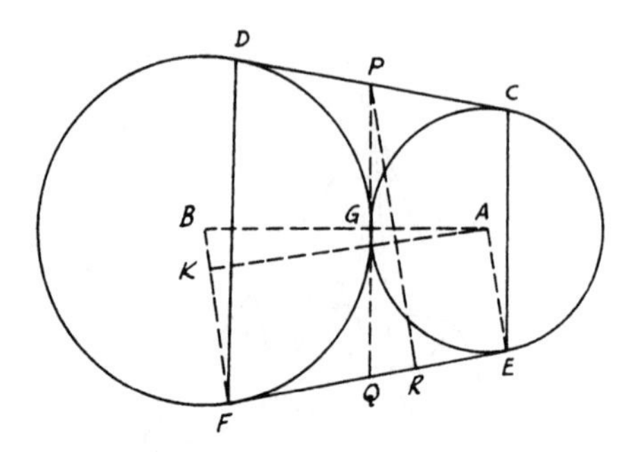

FIGURE 121

$= 4\,rr'/(r + r')^2$ or $PR = 4\,rr'/(r + r')$. Hence trapezoid $CEFD = PR \cdot EF = 8\,rr'\sqrt{rr'}/(r + r')$.

4.22. *From a point P in the base BC of a triangle ABC, PM, PN are drawn perpendiculars to AB, AC respectively and the parallelogram AMNR is completed. Show that R lies on a fixed straight line.*

CONSTRUCTION: Draw BE, CD $\perp$s AC, AB respectively. Complete the $\square$s $ABES$, $ADCQ$. Join QR, QS and draw $CFG \parallel QR$, meeting NR, ES in F, G. Then QS is the fixed line on which R lies (*Fig.* 122).

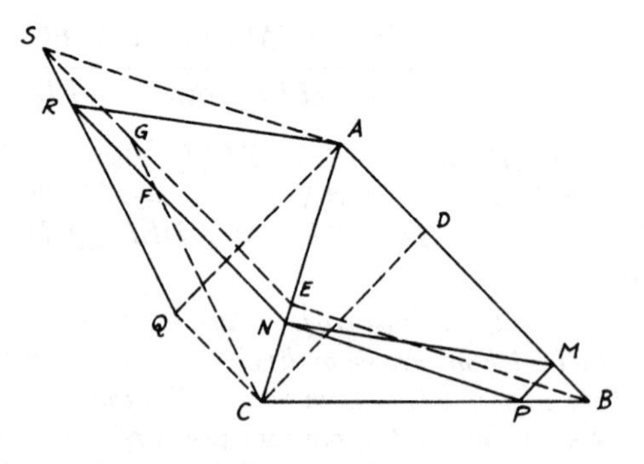

FIGURE 122

Proof: If it is shown that QR coincides with QS, then QRS is one line. Now, $RF = QC = AD$. $\therefore$ $FN = DM$ (since $RN = AM$). $\because$ $GE/FN = EC/NC = BC/CP = BD/DM$, $\therefore$ $GE = BD$. Hence $SG = AD = QC$. $\therefore$ $SGCQ$ is a $\square$. $\therefore$ $CG \parallel QS$. Since it is also $\parallel QR$ (by construction), $\therefore$ QR coincides with QS. But, BE, CD are fixed alternates of $\triangle ABC$. Hence QRS is also fixed.

4.23. *Circles are described on the sides of a quadrilateral ABCD, whose diagonals are equal, as diameters. Prove that the four common chords of pairs of circles on adjacent sides form a rhombus.*

CONSTRUCTION: Let $ABCD$ be the original quadrangle and $EFGH$ be the quadrangle formed by the intersection of the common chords $BPEF$, $ASGF$, $ECQH$, $GDRH$. Join PQ, QR, RS (*Fig.* 123).

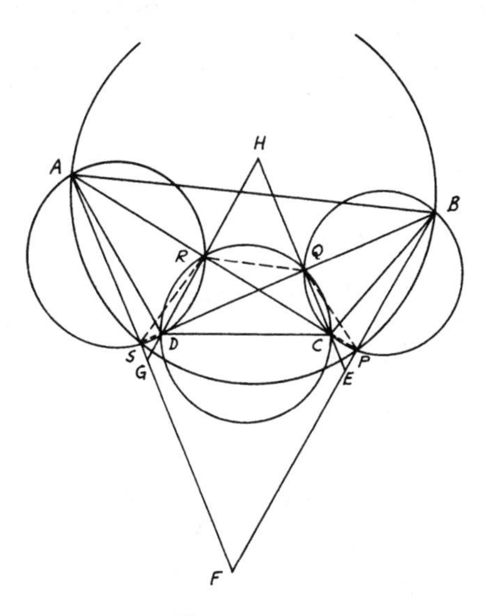

FIGURE 123

Proof: Since ⊙s on AB, BC as diameters intersect on AC ($\because \angle APB = \angle CPB =$ right), then P lies on AC or produced. Similarly, for ⊙s taken in pairs, their points of intersection lie on AC, BD. $\because$ quadrangle $ARDS$ is cyclic, $\therefore \angle QSR = \angle CAD$. Similarly, quadrangle $QCDR$ is cyclic. $\therefore \angle SQR = \angle ACD$. Hence △s QRS, CDA are similar. $\therefore CD/AC = QR/QS$. Also, △s PQR, BCD are similar. $\therefore BD/CD = PR/QR$.

Multiplying yields $BD/AC = PR/QS$. $\because BD = AC$, $\therefore PR = QS$. Since also △s HCR, HDQ are similar, $\therefore HD/QD = HC/CR$. But $\angle HQD = \angle QSA =$ right. $\therefore HQ \parallel SG$. Similarly, $HR \parallel EP$. $\therefore HD/QD = HG/QS$ and $HC/CR = HE/PR$. $\therefore HG/QS = HE/PR$. $\therefore HG = HE$. $\because HE \parallel GF$ ($\perp SB$), $\therefore EFGH$ is a rhombus. Hence a ⊙ can be inscribed inside the rhombus $EFGH$.

4.24. *Given the vertical angle of a triangle, the ratio of the radii of the circles inscribed in the two triangles formed by the line bisecting the vertical angle and* *meeting the base, and the distance between the centers of these circles, construct the triangle.*

CONSTRUCTION: Let BAC be the given vertical angle of $\triangle$. Bisect $\angle BAC$ by AP and $\angle$s BAP, CAP by AQ, AR. On AQ, AR take the distances AD, AE with the same given ratio of the radii. Between AD, AE, take the distance MN = given distance between the centers of inscribed $\odot$s and $\parallel DE$. This is done by taking EL = this distance on DE and drawing $LM \parallel AR$. On MN as diameter, construct a semicircle to cut AP in F. Then draw FB, FC making $\angle MFB = \angle MFA$ and $\angle NFC = \angle NFA$ (*Fig.* 124). Therefore ABC is the required $\triangle$.

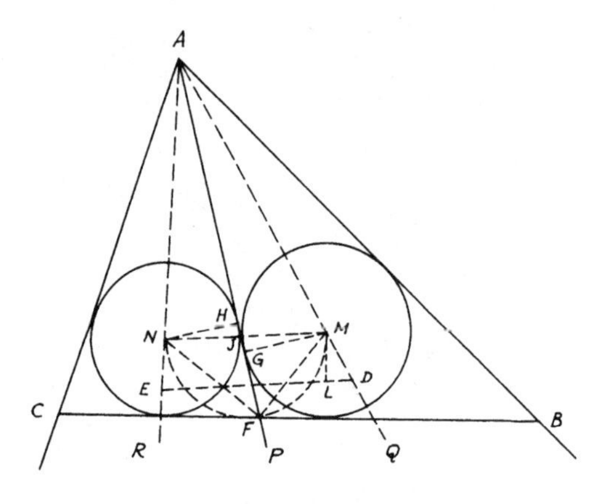

FIGURE 124

Proof: Draw the two $\odot$s with centers M, N and radii MG, NH (which are $\perp$s AP). Now, $\triangle$s MGJ, NHJ are similar, and AJ bisects $\angle MAN$. $\therefore$ $MJ/JN = MG/NH = AM/AN = AD/AE$ = given ratio. Since FM, FN bisect $\angle$s AFB, AFC, $\therefore$ M, N are the centers of inscribed $\odot$s in $\triangle$s AFB, AFC and ratio of their radii $MG : NH$ = given ratio. Again, $\angle MFN$ = right. $\therefore$ $\angle AFB + \angle AFC$ = 2 right angles. Hence BFC is a straight line touching $\odot$s M, N with MN = given distance.

4.25. *$ABCD$ is a quadrilateral inscribed in a circle. BA, CD produced and AD, BC produced meet in E, F. If G, H, K are the middle points of BD, AC, EF and M is the intersection of BD, AC, show that (i) $AC/BD = 2\,HK/EF = EF/2\,GK$; (ii) $GH/EF = (BD^2 - AC^2)/2\,BD \cdot AC$. If also from M a line parallel to BC is drawn meeting AB, CD, AF, EF in P, Q, R, S, prove that $RS^2 = RP \cdot RQ$.*

CONSTRUCTION. Join FH and produce it to N such that $FH = HN$; then join AN, EN (*Fig.* 125).

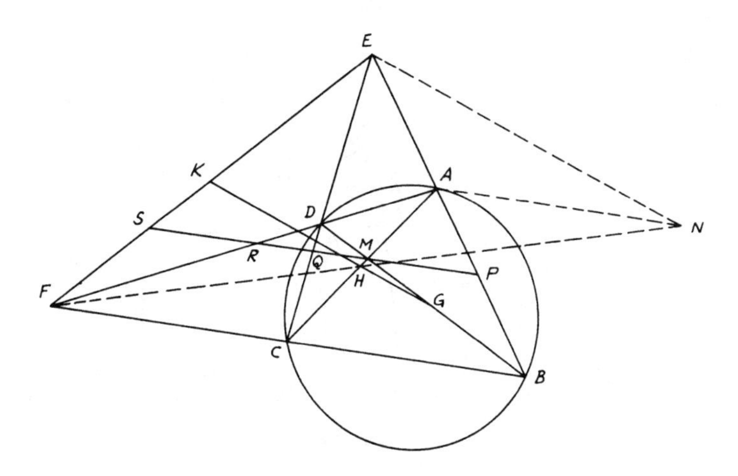

FIGURE 125

Proof: (i) From congruence of $\triangle$s AHN, CHF : $AN =$ and $\parallel CF$. Hence $\angle NAE =$ supp. of $\angle NAB =$ supp. of $\angle ABC = \angle ADC = \angle EDF$. In similar $\triangle$s EAC, EDB : $AC/BD = AE/ED$. Also in similar $\triangle$s FDB, FCA : $AC/BD = CF/DF$. $\therefore$ $AE/ED = CF/DF$. Hence $ED/DF = AE/CF = AE/AN$. Therefore, $\triangle$s ANE, DFE are similar (Th. 4.86). $\therefore$ $AE/ED = EN/EF$. $\therefore$ $AC/BD = AE/ED = EN/EF = 2\,HK/EF$. Similarly, $AC/BD = EF/2\,GK$. Therefore, $AC/BD = 2\,HK/EF = EF/2\,GK$.

(ii) Since the mid-points of the three diagonals of a complete quadrangle are collinear (Problem 2.16) and $\because$ $AC/BD = 2\,HK/EF = EF/2\,GK$,

$$\therefore \quad \frac{1}{2}\left(\frac{BD}{AC} - \frac{AC}{BD}\right) = \frac{1}{2}\left(\frac{2\,GK}{EF} - \frac{2\,HK}{EF}\right) \text{ or } \frac{BD^2 - AC^2}{2\,BD \cdot AC} = \frac{GH}{EF}.$$

(iii) $SQ/PQ = CF/BC$. Also, $QR/QM = CF/BC$. $\therefore$ $CF/BC = (SQ - QR)/(PQ - QM) = RS/PM$. Since $CF/BC = RM/PM$ (in $\triangle ABF$), $\therefore$ $RM = RS$. $\because RQ/RM = RM/RP = FC/FB$, $\therefore$ $RM^2 = RP \cdot RQ = RS^2$.

4.26. *If D, D′ and E, E′ and F, F′ are the points where the bisectors of the interior and exterior angles of a triangle ABC meet the opposite sides BC, CA, AB or produced in pairs, show that (i)* $1/DD' = 1/EE' + 1/FF'$;

(*ii*) $a^2/DD' = b^2/EE' + c^2/FF'$, *where* a, b, c *are the lengths of the sides* BC, BA, *respectively* (Fig. 126).

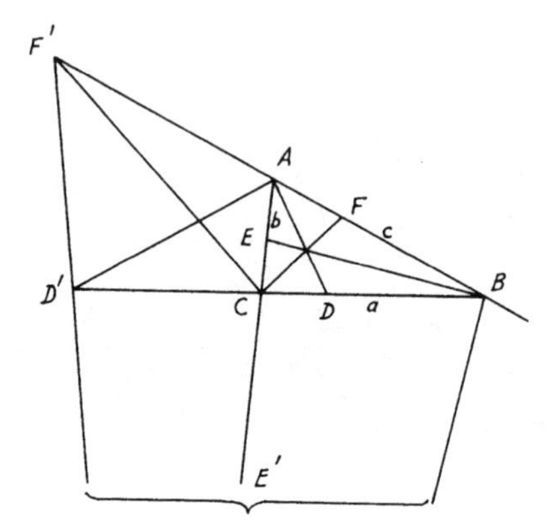

FIGURE 126

Proof: (i) $BD : C = CD : b$ [Th. 4.90(i)]. $\therefore$ $DB + CD : (b + c) = CD : b$. $\therefore$ $CD = ab/(b + c)$. Similarly, $CD' = ab/(c - b)$. Hence $CD + CD' = DD' = 2\,abc/(c^2 - b^2)$. Likewise, it can be shown that $EE' = 2\,abc/(c^2 - a^2)$ and $FF' = 2\,abc/(a^2 - b^2)$. Assuming $c > a$ and $a > b$ and considering counterclockwise rotation to be (+), then CB, BA, AC or $D'D$, FF', EE' have a (+) sign. Therefore, DD' is a (−) quantity. Hence

$$-\frac{1}{DD'} + \frac{1}{EE'} + \frac{1}{FF'} = \frac{1}{2abc}\{b^2 - c^2 + c^2 - a^2 + a^2 - b^2\} = 0.$$

$\therefore$ $1/DD' = 1/EE' + 1/FF'$.

(*ii*) Similarly,

$$-\frac{a^2}{DD'} + \frac{b^2}{EE'} + \frac{c^2}{FF'} = \frac{1}{2abc}\{a^2b^2 - a^2c^2 + b^2c^2 - b^2a^2 + c^2a^2 - c^2b^2\} = 0.$$

$\therefore$ $a^2/DD' = b^2/EE' + c^2/FF'$.

4.27. *G, H are the middle points of the diagonals BD, AC of a quadrilateral ABCD. If GH produced from both sides meets the sides AB, BC, CD, DA in E, F, J, K respectively, show that* (*i*) $AE : EB = FC : FB = CJ : JD = AK : KD$; (*ii*) $HE : HF = HJ : HK = GE : GK = GJ : GF$ *and if a circle with center M can be inscribed in the quadrilateral ABCD, then* (*iii*) $ME : MJ = AB : CD$; (*iv*) $MF : MK = BC : AD$.

CONSTRUCTION: Draw the $\perp$s AN, BO, CL, DT on transversal $FEGHJK$ (*Fig.* 127).

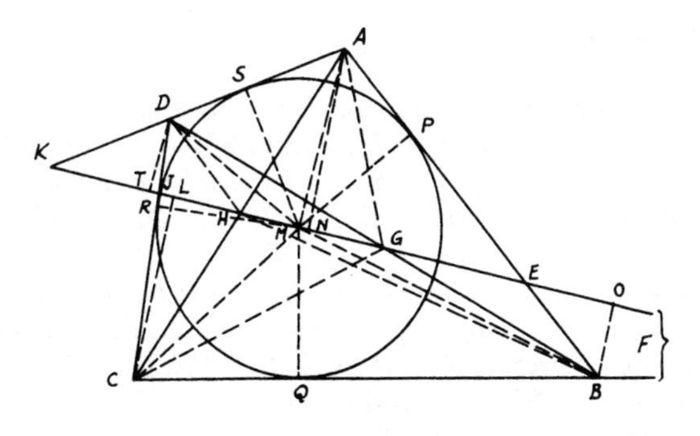

FIGURE 127

Proof: (i) Since G is the mid-point of BD, then $\triangle$s BOG, DGT are congruent. $\therefore BO = DT$. Similarly, $AN = CL$. Now, $\triangle$s BOE, ANE are similar. $\therefore AE/BE = AN/BO$. Similarly, $FC/FB = CL/BO = AN/BO$, $CJ/JD = CL/DT = AN/BO$, and $AK/KD = AN/DT = AN/BO$. Hence $AE/BE = FC/FB = CJ/JD = AK/KD$.

(ii) Suppose that the previous ratios are denoted (a). Join AG, CG, BH, HD. Now,

$$\frac{HE}{HF} = \frac{\triangle BEH}{\triangle BFH} = \frac{\triangle BEH\ \triangle BCH}{\triangle BFH\ \triangle BFH} = \frac{BE\ BC}{BA\ BF} = \frac{a-1}{a+1} \quad (\because \triangle ABH = \triangle BCH).$$

Similarly, $HJ/HK = GE/GK = GJ/GF = (a-1)/(a+1)$. Hence these ratios are equal.

(iii) Let the incircle touch the sides AB, BC, CD, DA in P, Q, R, S. Join MA, MB, MC, MD, MP, MQ, MR, MS. $\because \triangle BEM/\triangle DJM = ME\cdot BO/MJ\cdot DT = ME/MJ$. Similarly, $\triangle AEM/\triangle CJM = ME/MJ$. Hence

$$\frac{\triangle BEM + \triangle AEM}{\triangle DJM + \triangle CJM} = \frac{\triangle AMB}{\triangle DMC} = \frac{ME}{MJ} = \frac{MP \cdot AB}{MR \cdot CD} = \frac{AB}{CD}$$

(since $MP = MR$).

(iv) In a similar way to (iii),

$$\frac{\triangle FMC - \triangle BFM}{\triangle MAK - \triangle MDK} = \frac{\triangle BMC}{\triangle AMD} = \frac{MF}{MK} = \frac{MQ \cdot BC}{MS \cdot AD} = \frac{BC}{AD}$$

(since $MQ = MS$).

4.28. *If O is the center of the circumscribed circle of the triangle ABC, I is the center of the inscribed circle, I_a is the center of the escribed circle opposite to A, R, r, r_a are the radii of these three circles in order, and N is the center of the nine-point circle, then (i) $OI^2 = R^2 - 2\,Rr$; (ii) $OI_a{}^2 = R^2 + 2\,Rr_a$; (iii) $IN = \frac{1}{2}R - r$; (iv) $I_aN = \frac{1}{2}R + r_a$; (v) $R^2 - OG^2 = \frac{1}{9}\{AB^2 + BC^2 + CA^2\}$, where G is the centroid of the triangle ABC.*

CONSTRUCTION: (i) Join AII_a cutting the $\odot O$ in D. DO produced meets the circumference of $\odot O$ in E. Join CE, CD, IF (F point of contact of $\odot I$ and AC) and produce OI from both sides to meet the circumference in M, X (*Fig.* 128).

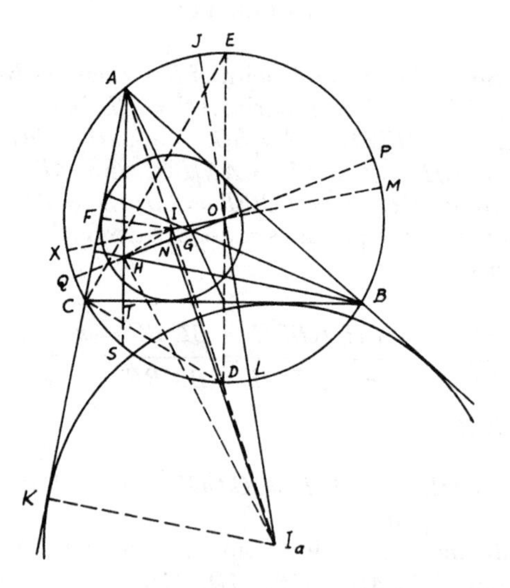

FIGURE 128

Proof: $\triangle$s AIF, EDC are similar. $\therefore$ $AI/DE = IF/CD$. $\therefore AI \cdot CD = IF \cdot DE$. $\because$ $CD = DI$ (see Problem 3.10), $\therefore$ $AI \cdot DI = 2\,Rr$. Since $AI \cdot DI = MI \cdot IX = OM^2 - OI^2$, $\therefore$ $OI^2 = R^2 - 2\,Rr$.

(ii) Produce I_aO to meet the circumference of $\odot O$ in J, L and join I_aK (K point of contact of $\odot I_a$ and AC produced). Again, $\triangle$s AI_aK,

EDC are similar. $\therefore$ $AI_a/DE = KI_a/CD$. $\therefore$ $AI_a \cdot CD = KI_a \cdot DE = 2\,Rr_a$. $\because$ $CD = DI_a$, $\therefore$ $AI_a \cdot DI_a = I_aJ \cdot I_aL = OI_a^2 - R^2$. $\therefore$ $OI_a^2 = R^2 + 2\,Rr_a$.

(iii) H is the orthocenter of $\triangle ABC$. Join MH, OH and produce the latter to meet $\odot O$ in P, Q. Produce alt. AT to meet $\odot O$ also in S. $\because$ N is the middle point of OH [Th. 3.77(i)], $\therefore$ in $\triangle OIH$, $OI^2 + HI^2 = 2\,IN^2 + 2\,NH^2$. $\because$ $OI^2 = R^2 - 2\,Rr$ and $HI^2 = 2\,r^2 - AH \cdot HT$, $\therefore$ $2\,HI^2 = 4\,r^2 - AH \cdot HS$ and $2\,NH^2 = \frac{1}{2}\,OH^2 = \frac{1}{2}\,R^2 - AH \cdot HT$. $\therefore$ By adding, $2\,(R^2 - 2\,Rr) + 4\,r^2 - AH \cdot HS = 4\,IN^2 + R^2 - AH \cdot HS$, from which $IN = \frac{1}{2}\,R - r$.

(iv) Join HI_a. In $\triangle OI_aH$, $OI_a^2 + HI_a^2 = 2\,I_aN^2 + 2\,NH^2$. $\therefore$ $2(R^2 + 2\,Rr_a) + 4r_a^2 - AH \cdot HS = 4\,I_aN^2 + R^2 - AH \cdot HS$, from which $I_aN = \frac{1}{2}\,R + r_a$.

(v) G is the point of trisection of OH [Th. 3.77(ii)]. Since $AO^2 + BO^2 + CO^2 = AG^2 + BG^2 + CG^2 + 3\,OG^2$ (easily proved), $\therefore$ $3\,R^2 - 3\,OG^2 = AG^2 + BG^2 + CG^2 = \frac{1}{3}\,(AB^2 + BC^2 + CA^2)$. Hence $R^2 - OG^2 = \frac{1}{9}\,\{AB^2 + BC^2 + CA^2\}$.

4.29. *AA', BB', CC' are the altitudes of a triangle ABC. D, E, F are the centers of the inscribed circles in the triangles $AB'C'$, $BC'A'$, $CA'B'$. If the inscribed circle in the triangle ABC touches BC, CA, AB in L, M, N respectively, show that the sides of the triangle DEF are equal and parallel to those of triangle LMN.*

Construction: Produce LO to meet AB in G (O being the incenter of $\triangle ABC$). Join ADO, BEO, CFO, NE, MF, EA' (ADO, etc., are straight lines, since they bisect $\angle$s A, etc.) (*Fig.* 129).

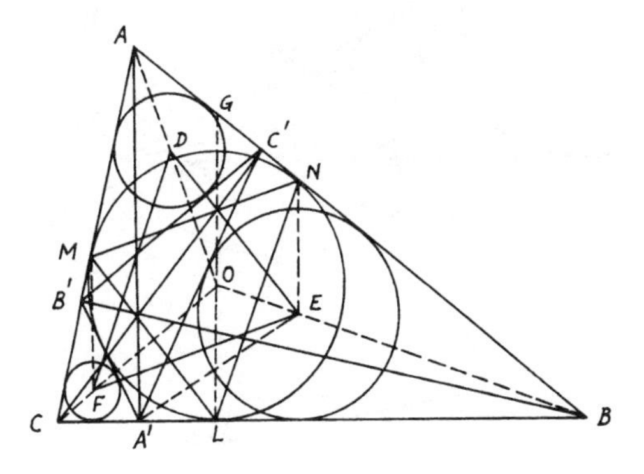

Figure 129

Proof: Quadrangle $AC'A'C$ is cyclic. $\therefore$ $\angle C'A'B = \angle A$. Since EA' bisects $\angle C'A'B$ and ADO bisects $\angle A$, $\angle EA'B = \angle BAO = \frac{1}{2}\,\angle A$.

Since, also, BEO bisects $\angle B$, $\therefore$ $\triangle$s ABO, $A'BE$ are similar. $\therefore$ $BA'/BA = BE/BO$. $\because$ $LOG \parallel AA'$ (both being $\perp BC$), $\therefore$ $BL/BG = BA'/BA$. $\because$ $BL = BN$, $\therefore$ $BN/BG = BA'/BA = BE/BO$. $\therefore$ $NE \parallel LOG$. In $\triangle BOG$, $\therefore$ $BN/BG = NE/GO = BL/BG = LO/GO$ (BEO bisects $\angle B$). $\therefore$ $NE = LO$. Hence $NE =$ and $\parallel LO$. Similarly, $MF =$ and $\parallel LO$ or NE. Therefore, $NEFM$ is a ▱. $\therefore$ $FE =$ and $\parallel$ MN. Similarly with DF, NL and DE, ML.

4.30. *In a triangle ABC, Q is the point of intersection of AF and BN where F and N are the points of contact of the escribed circles opposite A and B with BC and AC respectively, I the center of the inscribed circle, D the middle point of BC, G the centroid, H the orthocenter, O the center of the circumscribed circle. If DI is produced to meet AH in E, show that (i) AE = radius of inscribed circle; (ii) AQ is parallel and equal to 2 ID; (iii) Q, G, I are collinear and, QG = 2 IG; (iv) H, Q, O, I are the vertices of a trapezoid one of whose parallel sides is double the other and whose diagonals intersect in G.*

CONSTRUCTION: (i) Let $\odot$s I, M touch AC, BC or produced in K, L and J, F. Join IK, ML, IJ, MF, AIM. Produce JI to meet AF in X (*Fig.* 130).

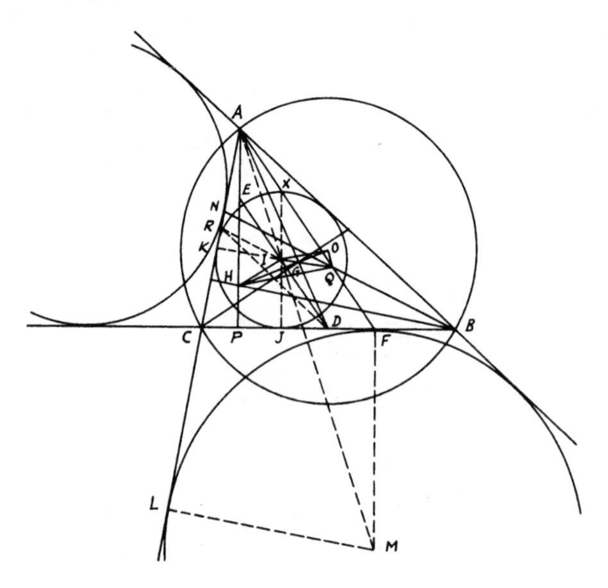

FIGURE 130

Proof: In $\triangle AML$, $\because$ $IK \parallel ML$, $\therefore$ $AI/AM = IK/ML$, since $MF \parallel JX$ ($\perp BC$). $\therefore$ In $\triangle AMF$, $AI/AM = IX/MF$. $\therefore$ IX/MF

$= IK/ML$. $\because$ $MF = ML$, $\therefore$ $IX = IK$. Hence X lies on the inscribed $\odot I$. $\because$ $BF = CJ$ and D is the mid-point of BC, $\therefore$ D is also the mid-point of FJ.

Also I is the mid-point of JX. $\therefore$ $DI \parallel AF$. But $JX \parallel AHP$. $\therefore$ $AXIE$ is a $\square$. $\therefore$ $AE = IX =$ radius of $\odot I$.

(ii) Bisect AC in R and join IR, RD. $\because$ $DI \parallel AF$ and similarly $RI \parallel BN$ and since $RD \parallel AB$, $\therefore$ $\triangle$s IRD, QBA are similar. Therefore, $ID/AQ = RD/AB = \frac{1}{2}$ or AQ is $\parallel$ and $= 2\,ID$.

(iii) Join AD cutting QI in G. $\triangle$s AQG, DIG are similar. $\therefore$ $AQ/DI = AG/DG = 2$ [from (ii)]. Hence $AG = 2\,DG$, but AD is a median in $\triangle ABC$. $\therefore$ G is the centroid and Q, G, I are collinear. Since also $AQ/DI = QG/IG = 2$, $\therefore$ $QG = 2\,IG$.

(iv) According to [Th. 3.77(ii)], O, G, H are collinear and $HG = 2\,GO$. Since $QG = 2\,IG$, $\therefore$ $\triangle$s GQH, GIO are similar (Th. 4.86). $\therefore$ $HQ = 2\,IO$ and $\parallel$ to it, since the diagonals QI, HO intersect in G. Hence $HQOI$ is the required trapezoid.

Miscellaneous Exercises

1. On the sides AB, AC of a triangle ABC, points D, E are taken such that AD is to DB as CE to EA. If the lines CD, BE intersect in F, the triangle BFC will be equal in area to the quadrilateral $ADFE$.
2. If O is the center of the inscribed circle of the triangle ABC, and AO meets BC in D, prove that $AO : OD = (AB + AC) : BC$.
3. From any point on the circumference of a circle perpendiculars are drawn to two tangents to the circle and their chord of contact. Prove that the perpendicular on the chord is a mean proportional between the other two perpendiculars.
4. The inscribed circle touches the side BC of the triangle ABC in D. An escribed circle touches BC in E. Show that the foot of the perpendicular from E on AD lies on the escribed circle.
5. A straight line is drawn from a vertex of a triangle to the point where the escribed circle touches the base. Show that if tangents be drawn to the inscribed circle at the points of intersection, one of them will be parallel to the base.
6. $ABEC$ is a straight line harmonically divided; i.e., $AB : AC = BE : EC$. If through E a straight line is drawn meeting parallels through B, C in D and Q, and if AQ cuts DB produced in P, show that $BP = BD$.
7. If a circle PDG touches another circle ABC externally in D and a chord AB extended in P, and if CE is perpendicular to AB at its middle point E and on the same side of PAB as the circle PDG, prove that the rectangle contained by CE and the diameter of $PDG = PA \cdot PB$.

8. ABC, DEF are two straight lines AD, BE, CF being parallel to one another. Prove that if AF passes through the middle point of BE, then CD will also pass through that point.

9. $ABCD$ is a rectangle inscribed in a circle. AD is produced to any point E. EC is joined and produced to cut the circle and AB produced in F, G respectively. Prove that $AG \cdot AE : FB \cdot FD = EG^2 : BD^2$.

10. If AD, BE, CF are any three concurrent straight lines drawn from the vertices of a triangle ABC to the opposite sides and M is their point of concurrency, then $DM/DA + EM/EB = FM/FC = 1$.

11. DE is any straight line parallel to the side AC of a triangle ABC meeting BC, AB in D, E. If a point F is taken on AC such that the triangles ABF, BED are equal and FG is drawn parallel to AB meeting BC in G, show that BD is a mean proportional between BC, BG. (Join AG, GE, AD and show that $GE \parallel AD$.)

12. $ABCD$ is a quadrilateral inscribed in a circle and E is the intersection of the diagonals. Show that $BE : DE = AB \cdot BC : AD \cdot CD$.

13. If a circle is drawn passing through the corner A of a parallelogram $ABCD$ and cutting AB, AC, AD in P, Q, R respectively, then $AQ \cdot AC = AP \cdot AB + AR \cdot AD$. (Join PR, QR and draw CG making with AD produced $\angle DGC = \angle AQR$. Another solution apply Ptolemy's theorem to $APQR$.)

14. Circles are drawn on the sides of a right-angled triangle ABC at A, as diameters. From A any transversal is drawn cutting circles on AB, BC, CA in F, G, H. Show that $FG = AH$.

15. ABC is a triangle and lines are drawn through B and C to meet the opposite sides in E, F. If BE, CF intersect in a point on the median from A, show that EF is parallel to BC.

16. Show that if on the sides of a right-angled triangle ABC, similar triangles are described so that their angles opposite to the sides AB, BC, CA are equal, then the triangle on the hypotenuse is equal to the sum of the other two triangles. (Let $\angle A$ = right angle, ABD, BCE, ACF be the equiangular $\triangle$s. Draw $AG \perp BC$ and join GE. Then show that $\triangle ABD = \triangle BGE$.)

17. In the triangle ABC, AD, BE, CF are the altitudes and EM, FN are the perpendiculars from E, F on BC. Show that $\triangle CED : \triangle BFD = CM : BN$.

18. Show that the rectangle contained by the perpendiculars from the extremities of the base in a triangle to the external bisector of the vertex angle is equal to the rectangle contained by the perpendiculars from the middle point of the base to the same external bisector and to the internal bisector of the vertex angle.

19. The side $AKLB$ of a rectangle $ABCD$ is three times the side AD. If K, L are the points of trisection of AB, and BD meets KC in R, prove that C, R, L, B lie on a circle.

20. $ABCD$ is a quadrilateral inscribed in a circle. If the bisectors of the angles CAD, CBD meet in G, show that $AG : BG = AD + AC : BD + BC$.

21. $ABCD$ is a trapezoid whose parallel sides CD, AB have a ratio of $2 : 5$. If DE, CL are drawn parallel to BC, AD meeting AB in E, L and DE, cutting AC, CL in H, G, show that $\triangle CGH = 8/105$ trapezoid $ABCD$.

22. Show that the lines joining the vertices of a triangle to the points of contact of the inscribed circle with the opposite sides are concurrent and that these lines are bisected by the lines joining the middle points of the opposite sides to the center of the incircle.

23. ABC is a triangle inscribed in a circle. If the altitudes AD, BE, CF are produced to meet the circle in X, Y, Z respectively, show that $AX/AD + BY/BE + CZ/CF = 4$. (Join BX and use Exercise 10.)

24. If a, b, c are the lengths of the sides of a triangle ABC and if the internal bisectors of the angles A, B, C meet BC, CA, AB in X, Y, Z and assuming s is half the perimeter of the triangle, prove that (a) $AX^2 = bc\{1 - [a^2/(b + c)^2]\}$; (b) $AX \cdot BY \cdot CZ = 8abc \cdot s \cdot \triangle ABC/(a + b)(b + c)(c + a)$; (c) $\triangle XYZ/\triangle ABC = 2\,abc/(a + b)(b + c)(c + a)$.

25. Find the radius of the circle inscribed in a rhombus whose diagonals are $2a$ and $2b$.

26. M, N are the centers of two circles intersecting in A, B. From any point C on the circumference of either one of the circles, a tangent CD is drawn to the other. Show that $(CD^2 : CA \cdot CB)$ is constant. (Let c be on circle N, CD tangent to circle M. Produce CA to meet circle M in E, then join EB, AB, MN, BM, BN.)

27. Divide a given arc of a circle into two parts so that the chords of the parts are in a given ratio.

28. Through two fixed points on the circumference of a circle draw two parallel chords which will be to each other in a given ratio.

29. ABC is a triangle inscribed in a circle. From A straight lines AD, AE are drawn parallel to the tangents at B, C respectively, meeting BC produced if necessary in D, E. Prove that $BD : CE = AB^2 : AC^2$.

30. In Problem 2.8, show that AM is a mean proportional between MC, BN.

31. ABC is an equilateral triangle inscribed in a circle and D any point on the circumference. If BD, AD produced meet AC, CB or produced in E, F and DC cuts AB in G, prove that (a) triangles ABF, CDF, BAE, CDE are similar; (b) $BG : CE = CG^2 : BE^2$.

32. Prove by ratio and proportion that the middle points of the three diagonals of a complete quadrilateral are collinear. (See Problem 2.16. Complete ▱s $AECK$, $EBLD$. Join EK, EL, DL, FL, LK. Prove that KLF is a straight line. Hence GHJ is one line $\parallel KLF$.)

33. DM, DR are two tangents to a circle from a point D. From D a line is drawn parallel to another tangent from any point A on the circle, meeting AM, AR produced in B, C. Show that $BMRC$ is cyclic, and find the center of its circle.

34. In an equilateral triangle draw a straight line parallel to one of the sides so as to divide the triangle into two parts whose areas are proportional to the squares on lines equal to their perimeters.

35. AB is a diameter of a semi-circle whose center is O. AO is bisected in C and on AC, CB as diameters two semi-circles are inscribed in the first one. If DE is a common tangent to the smaller semi-circles and produced to meet BA produced in M, show that $AC = 2\ AM$.

36. CD, CE are tangents to a circle from any point C. AB is a chord in the circle bisected by DE. If AG, AH, BM, BN are perpendiculars to both tangents CD, CE from A, B, show that $AG \cdot AH = BM \cdot BN$. (Use Exercise 3.)

37. ABC is a triangle. Two circles are described passing through B, C such that one touches AB in B and the other touches AC in C. If the circles cut AB, AC or produced in F, G, show that (a) $BG \parallel CF$; (b) $AG : AF = AB^3 : AC^3$.

38. (a) Construct a triangle given the base, the area, and the sum of the sides; (b) construct a triangle given the base, the area, and the difference of the sides. (Use Problem 3.25.)

39. $ABCD$ is a quadrilateral and F, G, H are three points on AD, BD, CD respectively such that $AF : FD = BG : GD = CH : HD$. If M, N, L are the middle points of AB, AC, BC respectively, prove that FL, GN, HM are concurrent. (Join the sides of the triangles FGH, MNL and show that they are homothetic.)

40. A straight line bisects the base BC of a triangle ABC, passes through the center of the inscribed circle, and meets at P a straight line drawn through A parallel to the base. Show that $AP = \frac{1}{2}(AB - AC.)$

41. A circle cuts the sides BC, CA, AB of a triangle ABC in six points A', A'', B', B'', C', C'' respectively and the perpendiculars to the respective sides at three of these points are concurrent. Show that those erected at the other three points are also concurrent.

42. From the vertex A of a triangle ABC, AD and AE are drawn to the base making with AB, AC two equal angles. Show that $AB^2 : AC^2 = BD \cdot BE : CD \cdot CE$.

43. Two straight lines BP, CQ are drawn from B, C of a triangle ABC to meet the opposite sides in P, Q and intersect on the altitude AD in E. Show that AD bisects the angle PDQ.

44. From a given point on the circumference of a given circle, draw two chords so as to be in a given ratio and to contain a given angle.

45. A point D is taken in the side AB of a triangle ABC, and DC is drawn. It is required to draw a straight line EF parallel to BC and meeting AB, AC in E, F so that the quadrilateral $EBCF$ may be equal to the triangle DBC. (In AB take AE the mean property between AB, AD and draw $EF \parallel BC$.)

46. Divide a triangle into any number of equal parts by straight lines parallel to the base.

47. $ABCD$ is a parallelogram; APQ is drawn cutting BC and DC produced in P, Q. If the angle ABP' be made equal to the angle ADQ', $BP' = BP$ and $DQ' = DQ$, prove that the angles PBP', QDQ', $P'AQ'$ are all equal and that $AP' : AQ' = AP : AQ$.

48. If two of the sides of a quadrilateral are parallel, show that the difference of the squares on the two diagonals is to the difference of the squares on the non-parallel sides as the sum of the lengths of the parallel sides is to their difference.

49. $ABCD$ is a rhombus and any straight line is drawn from C to cut AB, AD produced in F, G. Show that $1/AB = 1/AF + 1/AG$.

50. D, E are the points of intersection of two circles of which one is fixed and the other passes always through fixed points A, B. Prove that the ratio $(AD \cdot AE : BD \cdot BE)$ is constant. (Produce AD, BE to cut a fixed circle in L, R. Join LE, RD.)

51. AD, BE, CF, the perpendiculars from the vertices of a triangle on the opposite sides, intersect in O. Prove that $AO \cdot AB = AE \times$ diameter of the circumscribed circle and $OB \cdot OC = OD \times$ diameter of the circumscribed circle.

52. Construct a quadrilateral $ABCD$ given the four sides and the area. [Produce BC to E such that $BC/BE = AD/AB$. Draw CG, $EF \perp$s AB, and $CH \perp AD$. Prove that $2\,AD(BF + DH) = AB^2 + BC^2 - AD^2 - DC^2 =$ given. $\therefore$ $(BF + DH)$ is given. Also, $2 \square ABCD = AD(CH + EF)$. $\therefore$ $(CH + EF)$ is given, also BE is given.]

53. AB, AC are the sides of a regular pentagon and a regular decagon inscribed in a circle with center O. The angle AOC is bisected by a straight line which meets AB in D. Prove that the triangles ABC, ACD are similar, also the triangles AOB, DOB. Thence prove that $AB^2 = AC^2 + AO^2$.

54. From the vertex of a triangle draw a line to the base so that it may be a mean proportional between the segments of the base. (Describe a circle about given $\triangle ABC$ and find its center O. On OA describe a semicircle ODA cutting BC in D. Draw chord ADE and show that $AD^2 = AD \cdot DE = BD \cdot DC$.)

55. Two circles touch externally in C. If any point D is taken without them such that the radii AC, BC subtend equal angles at D and DE, DF are tangents to the circles, then $DE \cdot DF = DC^2$.

56. Two triangles which have one angle of the one equal to one angle of the other, and are to one another in the ratio of the squares of the sides opposite those angles, are similar to each other.

57. Two circles touch in O, and a straight line cuts one in A, B, the other in C, D. Show that $OB \cdot OC : OA \cdot OD = BC : AD$. (Draw the common tangent OK meeting AB in K and CE, $AF \perp BO$, DO respectively.)

58. From the vertex A of a triangle ABC inscribed in a circle, a tangent AD is drawn to touch the circle at A and meet the base BC produced in D. From D another tangent DE is drawn touching the circle at E. BML is drawn parallel to AD meeting AE, AC or produced in M, L. Show that $BM = ML$.

59. E is the middle point of the side BC of a square $ABCD$, O is the intersection of its diagonals, F the middle point of AE, G the centroid of the triangle ABE. If OG cut AE in H, prove that the square $ABCD = 8\, AE \cdot FH$.

60. AB, the diameter of a circle, is trisected in C, D. PCQ, PDR are two chords of the circle and QR meets AB produced in N. Prove that $PC^2 : PD^2 = NR : NQ$. (From Q draw $QF \parallel PDR$ meeting AB in F.)

61. Construct a triangle having given the radius of its circumscribed circle and the radii of two of the four circles touching the sides.

62. ABC is a triangle. Find a point D on AB such that if a parallel DE to the base BC be drawn to meet AC in E, then $DE^2 = BD^2 + CE^2$.

63. Construct a triangle on a given base and with a given vertical angle such that the base may be a mean proportional between the sides. Show that the problem is possible only when the given angle is not greater than 60°.

64. Show that any two diagonals of a regular pentagon cut each other in extreme and mean ratio.

65. ABC is a triangle inscribed in a circle. If the bisector of the vertex angle BAC meets the base BC in D and the circumference in E and is bisected in D, show that $AB^2 = 2\, BD^2$ and $AC^2 = 2\, CD^2$.

66. Construct a triangle given the base and the vertex angle so that the rectangle contained by the sides may be a maximum.

67. Construct a triangle given the base, the vertical angle, and the ratio between the perimeter and the altitude to that base.

68. A circle is described about an isosceles triangle. Prove that the distance of any point in the arc subtended by the base opposite the vertex from the vertex bears a constant ratio to the sum of its distances from the other two vertices.

69. Through a given point P in the base BC of a triangle ABC, draw a straight line to cut the sides AB, AC in R, Q respectively so that BR will be equal to CQ.

70. Three circles touch each other internally at the same point. Prove that the tangents drawn from any point on the largest circle to the other two circles bear to one another a constant ratio.

71. Prove that the side of a regular polygon of twelve sides inscribed in a circle is a mean proportional between the radius of the circle and the difference between the diameter of the circle and the side of an equilateral triangle inscribed in the circle. (Let ABC be the equilateral triangle inscribed in circle O. Draw diameter AOD and join BD. Bisect arc BD in F and join FB, FO, BO. FB will be the side of a twelve-sided regular polygon.)

72. ABC is a triangle; the sides AB, AC are cut proportionally in the points D, E. From any point P in BC two straight lines PQ and PR are drawn meeting AB or AB produced in Q, R, and always intercepting a portion QR which is equal to AD. Also, PQ', PR' are drawn meeting AC or AC produced in Q', R', and always intercepting a portion $Q'R'$ which is equal to AE. Show that the sum of the areas PQR, $PQ'R'$, is constant.

73. Prove that the straight line drawn from a vertex of a triangle to the center of the inscribed circle divides the line joining the orthocenter to the center of the circumscribed circle into segments, which are in the ratio of the perpendicular from the center of the circumscribed circle on the opposite side of the triangle to the radius of the nine-point circle.

74. Three circles have a common chord, and from any point in one, tangents are drawn to the other two. Prove that the ratio of these tangents is constant.

75. From the point of contact of two circles which touch internally are drawn any two chords at right angles, one in each circle. Prove that the straight line joining their other extremities passes through a fixed point. (Let $\odot ABE$ with center O touch $\odot ADC$ with center O' internally at A. Let AEC, AB be chords $\perp$ each other. Produce AB to D; BE, DC are diameters, and let BC cut OO' in G. Show that G is a fixed point.

76. AD is drawn perpendicular to the hypotenuse BC of a right-angled triangle ABC. On BC, AB similar triangles BEC, AGB are similarly described so that the angles CBE, ABG are equal; DE is drawn. Prove that the triangles ABG, BDE are equal.

77. From a point T a tangent is drawn to each of two concentric circles, and through the common center C, CRR' is drawn parallel to the bisector of the exterior angle at T, meeting the tangents in R, R'. Show that the ratio of CR to CR' is independent of the position of T.

78. ABD is the diameter of a semi-circle ACD, and ABC is a right angle. E any point on the chord AC inside the semi-circle is joined to B, and CF is drawn cutting AD in F and making the angle BCF equal to the angle ABE. Prove that $AE : EC = BF : BD$.

79. Show how to draw through a given point in a side of a triangle a straight line dividing the triangle in a given ratio. (In BA take BD so that $BD : AB$ in the given ratio. Let D be between B and P the given point in AB. Draw $DE \parallel CP$; DE will be the required line.)

80. Two parallelograms $ABCD$, $A'BC'D'$ have a common angle at B. If AC' and DD' meet in O, prove that $OD' : OD =$ fig. $A'BC'D'$: fig. $ABCD$.

81. A, B are the centers of two circles and $DEFG$ is a transversal cutting the circles A, B in D, E and F, G respectively such that the ratio $DE : FG$ is fixed. Two tangents DP, GP are drawn to the circles A, B respectively. Show that $DP : GP$ is a constant ratio.

82. Given the three altitudes of a triangle, construct the triangle.

83. Construct a triangle equal to a given triangle and having one of its angles equal to an angle of the triangle and the sides containing this angle in a given ratio.

84. Two circles intersect in A, B. The chords BC, BD are drawn touching the circles at B, and the points D, C are joined to A. Prove that AC, AD are to each other in the ratio of the squares of the diameters of the circles.

85. Two fixed circles touch externally at A, and a third passes through A and cuts the other two orthogonally in P, Q. Prove that the straight line PQ passes through a fixed point.

86. If the base of a triangle be a mean proportional between the sides, prove that the bisectors of the angles at the base will cut off on the sides segments measured from the vertex such that their sum is equal to the base.

87. If perpendiculars be drawn from any point on the circumference of a circle to the sides of an inscribed quadrilateral, the rectangle contained by the perpendiculars on two opposite sides is equal to the rectangle contained by the other two perpendiculars.

88. Prove that any straight line drawn from the orthocenter of a triangle to the circumference of the circumscribing circle is bisected by the nine-point circle.

89. The opposite sides AB, DC of a quadrilateral $ABCD$ are divided in a given ratio at E and F so that $AE : EB = DF : FC$, and the other pair of opposite sides BC, AD are divided at G, H in another given ratio so that $BG : GC = AH : HD$. Show that the point of intersection of EF and GH divides GH in the first of the given ratios and EF in the second.

90. Construct a triangle similar to a given triangle and having its vertices on three given parallel lines.

91. Divide a quadrilateral in a given ratio by a straight line drawn from a given point in one of its sides. (Let $ABCD$ be the given quadrilateral and P given point in CD. Convert $ABCD$ into an equal triangle in area through P by drawing $CE \parallel PB$ and $DF \parallel PA$ meeting AB produced in E, F. $\therefore$ $\triangle PEF = \square ABCD$ [see Problem 2.4(ii)]. Divide EF in the given ratio by point Q. Hence PQ is the required line.)

92. $ABCD$ is a quadrilateral of which the sides AB, DC meet in P and the sides AD, BC in Q. Prove that $PA \cdot PC : PB \cdot PD = QA \cdot QC : QB \cdot QD$. (Draw $BE \parallel AQ$ meeting PC in E and $DF \parallel AP$ meeting CQ in F.)

93. Two escribed circles of the triangle ABC are drawn, one touching AB, AC produced in D, E respectively, the other touching BA, BC produced in F, G. Through D the diameter DH is drawn. Prove that (a) $AF = BD$; (b) HC produced passes through F. (Use Problem 4.30.)

94. Construct a triangle given the base, the difference between the base angles, and the rectangle contained by the sides.

95. Divide a triangle by a straight line drawn through a given point into (a) two equal parts; (b) two parts of which the areas have a given ratio. [(a) Let D be the given point inside $\triangle ABC$. Join AD and make AE subtend $\angle BAE = \angle CAD$. Take AE such that $AD \cdot AE = \frac{1}{2} AB \cdot AC$. Join DE and construct on it an arc of $\odot$ subtending $\angle = \angle EAC$. If this arc cuts AB in F, F', then two solutions are possible by joining FD, $F'D$.]

96. ABC is a triangle whose sides AB, BC, CA are in ratio of $4 : 5 : 6$. Show that $\angle B = 2 \angle C$.

97. ABC is a triangle inscribed in a circle. If through A another circle is described cutting the first one, AB, AC in E, D, G, show that $AB : AC = EB + EG : EC + ED$. (Join AE and draw $CM \parallel AE$.)

98. Construct an isosceles triangle having given its vertex angle and the area.

99. ABC is a right-angled triangle with the right angle at A. If AD, AE are the altitude and bisector of the right angle and F is the middle point of BC, show that $(AB + AC)^2 : AB^2 + AC^2 = FD : FE$.

100. Construct a triangle having given the area and angles. [Let the given area of a triangle be a^2. Construct any triangle $AB'C'$ similar to the required triangle, the angles being given. Produce AB' to D' so that $B'D' = \frac{1}{2} C'D$ the altitude of $\triangle AB'C'$. On AD' as diameter, describe a semi-circle which is cut by perpendicular $B'E$ to AD' in E. Produce $B'E$ to F such that $B'F = a$ (side of square equal in area to triangle). AE produced meets FR which is $\perp B'F$ in R. RB is $\perp AD'$ produced and $BC \parallel B'C'$. Hence ABC is the required triangle.]

101. ABC is an equilateral triangle and D any point in BC. Show that if AL be drawn perpendicular to BC and BM, CN perpendicular to AD, then $AL^2 = BM^2 + CN^2 + BM \cdot CN$.

102. O is a point inside a triangle ABC. Lines are drawn from the middle points of BC, CA, AB parallel to OA, OB, OC respectively. Prove that they meet in a point O' and that, whatever be the position of O, OO' passes through a fixed point and is divided by it in the ratio of $2:1$.

103. In *Fig.* 59, prove that EK produced passes through one of the points of trisection of BH.

104. H_1, H_2, H_3, H_4 and G_1, G_2, G_3, G_4 are the orthocenters and centroids of the four triangles BCD, ACD, ABD, ABC formed by the cyclic quadrilateral $ABCD$. Show that the two figures $ABCD$ and $H_1H_2H_3H_4$ are congruent, also that $G_1G_2G_3G_4$ and $H_1H_2H_3H_4$ are similar.

105. Two equal circles having centers A, B touch at C. A point D, in AB produced, is the center of a third circle passing through C. Take a common tangent (other than that at C) to the circles whose centers are A and D, and let P and Q be the points of contact. Draw the line CQ cutting the circle with center B in M and produce it.to meet in N the tangent to this circle at E which is diametrically opposite to C. Show that $EN = PQ$ and $CM = QN$.

106. If two semi-circles are on opposite sides of the same straight line, and the radius of the greater is the diameter of the less, draw the greatest straight line perpendicular to the diameter and terminated by the circles. (Let ADB, AEC be the semi-circles, $AC = CB$. Bisect AC in F and trisect FC in G so that $GC = 2\ FG$. Draw $DGE \perp AC$. Hence DE is the greatest line.)

107. ABC, $A'B'C'$ are two similar triangles. In BC, CA, AB points D, E, F are taken and $B'C'$, $C'A'$, $A'B'$ are divided in D', E', F' similarly to BC, CA, AB respectively so that $BD : DC = B'D' : D'C'$, etc. Prove that the triangles DEF, $D'E'F'$ are similar and that if the straight lines drawn through D, E, F at right angles to BC, CA, AB respectively are concurrent, so also are the corresponding straight lines drawn from D', E', F'.

108. $ABCD$ is a quadrilateral of which the angles A and B are right angles. A point L is taken in AB such that $AL : LB = AD : BC$. Show that $LD \cdot DC + LP^2 = 2\ AD \cdot BC + PD^2$, where P is the middle point of DC.

109. Two circles touch each other internally at O and a chord $ABCD$ is drawn. The tangent at A intersects the tangents at B and C in G, E. The tangent at D intersects the tangents at B and C in F, H. Prove that OA bisects the angle GOE and that $EFGH$ can be inscribed in a circle which touches both given circles at O.

110. Construct a circle which will cut three straight lines at given angles. (Let AB, BC, CA forming $\triangle ABC$ be the given lines. With any point O as center describe a circle. Draw radius $OM \perp$ the direction of BC; make $\angle MOD = \angle MOE =$ angle at which the required circle is to cut BC. $\because$ DE is $\perp OM$, $\therefore$ DE is $\parallel BE$ and $\therefore$ the angle which tangent at D

makes with $DE = \frac{1}{2} \angle DOE$. $\therefore$ $\odot DME$ cuts DE at the same angle at which the required circle is to cut BC. Similar chords FG, HK of circle DME may be found $\parallel$ BA, AC and cutting circle DME at the same given angles.)

CHAPTER 5

LOCI AND TRANSVERSALS

Definitions and Theorems

Loci

Definition: If any and every point on a line, part of a line, or group of lines whether straight or curved satisfy an assigned condition and no other point does so, then that line, part of a line, or group of lines is called the locus of the point satisfying that condition.

Among the most important loci are:

5.1. *The locus of a point at a given distance from a given point is the circumference of the circle having the given point as center and the given distance as radius.*

5.2. *The locus of a point at a given distance from a given straight line is a pair of straight lines parallel to the given line, one on each side of it.*

5.3. *The locus of a point equidistant from two given points is the straight line bisecting at right angles the line joining the given points.*

5.4. *The locus of a point equidistant from two given intersecting lines is the two bisectors of the angles formed by the two given lines.*

5.5. *The locus of a point from which tangents of given length or subtending a given angle are drawn to a given circle is another circle concentric with the given one.*

5.6. *The locus of a point which subtends a given angle at a given line is an arc of a circle passing through the ends of the given line. If this is a right angle then the locus will be the circle on the given line as diameter.*

Some of the most frequently used loci are:

5.7. *The locus of the mid-points of the chords of given length drawn in a given circle is a circle concentric with the given one and touching the chords at these points.*

5.8. *The locus of a point the sum of the squares of whose distances from two given points is constant is a circle with the mid-point between the given points as center.*

5.9. *The locus of a point the difference of the squares of whose distances from two given points is constant is a straight line perpendicular to the line joining the two given points.*

5.10. *The locus of a point the ratio of whose distances from two fixed points is constant is a circle, called the* Apollonius circle.

TRANSVERSALS

5.11. *Ceva's theorem: Any three straight lines drawn through the vertices of a triangle so as to intersect in the same point either inside or outside the triangle divide the sides into segments such that the product of three non-adjacent segments is equal to the product of the other three.*

5.12. *Conversely, if three straight lines drawn through the vertices of a triangle cut the sides themselves, or one side and the other two produced, so that the product of three non-adjacent segments is equal to the product of the others, the three lines are concurrent.*

5.13. *Menelaus' theorem: If the sides or sides produced of a triangle be cut by a transversal, the product of three non-adjacent segments is equal to the product of the other three.*

5.14. *Conversely, if three points be taken on two sides and a side produced of a triangle, or on all three sides produced, such that the product of three non-adjacent segments is equal to the product of the others, the three points are collinear.*

Solved Problems

5.1. *Find the locus of a point the sum of whose distances from two given intersecting straight lines is equal to a given length.*

ANALYSIS: Let AOB, COD be the given intersecting straight lines and X the given length (*Fig.* 131). Between OA, OC place the straight

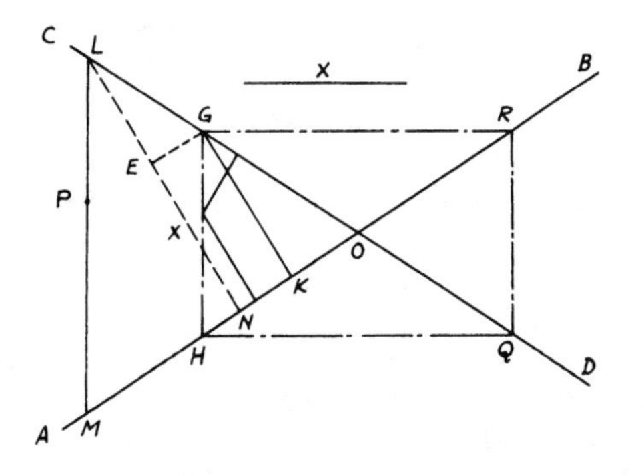

FIGURE 131

line $GK = X$ and $\perp$ to OA. This is done by taking $EN = X$ on any line $LN \perp OA$ and drawing $EG \parallel OA$. Then $GK = EN = X$. Cut off

$OH = OG$ and join GH. Hence in the isosceles $\triangle OGH$, the sum of the $\perp$ distances of any point in GH from OA, OC is equal to GK (see Problem 1.20). Therefore, every point in GH satisfies the required condition.

Also, no point within the angle AOC not in GH has the sum of its distances from OA, OC equal to X.

Proof: Take any such point P. Through P draw $LPM \parallel GH$; draw $LN \perp AO$. Then the sum of the $\perp$distances of P to OA, $OC = LN$. But LN is not equal to GK or X, since $LNKG$ cannot be a rectangle because NK meets LG and O. And if we take OR, OQ each equal to OG and join GR, RQ, QH, it can be shown in the same way that every point on the lines GR, RQ, QH (and no other) has the sum of its $\perp$ distances from AB, CD equal to X. Therefore the perimeter of the rectangle $GHQR$ is the required locus.

5.2. *ABCD is a quadrilateral and P is a point inside it such that the sum of the squares on PA, PB, PC, PD is constant. Show that the locus of P is a circle and find its center.*

CONSTRUCTION: Bisect AB, BC, CD, DA in E, F, G, H respectively. Join EG, FH to intersect in O. Then O is the center of the locus $\odot$ of P whose radius is OP (*Fig.* 132).

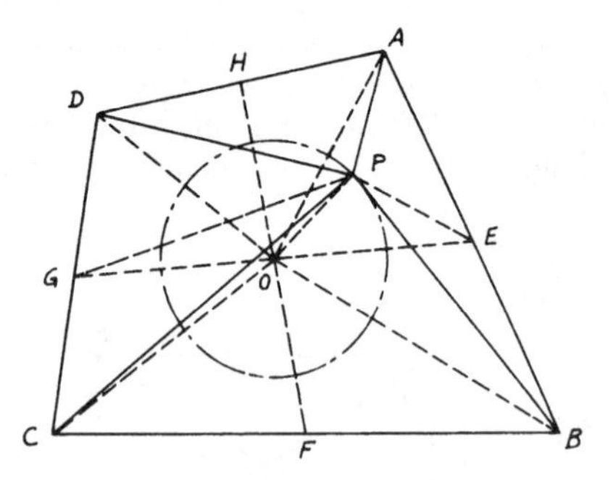

FIGURE 132

Proof: Join PE, PG, OA, OB, OC, OD. O is the mid-point of EG, FH (see Problem 1.14). $AP^2 + PB^2 = 2\,(PE^2 + AE^2)$ and $PC^2 + PD^2 = 2\,(PG^2 + DG^2)$. $\therefore AP^2 + PB^2 + PC^2 + PD^2 = 2\,(PE^2 + PG^2 + AE^2 + DG^2) = 4\,(PO^2 + OG^2) + 2\,(AE^2 + DG^2) = 4\,PO^2 + OA^2 + OB^2 + OC^2 + OD^2 =$ constant. Since O is the mid-point of EG, FH and hence fixed and OA, OB, OC, OD are fixed lengths, $\therefore PO$ is fixed in length. $\therefore$ Locus of P is $\odot$ with O as center and PO as radius.

5.3. *From any point P on the circumference of a circle circumscribing a triangle ABC perpendiculars PD, PE are let fall on the sides AB, BC. Prove that the locus of the center of the circle circumscribing the triangle PDE is a circle.*

CONSTRUCTION: Join PB and bisect it in M, which will be the center of ⊙ circumscribing $\triangle PDE$. Join OB and with OB as diameter draw a ⊙ which is the required locus of M (*Fig.* 133).

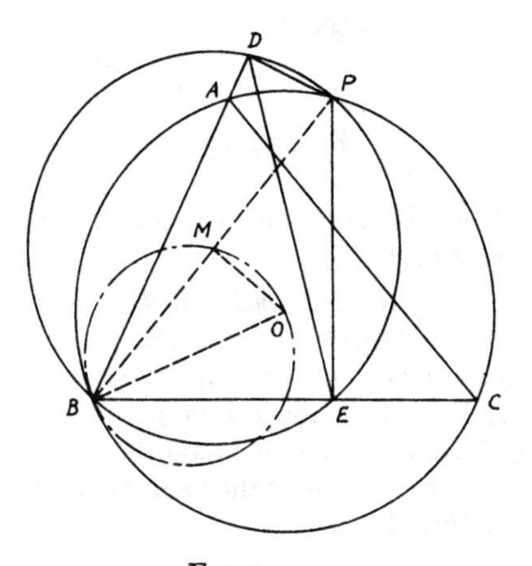

FIGURE 133

Proof: As will be seen later, DE is the Simson line of P with respect to $\triangle ABC$. Since $\angle PDB + \angle PEB = 2$ right angles, $\therefore$ quadrangle $PDBE$ is cyclic. Hence ⊙ circumscribing $\triangle PDE$ will pass through B. $\therefore$ PB is a diameter of ⊙PDE. Since the $\perp$ from O the circumcenter of $\triangle ABC$ bisects PB, $\therefore$ OM is $\perp$ PB. $\because$ OB is a radius of ⊙ABC and fixed in position and length, $\therefore$ the locus of M, the circumcenter of $\triangle PDE$, is a circle with OB as diameter.

5.4. *M and N are the centers of two circles which intersect each other orthogonally at A, B. Through A a common chord CAD is drawn to the circles M, N meeting them in C, D respectively. Find the locus of the middle point of CD.*

CONSTRUCTION: Join MN and on it as diameter construct a ⊙ which will be the locus of E the mid-point of CAD. Draw MG, NK $\perp$s to CAD and join ME, EN, MA, AN (*Fig.* 134).

Proof: Since ⊙s cut orthogonally, $\therefore$ MA is $\perp$ AN. $\therefore MN^2 = MA^2 + AN^2$. $\because$ E, G, K are the mid-points of CAD, CA, AD, $\therefore$ $CG + GE = AE + AD$. $\therefore$ $2\,GE + AE = AE + 2\,AK$. $\therefore$ $GE = AK$. Again, $MA^2 + AN^2 = MG^2 + AG^2 + KN^2 + AK^2 = (MG^2 + GE^2) + (KE^2 + KN^2) = EM^2 + EN^2 = MN^2$. $\therefore MEN$ is a right

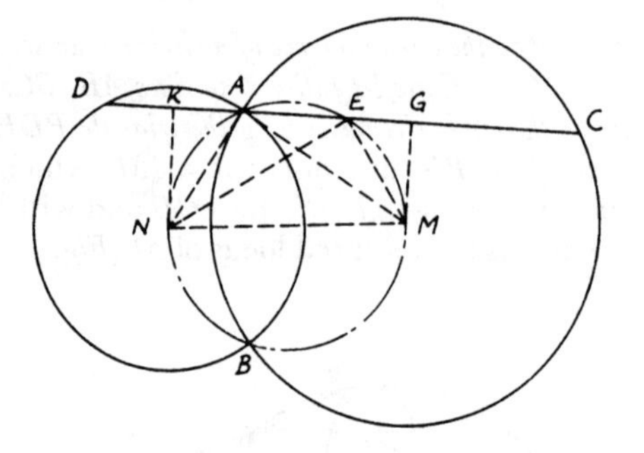

FIGURE 134

angle. ∵ MN is fixed, hence ⊙ on MN as a diameter is the locus of E for all locations of CAD.

5.5. *PQ is a chord in a fixed circle such that the sum of the squares on the tangents from P, Q to another fixed circle is always constant. Show that the locus of R the middle point of PQ is a straight line.*

CONSTRUCTION: Let M be the center of the ⊙, where PQ is a chord and PC, QD are the tangents to another ⊙ center N. From R draw $RS \perp$ to MN and this will be the locus of R. Join MP, MR, NP, NQ, NC, ND, NR (*Fig.* 135).

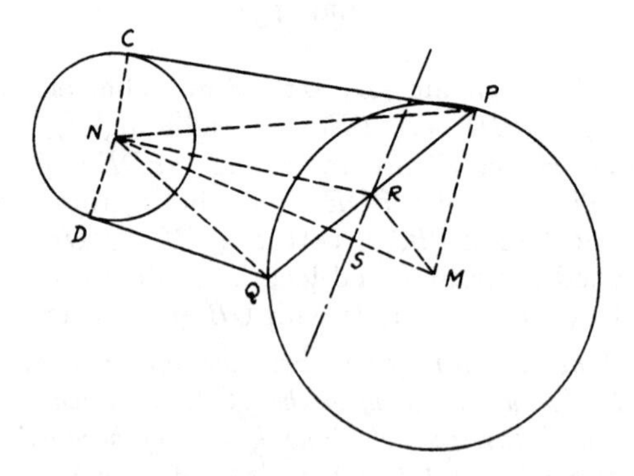

FIGURE 135

Proof: $PN^2 + NQ^2 = PC^2 + CN^2 + QD^2 + DN^2$. But $(PC^2 + QD^2)$ is constant and $CN = DN =$ fixed radii. ∴ $PN^2 + NQ^2 = \text{constant} = 2\ NR^2 + 2\ PR^2 = 2\ NR^2 + 2\ PM^2 - 2\ RM^2$. But PM is fixed also. ∴ $NR^2 - RM^2 = \text{constant}$. Since M, N are fixed centers, ∴ locus of R is the $\perp$ RS on MN (see Problem 5.9).

5.6. *If the rectangle ABCD can rotate about the fixed corner A such that B, D move along the circumference of a given circle whose center is O, find the locus of the remaining corner C.*

CONSTRUCTION: Join the diagonals AC, BD to intersect in E. Then the locus of C is a concentric ⊙ with O as center and OC as radius. Join OA, OD, OE (*Fig.* 136).

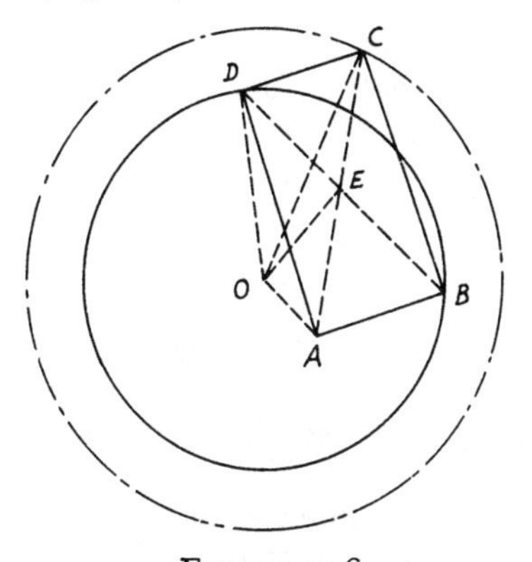

FIGURE 136

Proof: Since E is the mid-point of AC, BD, and OE is $\perp BD$, $\therefore$ $AO^2 + OC^2 = 2\,OE^2 + 2\,AE^2 = 2\,OE^2 + 2\,DE^2 = 2\,DO^2$. $\therefore$ $OC^2 = 2\,DO^2 - AO^2 =$ constant (since DO, AO are fixed). $\because$ O is a fixed center, $\therefore$ the locus of C is a concentric ⊙ with center O and radius $= \sqrt{2\,DO^2 - AO^2}$.

5.7. *A circle of constant magnitude passes through a fixed point A and intersects two fixed straight lines AB, AC in B, C. Prove that the locus of the orthocenter of the triangle ABC is a circle.*

CONSTRUCTION: Let H be the orthocenter of $\triangle ABC$, and O be the center of ⊙ABC. Join OC and drop $OD \perp BC$. With A as center and AH as radius construct a ⊙ to be the locus of H (*Fig.* 137).

Proof: Since the ⊙O is of constant magnitude and A, AF, AL are fixed in position, then BC is of constant magnitude. $\because$ OD is $\perp BC$, $\therefore$ D is the mid-point of BC. $\therefore$ $\angle DOC = \angle BAC$. Since $OD = CD \cdot \cot DOC = CD \cdot \cot BAC$ and $OD = \frac{1}{2} AH$ (see Problem 1.32), $\therefore$ $AH = 2\,CD \cdot \cot A = BC \cdot \cot A$. $\because$ BC is fixed in length and $\angle A$ is constant, $\therefore$ AH is a fixed length. But, since A is a fixed point, the locus of H is a ⊙ with A as center and $(BC \cdot \cot A)$ as radius.

5.8. *Through a fixed point O any straight line OPQ is drawn cutting a fixed circle M in P and Q. On OP, OQ as chords are described two circles touching*

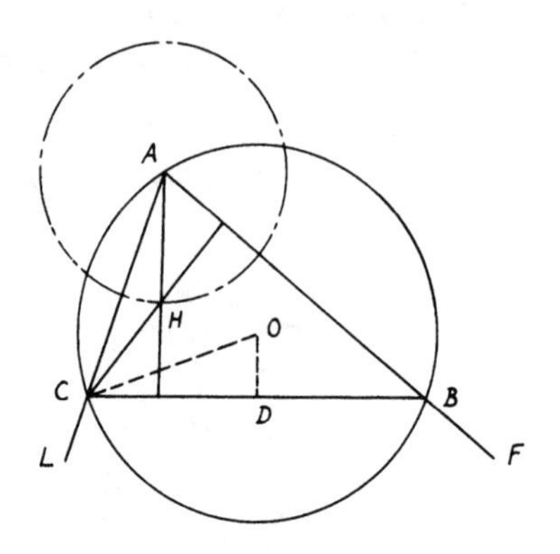

FIGURE 137

the fixed circle at P, Q. Prove that the two circles so described intersect on another fixed circle.

CONSTRUCTION: Let this point of intersection be S. Join OS, PS, QS and draw PR, QR tangents to the $\odot M$. Join RM and OM. On OM as diameter draw a $\odot$ which will be the locus of S (*Fig.* 138).

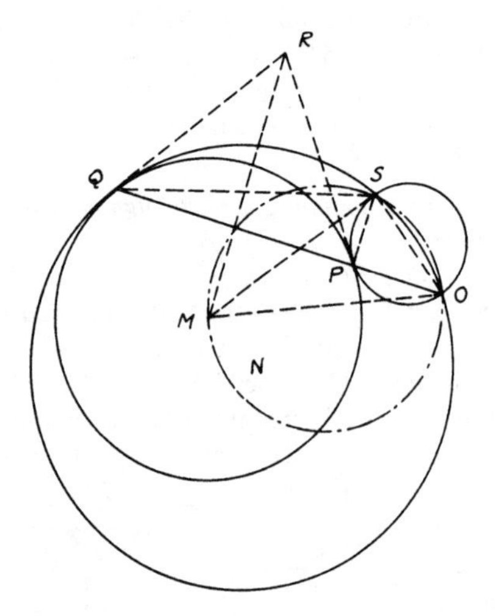

FIGURE 138

Proof: In $\triangle SOQ$, $\angle SOQ + \angle SQO + \angle OSQ = 2$ right angles. $\because$ RQ touches $\odot N$ passing through $\triangle OSQ$, $\therefore$ $\angle RQS = \angle SOQ$. $\therefore$

$\angle RQO + \angle OSQ = 2$ right angles. Since $\angle RQS = \angle RPS = \angle POS$, $\therefore$ $SPQR$ is a cyclic ⏢. But $PMQR$ is also cyclic. Then the figure $SPMQR$ is cyclic. Hence $\angle MSQ = \angle MRQ$. $\therefore$ $\angle RQO + \angle MRQ + \angle MSO = 2$ right angles. $\because$ $\angle RQO + \angle MRQ$ = right angle ($RM \perp PQ$), $\therefore$ $\angle MSO$ = right angle. But O, M are two fixed points. Therefore, ⊙ drawn on OM as diameter is the locus of S and is a fixed ⊙.

5.9. *Given a fixed circle and two fixed points A, B. From A, a line AC is drawn to intersect the circle in C. Produce AC to D so that AC = CD. E is the middle point of AB. If CB intersects DE in M, find the locus of M.*

CONSTRUCTION: Let O be the center of the fixed ⊙. Join DB, OB, OC. Draw $MN \parallel CO$ to cut OB in N. Then the locus of M is a circle with center N and radius MN (*Fig.* 139).

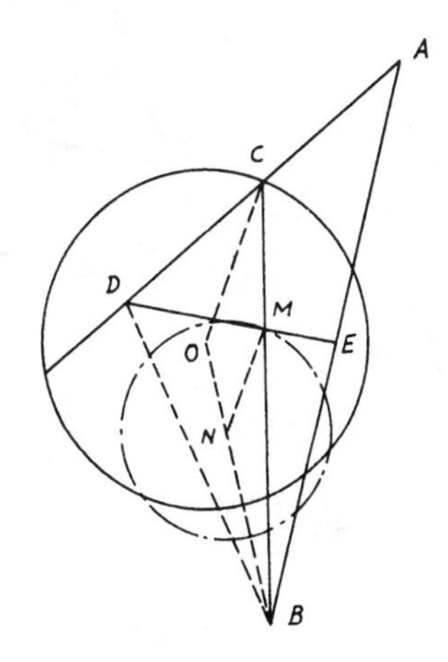

FIGURE 139

Proof: $\because$ C, E are the mid-points of AD, AB in $\triangle ADB$. $\therefore$ M is the centroid of $\triangle ADB$. $\therefore$ $BM = \frac{2}{3} BC$. $\because$ MN is $\parallel CO$ in $\triangle BOC$, $\therefore$ $BM/BC = MN/OC = \frac{2}{3}$. $\therefore$ $MN = \frac{2}{3} OC$, the fixed radius of ⊙O. Hence MN is of fixed length. Again, $BM/BC = BN/BO = \frac{2}{3}$. $\therefore$ $BN = \frac{2}{3} BO$. $\because$ B and O are both fixed and BN is of fixed length. $\therefore$ N is a fixed point on BO. Hence locus of M is a ⊙ with center N and radius $MN = \frac{2}{3}$ radius OC.

Note: If, in general, E divides AB into EB/AE = given ratio p and C divides AD into CD/AD = given ratio q, $\therefore$ in $\triangle ABC$: $(CD/DA)(AE/EB)(BM/MC) = 1$ (Th. 5.13). $\therefore$ $BM/MC = (DA/CD)$

$(EB/AE) = p/q$. $\therefore$ $BM/BC = p/(p + q) = MN/CO$. $\therefore$ $MN = CO\,(p/(p + q))$ and N is also fixed on OB. $\therefore$ Locus of M is $\odot N$ with MN as radius.

5.10. *The vertices of a triangle are on three straight lines which diverge from a point, and the sides are in fixed directions; find the locus of the center of the circumscribed circle.*

CONSTRUCTION: Let DEF, $D'E'F'$ be two $\triangle$s with their vertices on ODD', OEE', OFF'. Bisect DF in G and produce OG to meet $D'F'$ in G'. Let C, C' be the centers of the circumscribing $\odot$s on DEF, $D'E'F'$. Then OC, OC' will be the locus of C (*Fig.* 140).

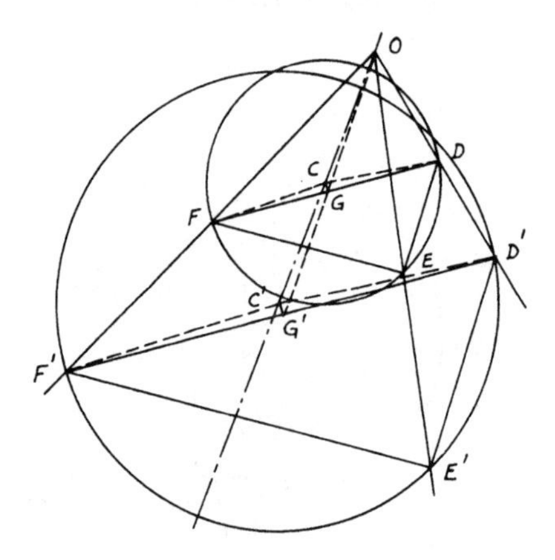

FIGURE 140

Proof: Since G is the mid-point of DF and $D'F'$ is $\parallel DF$, $\therefore$ G' bisects $D'F'$. $\because$ $\angle DEF = \angle D'E'F'$, $\therefore$ $\angle DCF = \angle D'C'F'$. $\therefore$ $\angle GCF = \angle G'C'F'$ and right angle CGF = right angle $C'G'F'$. $\therefore$ $CG/C'G' = FG/F'G' = OG/OG'$ and $\angle CGO = \angle C'G'O$. $\therefore$ $\angle GOC = \angle G'OC'$. $\therefore$ OCC' is a straight line and is the required locus.

Note: This is an explicit proof of a relationship that can also be developed by homothetic figures.

5.11. *Find the locus of a point moving inside an equilateral triangle such that the sum of the squares of its distances from the vertices of the triangle is constant.*

CONSTRUCTION: Let O be the center of the $\odot$ circumscribing the equilateral. $\triangle ABC$ and P is a point which satisfies the condition. The locus of P will be a $\odot$ with center O and OP as radius (*Fig.* 141).

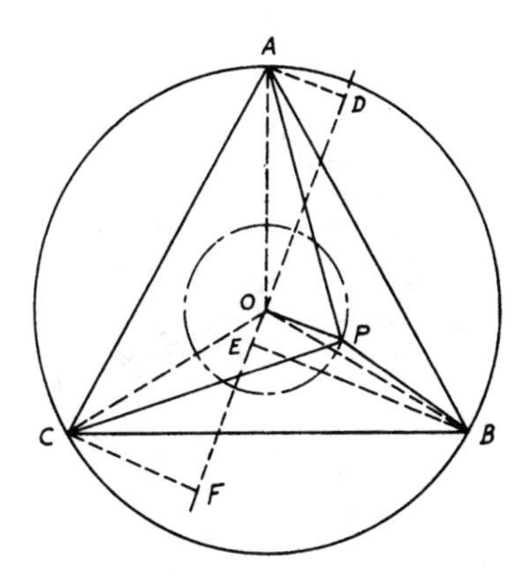

FIGURE 141

Proof: Join AO, BO, CO and draw $DOEF \perp OP$. From A, B, C the $\perp$s AD, BE, CF are drawn on $DOEF$. Now, $PB^2 = BO^2 + OP^2 - 2\,PO \cdot BE$. $PA^2 = AO^2 + OP^2 + 2\,PO \cdot AD$. $PC^2 = CO^2 + OP^2 + 2\,PO \cdot CF$. Since $AO = BO = CO = r$, then by adding, $PA^2 + PB^2 + PC^2 = 3r^2 + 3\,OP^2 + 2\,PO(AD + CF - BE) = \text{constant}$. $\because AD + CF = BE$ (see Problem 1.29), $\therefore OP^2 = \frac{1}{3}$ (construct $- 3r^2$) = constant. Hence, the locus of P is a circle with O as center and OP as radius.

5.12. *If a triangle ABC is similar to a given triangle and has one vertex A fixed, while another vertex B moves along a given circle, prove that the locus of the third vertex C is a circle.*

CONSTRUCTION: Suppose that M is the center of the $\odot$ on which B moves. Join MA and draw the lines MN, AN to make with MA $\angle NMA = \angle B$ and $\angle NAM = \angle CAB$. Then the locus of C is the $\odot$ with N as center and CN as radius (*Fig.* 142).

Proof: Join MB. $\because$ $\triangle$s AMN, ABC are similar, $\therefore AM/AB = AN/AC$. $\because \angle MAB = \angle NAC$, $\therefore$ $\triangle$s MAB, NAC are also similar. Hence $MB/NC = AB/AC$ = given ratio. But, since MB is a given radius, then NC is also given. Again, the $\triangle MNA$ has two of its vertices M, A fixed, and its angles are fixed because it is similar to $\triangle ABC$. $\therefore$ N is a fixed point. Therefore, $\odot N$ with radius $= MB \cdot AC/AB$ is the required locus of C.

5.13. *Two fixed straight lines AB, CD of given lengths meet, when produced, at a point O. P is a point such that the sum or difference of the areas of the triangles with P as vertex and AB, CD as bases is equal to the area of a*

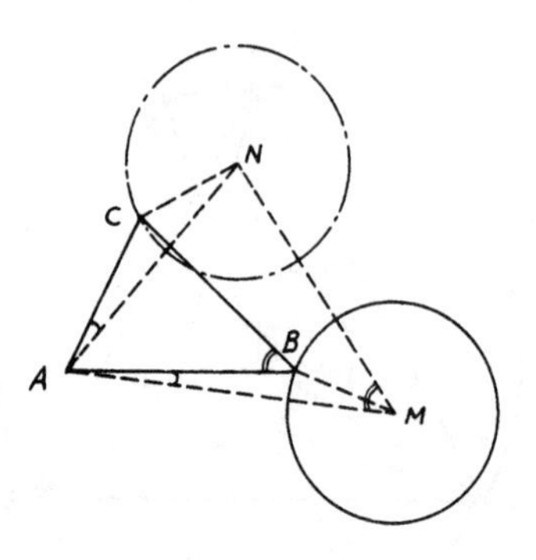

FIGURE 142

given triangle. Prove that P lies on a fixed straight line and construct this line for each case.

CONSTRUCTION: (i) For the case of a given sum of $\triangle$s PAB, PCD, produce AB, CD to meet in O. Take $OQ = AB$ and $OR = CD$, and join QR. Then the locus of P will be the line MPL drawn through $P \parallel QR$ (*Fig.* 143).

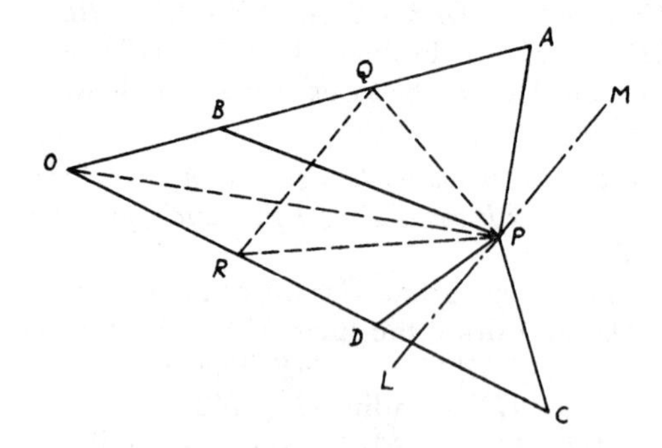

FIGURE 143

Proof: Since $\triangle APB = \triangle QPO$ and $\triangle PDC = \triangle PRO$, adding gives fig. $QPRO = \triangle APB + \triangle PDC =$ constant. $\because$ The area of $\triangle OQR$ is fixed (since AB, CD are fixed straight lines and O is fixed), $\triangle PQR$ is constant and since QR is fixed in direction, $\therefore$ the locus of P is a line through it $MPL \parallel QR$.

CONSTRUCTION: (ii) For the case of a given difference of $\triangle$s PAB, PCD, produce ABO to Q so that $OQ = AB$ and take OR on OC

$= CD$ and join QR. Then the locus of P is the line $MPL \parallel QR$ (*Fig.* 144).

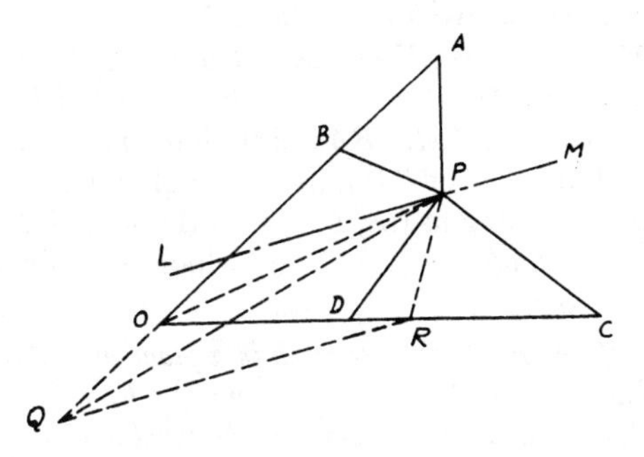

FIGURE 144

Proof: Similar to the first case. $\triangle PCD - \triangle APB = \triangle PRO - \triangle POQ =$ fig. $PRQO - \triangle OQR - \triangle POQ = \triangle PQR - \triangle OQR =$ constant. But, since $\triangle OQR$ is fixed, $\therefore$ $\triangle PQR$ is constant and also QR is fixed in direction. $\therefore$ The locus of P is a line $MPL \parallel QR$.

5.14. *From B, C the vertices of a triangle right-angled at A are drawn straight lines BF, CE respectively parallel to AC, AB and proportional to AB, AC. Find the locus of the intersection of BE and CF.*

CONSTRUCTION: Let P be a point on the locus. Draw EG, FH, $PK \perp BC$ and FL, $PM \perp AC$. Let PM meet BF, BC in N, R (*Fig.* 145).

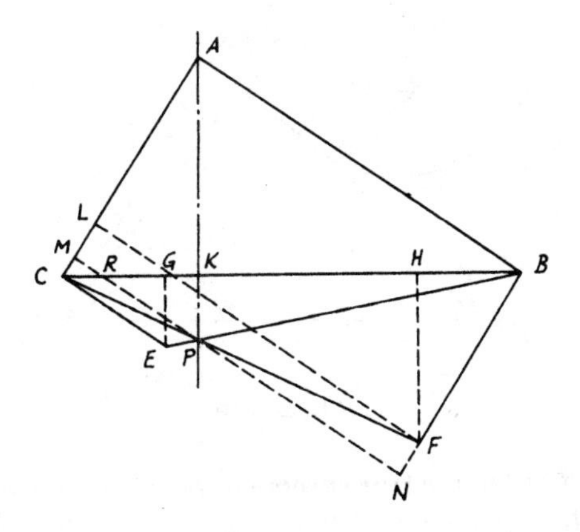

FIGURE 145

Proof: $\triangle$s ABC, HFB are similar. $\therefore$ $BH/BF = AC/BC$. But $BF/CE = AB/AC$ (hypothesis). Hence $BH/CE = AB/BC = CG/CE$ (since $\triangle$s ABC, GCE are also similar). $\therefore$ $BH = CG$. Also, $EG/CG = BH/HF$. $\therefore$ $EG/HF = EG^2/CG^2 = AC^2/AB^2$.

Again, $PM/AB = CP/CF = PK/FH$ and $AC/AM = BE/BP = EG/PK$. $\therefore$ $PM \cdot AC/AB \cdot AM = EG/FH = AC^2/AB^2$. $\therefore$ $PM/AM = AC/AB$. Hence $\triangle$s PAM, CBA are similar (Th. 4.86). $\therefore$ $\angle PAM = \angle ABC$. $\therefore$ The locus of P is a line through A making with CA or CA produced (according as CE, BF lie on the same side or on opposite sides of BC) an angle $= \angle ABC$ and which coincides with $\perp$ from A to BC.

5.15. *Four points A, B, A′, B′ are given in a plane with AB different from A′B′. Prove that there are always two positions of a point C in the plane such that the triangles CAB, CA′B′ are similar, the equal angles being denoted by corresponding letters.*

Construction: Draw AA', BB' and divide them internally and externally in the given ratio ($AB/A'B'$) in P, Q and M, N respectively. On PQ, MN as diameters describe two $\odot$s to intersect in C, C' which are the required points (*Fig.* 146).

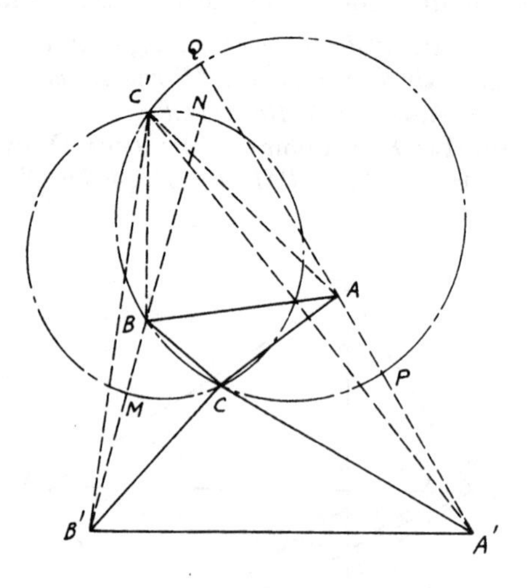

Figure 146

Proof: Join C, C' to the four corners A, B, A', B'. Since $\odot$ on PQ as diameter is the Apollonius $\odot$ of C in $\triangle CAA'$, $\therefore$ $AC/A'C = AP/A'P = AB/A'B'$. Similarly, $BC/B'C = AB/A'B'$. Hence $AB/A'B' =$

$AC/A'C = BC/B'C$. Therefore, $\triangle$s ACB, $A'CB'$ are similar. Similarly, $\triangle$s $AC'B$, $A'C'B'$ are similar. $\therefore$ There are two positions of C.

5.16. *The hypotenuse of a right-angled triangle is given. Find the locus of the mid-point of the line joining the outer vertices of the equilateral triangles described externally on its sides.*

CONSTRUCTION: Let BC be the hypotenuse of right-angled $\triangle ABC$. On CA, AB construct equilateral $\triangle$s CAD, AEB. On the same side as $\triangle ABC$, construct on BC equilateral $\triangle BFC$. Bisect BC, CF, FB, DE in G, H, K, P. Join GP, GH, GK, GD, GE, GA, GF. Therefore, the required locus of P is the $\odot$ described on HK as diameter (*Fig.* 147).

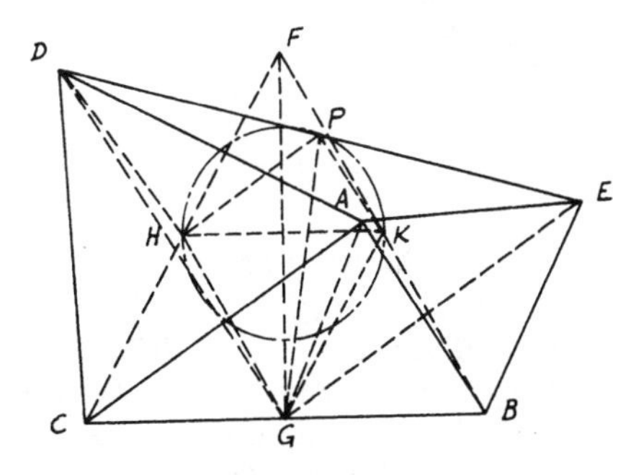

FIGURE 147

Proof: $\because GA = GC$ and $AD = DC$, $\therefore GD$ is $\perp AC$ and $\therefore \parallel AB$. Similarly, GE is $\perp AB$ and $\parallel AC$. $\because \angle GDC = \frac{1}{2} \angle ADC = \angle GFC$, $\therefore D$ lies on $\odot FGC$ whose center is H. $\therefore HD = HG$. $\because GD$ is $\parallel AB$ and $GE \parallel AC$, $\therefore \angle EGD$ is right. $\therefore GP = PD$ and PH is common to $\triangle$s GPH, DPH and $GH = HD$. $\therefore \angle GPH = \angle DPH$. Similarly, $\angle GPK = \angle KPE$. $\therefore \angle KPH$ is right. Since H, K are fixed points, $\therefore$ locus of P is a $\odot$ on diameter HK.

5.17. *The base and the vertex angle opposite to it in a triangle are given. Find the locus of the center of the circle which passes through the excenters of the three circles touching the sides of the triangle externally.*

CONSTRUCTION: Let O_1, O_2, O_3 be the centers of the excircles touching BC, CA, AB of $\triangle ABC$ in which BC and $\angle A$ are given. Draw the sides of $\triangle O_1$, O_2, O_3 and draw $\odot$ circumscribing $\triangle ABC$ cutting OO_1 and O_2O_3 in D, E (O being the incenter of $\triangle ABC$). Bisect DE in M, and produce OM to meet EP, which is $\perp O_2O_3$, in P the center of $\odot$ circumscribing $O_1O_2O_3$. Join AOO_1, BOO_2, COO_3 (*Fig.* 148).

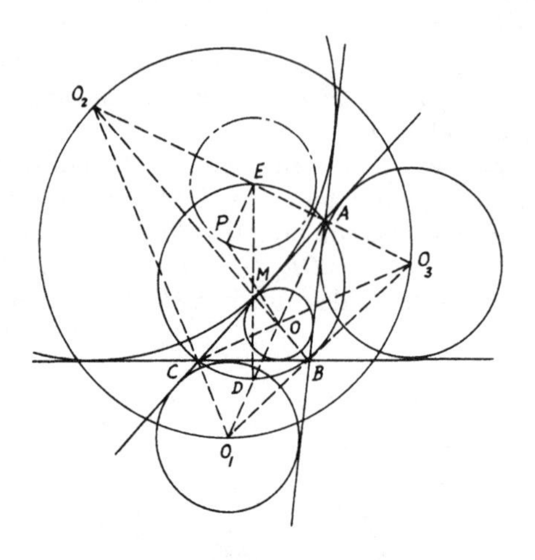

FIGURE 148

Proof: Since BC and $\angle A$ are given, $\therefore$ $\odot ABC$ is fixed. $\because$ $\triangle ABC$ is the pedal $\triangle$ of $\triangle O_1O_2O_3$, the fixed $\odot ABC$ is the nine-point $\odot$ of $\triangle O_1O_2O_3$, and D, E will be the mid-points of OO_1, O_2O_3. (O is also the orthocenter of $\triangle O_1O_2O_3$.) Since $\angle DAE$ = right angle, $\therefore$ DE is a diameter of the fixed $\odot ABC$, whose center M is also fixed. But D is a fixed point (being the mid-point of the fixed arc BC). $\therefore$ E is also a fixed point. $\because$ $OM = MP$ (see Problem 1.32), $\triangle$s OMD, PME are congruent. $\therefore$ $PE = OD$. But $\angle BOC$ = right + $\frac{1}{2}\angle A$ = fixed and BC is fixed. Then $\odot BOC$ is fixed and since it passes through O_1 ($BOCO_1$ is a cyclic quadrilateral) its radius OD is fixed. $\therefore$ PE is fixed in length. Therefore, the locus of P is a $\odot$ with E as center and PE as radius.

5.18. *Find the locus of a point such that the sum of the squares on the tangents from it to four given circles may be equal to a given square.*

CONSTRUCTION: Let A, B, C, D be four given circles and G a moving point, such that the sum of the squares of the tangents GH, GJ, GK, GL to these $\odot$s is given. Draw AB, CD and bisect them in P, Q respectively. Bisect PQ in R and join A, B, C, D to G (*Fig.* 149).

Proof: $GA^2 + GB^2 + GC^2 + GD^2 = (GH^2 + GJ^2 + GK^2 + GL^2) + (AH^2 + BJ^2 + CK^2 = DL^2)$ = constant, since the radii are given. Hence $2(GP^2 + PB^2 + GQ^2 + QC^2)$ = constant. $\because$ $(PB^2 + QC^2)$ is a fixed quantity, $\therefore$ $GP^2 + GQ^2$ = constant = $2(GR^2 + PR^2)$. But, since P, Q are two fixed points, $\therefore$ R is fixed and PR is

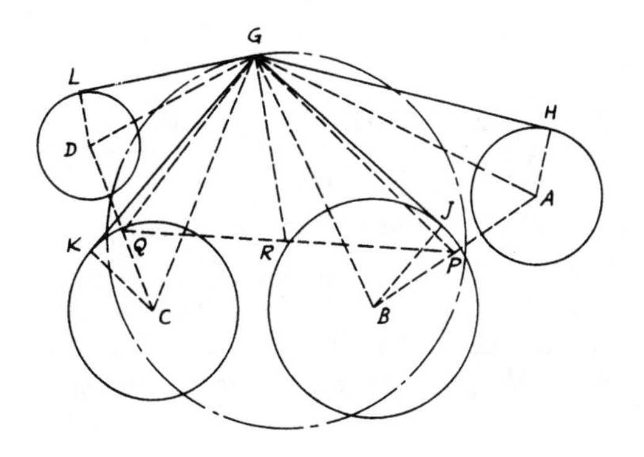

FIGURE 149

of constant magnitude. Hence GR is constant. Therefore, the locus of G is a $\odot$ having R as center and RG as radius.

5.19. *A line meets the sides BC, CA, AB of a triangle ABC at D, E, F. P, Q, R bisect EF, FD, DE. AP, BQ, CR, produced if necessary, meet BC, CA, AB at X, Y, Z. Show that X, Y, Z are collinear* (*Fig.* 150).

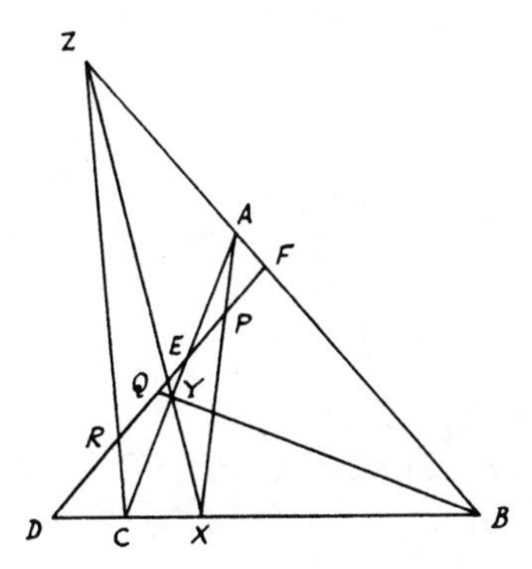

FIGURE 150

Proof:

$$\frac{\triangle ABX}{\triangle APF} = \frac{AX \cdot AB \sin BAX}{AP \cdot AF \sin BAX} = \frac{AX \cdot AB}{AP \cdot AF}.$$

$\triangle APE / \triangle ACX = AP \cdot AE / AC \cdot AX$. $\because$ $\triangle APF = \triangle APE$ (since PF

$= PE$), hence $\triangle ABX/\triangle ACX = BX/XC = AB\cdot AE/AC\cdot AF$. Similarly, $CY/YA = BC\cdot BF/BA\cdot BD$ and $AZ/ZB = CA\cdot CD/CB\cdot CE$. Multiplying yields

$$\frac{BX}{XC}\frac{CY}{YA}\frac{AZ}{ZB} = \frac{AB\cdot BC\cdot CA\; AE\cdot BF\cdot CD}{AC\cdot BA\cdot CB\; AF\cdot BD\cdot CE} = 1.$$

Therefore, X, Y, Z are collinear (converse, Menelaus' Th. 5.14).

5.20. *If P, Q, R are the points of contact of the inscribed circle with the sides BC, CA, AB of a triangle ABC and PQ, QR, RP produced meet AB, BC, CA in G, H, K, show that GHK is a straight line. If X, Y, Z are the mid-points of RG, PH, QK, show also that these points are collinear (Fig.* 151).

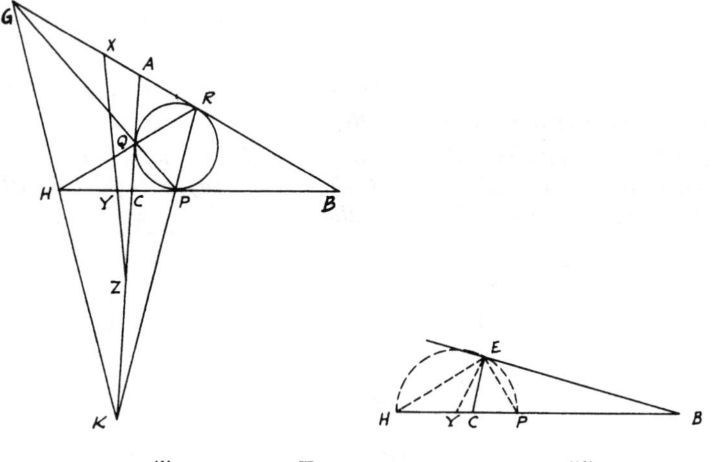

(i) FIGURE 151 (ii)

Proof: (i) $AG\cdot BP\cdot CQ/GB\cdot PC\cdot QA = 1$ (Th. 5.14). $\because$ $CP = CQ$, $\therefore$ $AG/GB = QA/BP$. Similarly, $BH/HC = BR/CQ$ and $CK/AK = CP/AR$. Multiplying gives $AG\cdot BH\cdot CK/GB\cdot HC\cdot KA = 1$. $\therefore$ GHK is a straight line.

(ii) Again, consider the line $HYCPB$ alone. Draw a $\odot$ on HP as diameter. Since $HC/BH = CQ/BR = CP/PB$, $\therefore$ $CP/HC = BP/BH$ and $CP/HC - CP = BP/BH - BP$ or $CP/BP = YC/YP$ (1) (Y is the mid-point of HP). $\because$ The $\odot$ on HP as diameter is the Apollonius $\odot$ of CB with respect to P, H, $\therefore$ any point E on this $\odot$ will yield the ratio $EC/EB = CP/BP$. $\because$ $\angle PEH$ = right angle and EP, EH are the internal and external bisectors of $\angle E$, $\therefore$ $\triangle$s YEC, YBE are similar. $\therefore$ $YE/YB = EC/EB = CP/BP$ (2). $\because$ $YP = YE$, hence from (1) and

(2), $\therefore$ $YC/YB = CP^2/BP^2$. Similarly, $XB/XA = BR^2/AR^2$. $AZ/ZC = AQ^2/CQ^2$. By multiplying, $\therefore$ $YC \cdot XB \cdot AZ/YB : XA \cdot ZC = 1$. Hence XYZ is also a straight line.

5.21. *The sides BC, CA, AB of a triangle ABC are cut by two lines in the points D, E, F and D′, E′, F′. Show that EF′, FD′, DE′ cut BC, CA, AB in three collinear points D″, E″, F″* (*Fig.* 152).

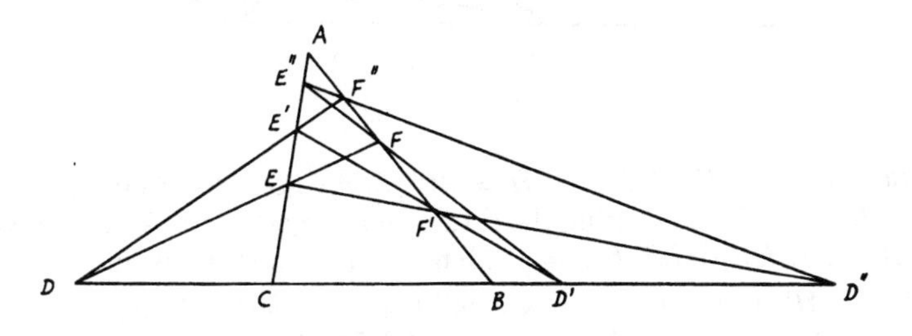

FIGURE 152

Proof: Since $EF'D''$ is a transversal of $\triangle ABC$, $\therefore$ $AE \cdot CD'' \cdot BF'/EC \cdot D''B \cdot F'A = 1$. Similarly with the other two transversals $E''FD'$, $DE'F''$:

$$\frac{BF \cdot AE'' \cdot CD'}{FA \cdot E''C \cdot D'B} = 1 \quad \text{and} \quad \frac{AE' \cdot CD \cdot BF''}{E'C \cdot DB \cdot F''A} = 1.$$

But, DEF, $D'E'F'$ are two transversals; $\therefore$ multiplying and using the ratios from the other transversals yields

$$\frac{CD'' \cdot BF'' \cdot AE''}{D''B \cdot F''A \cdot E''C} = 1.$$

Therefore, D'', F'', E'' are collinear.

5.22. *A transversal DEF cuts the sides BC, CA, AB of a triangle ABC in D, E, F respectively. If AD, BE, CF are joined and AG, BL, CH are drawn such that $\angle BAG = \angle CAD$, $\angle CBE = \angle ABL$, $\angle ACF = \angle BCH$, show that GHL is a straight line* (*Fig.* 153).

Proof:

$$\frac{\triangle ABG}{\triangle ACD} = \frac{AB \cdot AG \sin BAG}{AC \cdot AD \sin CAD} = \frac{AB \cdot AG}{AC \cdot AD} = \frac{BG}{DC}.$$

Likewise, $\triangle ACG/\triangle ABD = AC \cdot AG/AB \cdot AD = CG/BD$. Dividing gives $BD \cdot BG/DC \cdot CG = AB^2/AC^2$. Similarly, $CE \cdot CL/AE \cdot AL =$

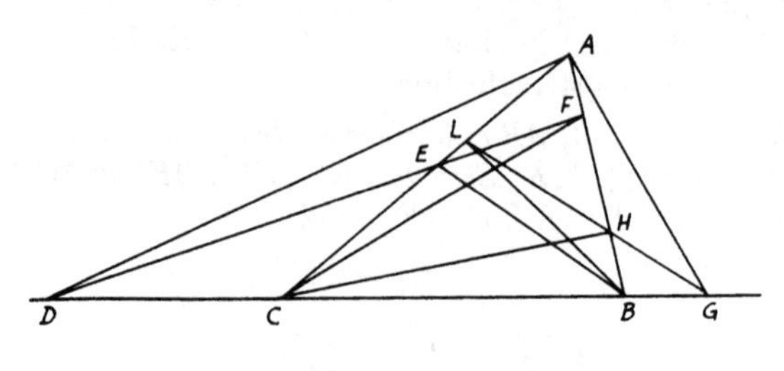

FIGURE 153

BC^2/AB^2 and $AF \cdot AH/BF \cdot BH = AC^2/BC^2$, since DEF is a transversal. $\therefore$ Multiplying and using the ratios from the other transversals gives $BG \cdot CL \cdot AH/GC \cdot LA \cdot HB = 1$. Hence GHL is a straight line.

5.23. *AB, CD, EF are three parallel straight lines. M, N, R are the intersections of pairs of lines AD, BC; AF, BE; CF, DE respectively. If X, Y, Z are the middle points of AB, CD, EF respectively, show that XR, YN, ZM are concurrent.*

CONSTRUCTION: Draw XR, YN, ZM, and also XY, YZ, ZX (*Fig.* 154).

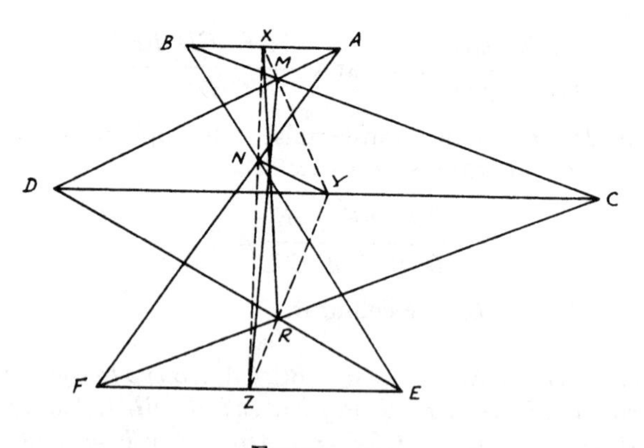

FIGURE 154

Proof: $\because$ AB is $\parallel$ CD and X is the mid-point of AB, $\therefore$ XM produced bisects CD. Hence XMY is one straight line. Likewise, YRZ, ZNX are also straight lines. By similarity, $XM/MY = AX/DY$, $YR/RZ = DY/ZE$, and $ZN/NX = ZF/AX$. Multiplying gives $XM \cdot YR \cdot ZN/MY \cdot RZ \cdot NX = 1$. Therefore, XR, YN, ZM are concurrent (Th. 5.12).

5.24. *On the sides BC, CA, AB of a triangle are taken the points X, Y, Z such that $BX = XC$, $CY = YA$, $AZ = 2\ ZB$. BY and CZ meet at P, AX and CZ at Q, AP and BQ at R, BP and CR at S. Show that $BY = 6\ SP$.*

CONSTRUCTION: Produce AR, BQ, CS to meet BC, CA, AB in D, E, F. Join DS, ZS (*Fig.* 155).

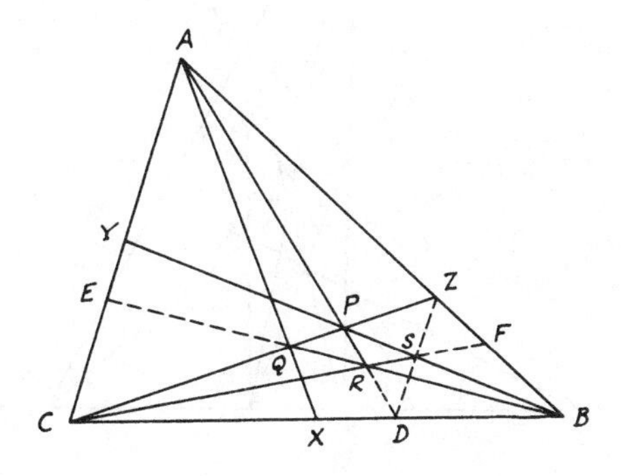

FIGURE 155

Proof: $\because AE \cdot CX \cdot BZ / EC \cdot XB \cdot ZA = 1$, and since $CX = XB$, $\therefore AE/EC = ZA/BZ = 2$. In $\triangle ACZ$, $ZB \cdot AY \cdot CP / BA \cdot CY \cdot PZ = 1$. $\because AY = CY$, $\therefore ZB/BA = PZ/CP = 1 : 3$. Also,

$$\frac{AY \cdot CD \cdot BZ}{YC \cdot DB \cdot ZA} = 1.$$

$\therefore CD/DB = ZA/BZ = 2$. Similarly, $AE \cdot CD \cdot BF / EC \cdot DB \cdot FA = 1$. $\therefore BF/FA = EC \cdot DB / AE \cdot CD = 1 : 4$. Hence $BF/AB = 1 : 5$ and $\because AB/BZ = 3$, $\therefore BF/BZ = 3 : 5$. $\therefore BF/FZ = 3 : 2$. Therefore, $ZP \cdot CD \cdot BF / PC \cdot DB \cdot FZ = 1$. $\therefore BP$, CF, DZ are concurrent at S. Hence ZSD is one straight line $\parallel AC$. $\therefore SP/PY = ZP/PC = 1 : 3$. $\therefore SP/SY = 1 : 4$. $\because SY/BY = AZ/AB = 2 : 3$, $\therefore SP/BY = 1 : 6$.

5.25. *If a transversal cuts the sides BC, CA, AB of a triangle ABC in P, Q, R respectively and if P′, Q′, R′ are the harmonic conjugates of P, Q, R with respect to B, C; C, A; A, B respectively, then AP′, BQ′, CR′ are concurrent. If X, Y, Z be the points of bisection of PP′, QQ′, RR′, then XYZ is a straight line.*

Note: Since P', Q', R' are the harmonic conjugates of P, Q, R with respect to B, C; C, A and A, B, then P, P' divide BC in the same ratio; i.e., $BP/CP = BP'/CP'$, and so on.

CONSTRUCTION: Draw PR', $R'Q'$, $P'Q$, $R'Q$, $P'R$, RQ' (*Fig.* 156).

Proof: $CP \cdot BR \cdot AQ / PB \cdot RA \cdot QC = 1$. Replacing equal ratios in the

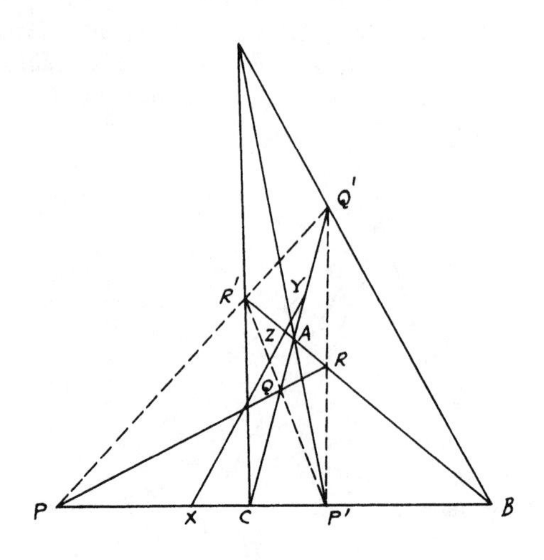

FIGURE 156

above quantity, $CP' \cdot BR' \cdot AQ'/P'B \cdot R'A \cdot Q'C = 1$. Hence AP', BQ', CR' are concurrent (Th. 5.12). Also, $CP' \cdot BR' \cdot AQ/P'B \cdot R'A \cdot QC = 1$. Therefore, $P'QR'$ is a straight line (Th. 5.14). Likewise, $PR'Q'$ and $P'RQ'$ are straight lines. Now, in the complete quadrilateral $QRQ'R'$, X, Y, Z are the mid-points of its diagonals PP', QQ', RR'. Hence they are collinear (see Problem 2.16).

5.26. *If AD, BE, CF are three concurrent lines drawn from the vertices of the triangle ABC and terminated by the opposite sides, then the diameter of the circle circumscribing triangle ABC will be equal to* $(AF \cdot BD \cdot CE) : \triangle DEF$.

CONSTRUCTION: Draw the $\perp$s BL, DR, CM to EF or produced; BN, CH to DR and AG to BC (*Fig.* 157).

Proof: Let $2r$ be the diameter of the circumscribing $\odot ABC$. $\because$ $BD/CD = ND/DH = (DR - LB)/(CM - DR)$, $\therefore BD \cdot CM + CD \cdot BL = DR\,(BD + CD) = BC \cdot DR$. Multiplying both sides by EF yields $\therefore BD \cdot \triangle CEF + CD \cdot \triangle BEF = BC \cdot \triangle DEF$. Hence $\triangle DEF = (BD \cdot \triangle CEF + CD \cdot \triangle BEF) : BC$. But $\triangle CEF/\triangle ACF = CE/AC$ and $\triangle ACF/\triangle ABC = AF/AB$. $\therefore \triangle CEF/\triangle ABC = CE \cdot AF/AC \cdot AB$. Similarly, $\triangle BEF/\triangle ABC = BF \cdot AE/AB \cdot AC$. Substituting gives $\triangle DEF = (BD \cdot CE \cdot AF \cdot \triangle ABC + CD \cdot BF \cdot AE \cdot \triangle ABC) : BC \cdot AC \cdot AB$. But $BD \cdot CE \cdot AF = CD \cdot BF \cdot AE$ (Th. 5.11) and $AB \cdot AC = AG \cdot 2r$ (Th. 4.100). Hence $BC \cdot AC \cdot AB = 2r \cdot BC \cdot AG = 4r \cdot \triangle ABC$. Therefore, $\triangle DEF = 2\,AF \cdot BD \cdot CE \cdot \triangle ABC/4r \cdot \triangle ABC$ or $2r = AF \cdot BD \cdot CE / \triangle DEF$.

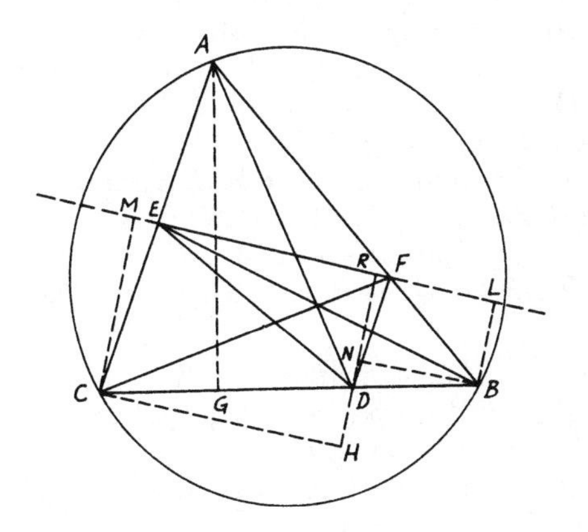

FIGURE 157

5.27. *D, E, F are the points of contact of the escribed circle opposite the vertex A with the sides BC, CA, AB of a triangle ABC. X, Y, Z and L, M, N are the similar points of contact of the other two escribed circles opposite to B, C with the same order of sides. If BY, CN intersect in G and BE, CF in H, show that A, G, D, H are collinear. If YN, ND, YD are also produced to meet BC, AC, AB in P, Q, R, then PQR is a straight line* (*Fig.* 158).

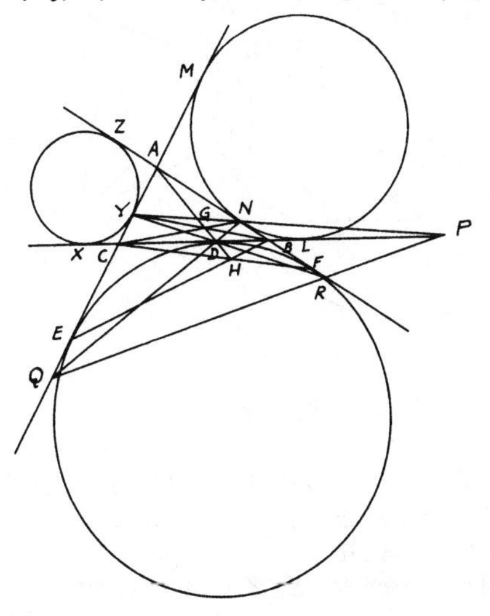

FIGURE 158

Proof: Let the perimeter of $\triangle ABC = 2s$ and a, b, c denote the sides BC, CA, AB. $BD = BF = AF - AB = s - c$. But $BD \cdot CY \cdot AN / DC \cdot YA \cdot NB = (s-c)\cdot(s-a)\cdot(s-b)/(s-b)\cdot(s-c)\cdot(s-a) = 1$. Hence AD, BY, CN are concurrent or A, G, D are collinear. Also, $BD \cdot CE \cdot AF/DC \cdot EA \cdot FB = (s-c)\cdot(s-b)\cdot s/(s-b)\cdot s\cdot(s-c) = 1$. Thus H lies on AGD produced. Since YNP is a transversal of $\triangle ABC$, $BP \cdot CY \cdot AN/PC \cdot YA \cdot NB = 1$. $\therefore$ $BP/PC = (s-c)/(s-b)$. Similarly, $CQ/QA = (s-a)/(s-c)$ and $AR/RB = (s-b)/(s-a)$. Hence $BP \cdot CQ \cdot AR/PC \cdot QA \cdot RB = 1$. Therefore, PQR is a straight line.

5.28. *In the triangle ABC, AD, BE, CF are the altitudes. If EF, FD, DE produced meet BC, CA, AB respectively in X, Y, Z, show that the centers of the circles ADX, BEY, CFZ are collinear.*

CONSTRUCTION: Let G, H, J be the centers of the ○s ADX, BEY, CFZ. Join XZ, YZ (*Fig.* 159).

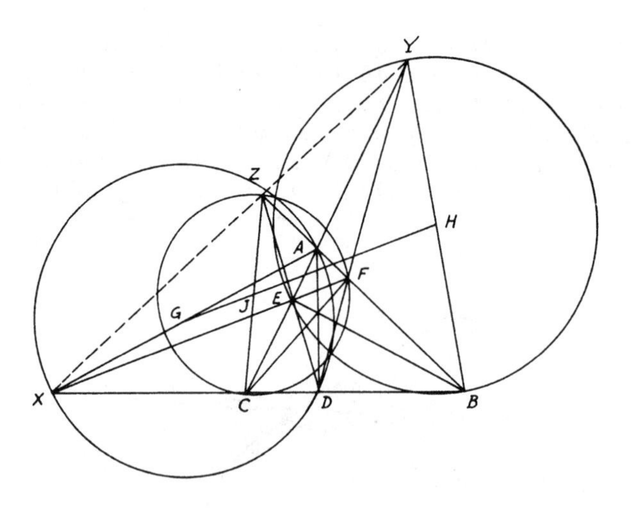

FIGURE 159

Proof: $\because$ $BD \cdot CE \cdot AF/DC \cdot EA \cdot FB = 1$ and $BX \cdot CE \cdot AF/XC \cdot EA \cdot FB = 1$, hence $BD/DC = BX/XC$. Likewise, $CY/YA = CE/EA$ and $AZ/ZB = AF/FB$. Multiplying yields $BX \cdot CY \cdot AZ/XC \cdot YA \cdot ZB = 1$. Therefore, XYZ is a straight line, since $\angle ADX$ = right. $\therefore$ G is the mid-point of AX the diameter of ⊙ADX. Similarly, H, J are the mid-points of BY, CZ. But, since G, H, J are the mid-points of the diagonals of the complete quadrilateral $ACXZ$, then they are collinear (see Problem 2.16).

5.29. *On the sides BC, CA, AB of a triangle are taken points X, X' and Y, Y' and Z, Z' such that X', Y, Z are collinear and also X, Y', Z and also X, Y, Z'. If $X'Z'$ cuts BY in Q and $X'Y'$ cuts CZ in R, show that QY' and RZ' meet on BC (Fig. 160).*

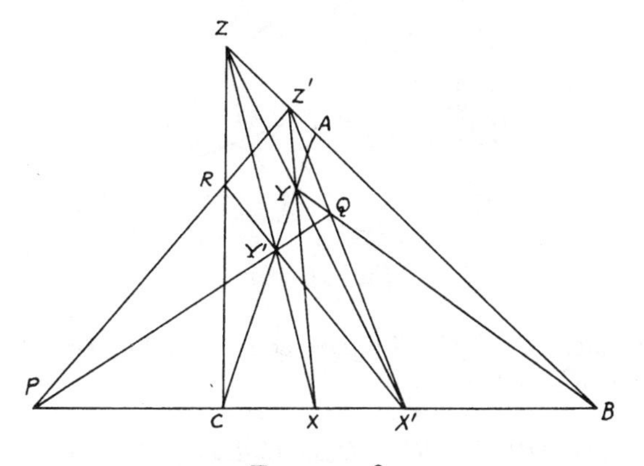

FIGURE 160

Proof: Produce QY' to meet BC in P. $Z'YX$ is a transversal to $\triangle BQX'$: $\therefore$ $QZ' \cdot BY \cdot X'X / Z'X' \cdot YQ \cdot XB = 1$. Likewise, in $\triangle YQY'$ and $\triangle YY'X'$, $Y'P \cdot YC \cdot QB / PQ \cdot CY' \cdot BY = 1$, and $X'R \cdot Y'C \cdot YZ / RY' \cdot CY \cdot ZX' = 1$. But, since in $\triangle YXX'$ $YZ/ZX' = XB \cdot YZ' / BX' \cdot Z'X$, then, multiplying the first three quantities and substituting in the fourth gives

$$\frac{QZ' \cdot Y'P \cdot X'R \cdot XX' \cdot QB \cdot YZ'}{Z'X' \cdot PQ \cdot RY' \cdot X'B \cdot YQ \cdot Z'X} = 1.$$

But $YZ' \cdot BQ \cdot XX' / Z'X \cdot QY \cdot X'B = 1$ in $\triangle YBX$ with $Z'QX'$ as transversal. Therefore, in $\triangle QY'X'$ the remainder $QZ' \cdot Y'P \cdot X'R / Z'X' \cdot PQ \cdot RY' = 1$. Therefore, $Z'RP$ is a straight line.

5.30. *A circle meets the sides BC, CA, AB of a triangle ABC at A_1, A_2; B_1, B_2; C_1, C_2. B_1C_1, B_2C_2 meet at X; C_1A_1, C_2A_2 at Y; and A_1B_1, A_2B_2 at Z. Show that AX, BY, CZ are concurrent.*

CONSTRUCTION: Join A_1B_2, B_1C_2, C_1A_2 and produce AX to meet B_1C_1 in D (*Fig.* 161).

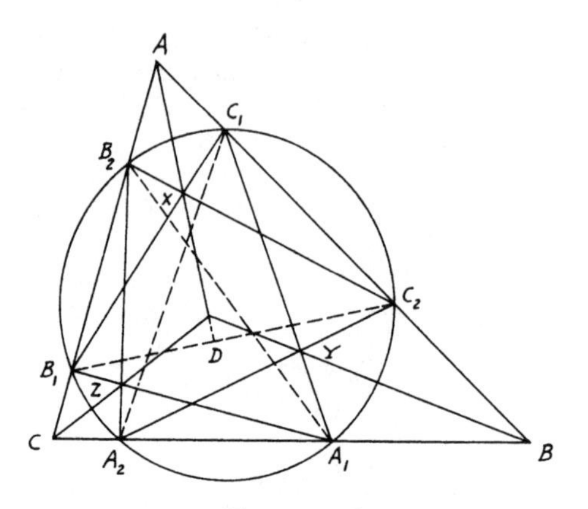

FIGURE 161

Proof: Since AX, B_1C_1, B_2C_2 are concurrent, $\therefore$ $DC_2 \cdot B_1B_2 \cdot AC_1/DB_1 \cdot AB_2 \cdot C_1C_2 = 1$, since

$$\frac{DC_2}{DB_1} = \frac{\triangle ADC_2}{\triangle ADB_1} = \frac{AC_2 \cdot AD \cdot \sin C_2AD}{AD \cdot AB \cdot \sin DAB_1} = \frac{AC_2 \cdot \sin C_2AD}{AB_1 \cdot \sin DAB_1},$$

and so on. Now substituting in the above quantity and simplifying,

$$\frac{\sin C_2AD \cdot \sin B_2B_1C_1 \cdot \sin B_1C_2B_2}{\sin DAB_1 \cdot \sin C_1B_1C_2 \cdot \sin B_2C_2C_1} = 1.$$

$$\therefore \frac{\sin BAX}{\sin XAC} = \frac{\sin C_1B_1C_2 \cdot \sin B_2C_2C_1}{\sin B_2B_1C_1 \cdot \sin B_1C_2B_2},$$

and so on. Hence

$(\sin BAX/\sin XAC) \cdot (\sin CBY/\sin YBA) \cdot (\sin ACZ/\sin ZCB)$

$$= \frac{\sin C_1B_1C_2 \cdot \sin B_2C_2C_1 \cdot \sin A_1C_1A_2 \cdot \sin C_2A_2A_1 \cdot \sin B_1A_1B_2 \cdot \sin A_2B_2B_1}{\sin B_2B_1C_1 \cdot \sin B_1C_2B_2 \cdot \sin C_2C_1A_1 \cdot \sin C_1A_2C_2 \cdot \sin A_2A_1B_1 \cdot \sin A_1B_2A_2}$$

and this is unity, since $\angle C_1B_1C_2 = \angle C_1A_2C_2$, and so on. Hence AX, BY, CZ are concurrent.

Miscellaneous Exercises

1. Any straight line is drawn cutting two fixed intersecting straight lines, and the two angles on the same side of it are bisected. Find the locus of the point of intersection of the bisecting lines.
2. Find the locus of a point the difference of whose distances from two given intersecting straight lines is equal to a given length.
3. P is a point inside a parallelogram $ABCD$ such that the area of the quadrilateral $PBCD$ is twice that of the figure $PBAD$. Show that the locus of P is a straight line.
4. Two given straight lines meeting in O are cut in P and Q by a variable third line. If the sum of OP, OQ be constant, show that the locus of the middle point of PQ is a straight line. [On OP, OQ take $OA = OB$ = half $(OP + OQ)$. AB is the required locus.]
5. Two sides AB, AD of a quadrilateral are given in magnitude and position and the area of the quadrilateral is given. Find the locus of the middle point of AC.
6. Given the base and the sum of the sides of a triangle. Find the locus of the feet of the perpendiculars let fall on the external bisector of the vertex angle from the ends of the base.
7. BAC is a fixed angle of a triangle and (a) the sum; (b) the difference of the sides AB, AC is given. Show that in either case the locus of the middle point of BC is a straight line.
8. A straight line XY moves parallel to itself so that the sum of the squares on the straight lines joining X and Y to a fixed point P is constant. Find the locus of the middle point of XY.
9. Show that the locus of a point such that the sum of the squares on its distances from three given points may be constant is a circle the center of which lies at the centroid of the triangle formed by the three given points.
10. Prove that the locus of a point such that the sum of the squares on its distances from the vertices of a quadrilateral is constant is a circle the center of which coincides with the intersection of the lines joining the middle points of opposite sides of the quadrilateral.
11. Through one of the points of intersection of two fixed circles with centers A and B, a chord is drawn meeting the first circle in P and the other in Q. Find the locus of the point of intersection of PA and QB. (The locus is an arc of a circle subtending the constant angle at the intersection of PA, QB.)
12. ABC is a triangle inscribed in a circle, P any point in the circumference of the circle, and Q is a point in PC such that the angle QBC is equal to the angle PBA. Show that the locus of Q is a circle.

13. ABC is a straight line, and any circle is described through A, B and meeting in P, P' the straight line bisecting AB at right angles. Find the locus of the points in which CP and CP' cut the circle.

14. Two circles intersect in A, B. Through A a chord PAQ is drawn to meet both circles in P, Q. Find the locus of the center of the circle inscribed in the triangle PBQ.

15. Find the locus of the center of the circle which bisects the circumferences of two circles given in position and magnitude. (The locus is the perpendicular on the line of centers of the two circles.)

16. A is a fixed right angle. Two equal distances AB, AC are taken on the sides of angle A. If C is joined to a fixed point D and BE is drawn perpendicular to CD, find the locus of E for all positions of B, C.

17. Two circles intersect in A and B and a variable point P on one circle is joined to A and B. PA, PB, produced if necessary, meet the second circle in Q and R. Prove that the locus of the center of the circle PQR is a circle.

18. AB is a fixed chord of a given circle. P is any point on the circumference. Perpendiculars AC and BD are drawn to BP and AP respectively. Find the locus of the middle point of CD.

19. Two circles touch and through the point of contact, a variable chord AB is drawn cutting the circles in A, B. Show that the locus of the middle point of AB is a circle. (Use the solution given in Problem 5.4.)

20. AB is a fixed chord of a circle and X any point on the circumference. Find the locus of the intersection of the other tangents from A and B to the circle drawn, with center X, to touch AB.

21. Two equal circles of given radius touch, each, one of two straight lines which intersect at right angles and also touch each other. Find the locus of their point of contact with each other.

22. Given a fixed straight line XY and a fixed point A. If B is a moving point on XY and C is taken on AB such that the rectangle $AC \cdot CB$ is constant, find the locus of C.

23. From a fixed point P, a straight line is drawn to meet the circumference of a fixed circle in Q. If PQ is divided in R such that $PR : PQ$ is constant, find the locus of R.

24. From a point P inside a triangle ABC perpendiculars PD, PE, PF are drawn to BC, CA, AB respectively. If the angle EDF is equal to A, prove that the locus of P is an arc of the circle passing through B, C and the center of the circle circumscribing ABC.

25. If the two circles do not intersect orthogonally in Problem 5.4, find the locus of E the middle point of CD when (a) the circles are not equal; (b) the circles are equal.

26. D is the middle point of a fixed arc AB of a given circle. From D any chord DE is drawn in the circle cutting AB in F. Show that the locus

of the center of the circle AEF is a straight line. (Join AD, DB. The locus is the line joining A to the center of circle AEF.)

27. Find the locus of the centers of the circles which pass through a given point and cut a given circle orthogonally.

28. If on each segment of a line, and on the same side of it, two equilateral triangles be described and their vertices joined to the opposite extremities of the line, the locus of the intersections of these lines is the circle circumscribing the equilateral triangle described on the other side of the line.

29. Through a fixed point which is equidistant from two parallel straight lines, a straight line is drawn terminated by the two fixed straight lines and on it as base is described an equilateral triangle. Show that the vertex of this triangle will lie on one of two straight lines.

30. A, B, C are three points not in the same line. D is any point on BC such that if AD is produced to E then $AD : AE = BD : BC$. Find the locus of E.

31. On the external bisector of the angle A of a triangle ABC, two points D, E are taken such that $AD \cdot AE = AB \cdot AC$. Prove that the locus of the point of intersection of DB and EC is an arc of a circle drawn on BC to subtend an angle = half angle A.

32. ABC is a triangle and D, E are two points on AB, AC respectively such that $BD \cdot BA + CE \cdot CA = BC^2$. Show that the locus of the point of intersection P of BE, CD is a circle. (Take point F in BC so that $BF \cdot BC = BD \cdot BA$ and prove that $ADPE$ is a cyclic quadrilateral, P being the required point for locus, which is a circle on BC subtending angle BPC.)

33. Given three points A, B, C on one straight line and D a moving point such that the angles ADB, BDC are always equal, find the locus of D.

34. AOB is a fixed diameter in circle with center O, and C is a moving point along the circumference. If D is taken on BC, such that the ratio of $BD : DC$ is constant, show that the locus of the point of intersection of OD, AC is a circle.

35. The middle points of all chords of a circle which subtend a right angle at a fixed point lie on a circle.

36. $ABCD$ is a square. From A, B two lines are drawn to meet CD or produced in E, F such that $CF \cdot DE = CD^2$ is always constant. What is the locus of the point of intersection of AE, BF?

37. Two circles intersect in A, B. Through A a line is drawn to meet the two circles in P, Q. If PQ is divided by R in a constant ratio, show that the locus of R is a circle.

38. The base BC and the vertex angle of the triangle ABC are given. From O the middle point of the base, OA is drawn and produced to P so that the ratio of $OP : OA$ is constant. Find the locus of P.

39. Find the locus of a point inside the vertex angle A of an isosceles triangle ABC, the distance of which from the base BC is a mean proportional between its distances from the equal sides. (This is the circle drawn on BC as a chord and touching AB, AC at B, C respectively.)

40. OAB is a straight line rotating about its fixed end O. A, B are two fixed points on the line and D is another fixed point outside it. If the parallelogram $DACB$ is completed, construct the locus of C.

41. P is a point in a segment of a circle described on a base AB, and the angle APB is bisected by PQ of such length that PQ is equal to half the sum of PA and PB. Prove that the locus of Q for different positions of P is a circle. (Let PQ meet the circle in E. E the mid-point of arc AB is a fixed point and $AE = EB$. Since $PE : PA + PB = AE : AB$, then $EQ : PQ = AE - AD : AD$, a construction ratio. Hence the locus of Q is a circle with center E.)

42. A triangle ABC is inscribed in a fixed circle, the vertex A being fixed and the side BC given in magnitude. If G be the centroid of the triangle ABC, show that the locus of G is a circle.

43. A square is described with one side always on a given line, and one corner always on another. Find the locus of the corner which lies on neither.

44. One end O of a straight line is fixed and the other P moves along a given straight line. If PQ be drawn at right angles to OP such that PQ/OP is constant, show that Q traces out a straight line. (Draw $OA \perp$ the line on which P moves. From AP cut off AB a fourth proportional to OP, PQ, OA. Then the locus of Q is a line from $B \perp OB$.)

45. ABC is a triangle inscribed in a circle and its vertex A is fixed. A point D is taken on BC such that the ratio $AD^2 : DB \cdot DC$ is constant. Show that the locus of D is a circle that touches circle ABC at A.

46. M is the center of a circle in which AB is a fixed diameter. Any point P is taken on AB produced and PQ is drawn tangent to the circle. If MD is drawn perpendicular to the bisector of the angle BPQ, show that when P moves along AB, D traces a straight line parallel to AB. (Join MQ and produce MD to meet PQ in R. Drop RL, DF $\perp$s AB.)

47. Two circles intersect in A, B and D, E are two points taken on their circumferences, such that the angle DAE is constant. If F divides DE in the fixed ratio of $DF : FE$, find the locus of F.

48. Two circles intersect in A and any straight line through A meets them again in P, Q. Show that the locus of a point which divides PQ in a constant ratio is a circle through the common points of the two circles.

49. Straight lines are drawn through the points B, C of a triangle ABC, making with AB and AC produced angles equal to those made by BC with AB and AC. These lines meet in A'. Prove that AA' passes through the center of the circle ABC and that as A moves round the circle, the locus of the orthocenter of $A'BC$ is a circle.

50. A is a fixed point inside a circle with center O. Any chord BAC is drawn through A and a semi-circle is constructed on BC as a diameter. If AD is drawn perpendicular to BC meeting the semi-circle in D, show that the locus of D is a circle which cuts circle O in H, G such that HG always passes through the fixed point A.

51. Find the locus of a point such that the triangles formed by joining it to the ends of two given straight lines are equal to each other.

52. Find the locus of the foot of the perpendicular from a fixed point on a chord of a given circle which subtends a right angle at the fixed point. (Let D be the fixed point and AB be the chord of the given circle with center O. Draw $DE \perp AB$ and join OD. Bisect OD in F and it can be proved that FE is constant and F is fixed. Hence the locus is a circle with center F.)

53. Two circles are described one of which passes through a fixed point A and has its center on a fixed line AB, and the other passes through a fixed point C and has its center on a fixed line CD parallel to AB. If the two circles touch, find the locus of their point of contact.

54. AB is a fixed chord of a circle subtending a right angle at the center. PQ is a variable diameter. Prove that the locus of the intersection of AP and BQ is a circle equal to the given circle.

55. A given straight line AB is trisected in C, D. Lines CP, DP are drawn through C, D inclined at a given angle. CP, DP are produced to E, F so that CP, DP are double of PE, PF respectively. Find the locus of the intersection of AF, BE.

56. From a fixed point D in the base BC of a given triangle ABC, any line DEF is drawn cutting AB, AC or produced in E, F respectively. The circles around DEB, DFC intersect in P. Find the locus of P.

57. A point moves such that the sum of the squares on its distances from the sides of a square is equal to twice the square. Show that the locus of this point is a circle the radius of which bears a constant ratio to the side of the square. What is the condition for the radius of this locus circle to be equal to the side of the square?

58. AOB is a fixed diameter of a circle with center O. AC is a variable chord and AC is divided at D so that the ratio of $AD : DC$ is constant. Find the locus of the intersection of OD, BC.

59. The squares on the two sides of a triangle on a given base are together equal to five times the area of the triangle. Prove that the locus of the vertex of the triangle is a circle.

60. From a given external point A a secant AMN is drawn to a given circle. Find the locus of the intersection of the circles which pass through A and touch the given circle in M, N, one externally, the other internally.

61. ABC is a right-angled triangle at A. On the sides AB, AC two squares $ABFG$, $ACHK$ are described. If FG, HK be produced to meet in P, show that, as the right-angle changes its position, the hypotenuse BC being

fixed, the locus of P will be a circle. (Draw BQ, $CR \perp BC$ meeting FG and HK or produced in Q, R. It can be proved that $BQ = CR = BC$. Hence Q, R are fixed points. Therefore, the locus of P is a semi-circle on diameter QR.)

62. In a triangle ABC, the base BC and the length of the line CD which divides the side AB in a fixed ratio at D are given. Find the locus of the vertex A of the triangle.

63. A, B, C, D are four points on a straight line. Find the locus of a point P outside this line which will subtend equal angles with AB, CD.

64. Find the locus of a point the difference of the squares on the distances of which from two opposite angular points of a square is equal to twice the rectangle contained by its distances from the other two angular points. (The locus is the circle circumscribing the square.)

65. A, B are the centers of two given circles of different diameters and M is a point which moves such that, if the two tangents MC, MD are drawn to circles A, B respectively, then $3\,MC^2 + MD^2$ is always constant. Prove that the locus of M is a circle the center of which lies on AB.

66. Show that in Problem 2.16 the line joining the middle points of the diagonals of a complete quadrilateral is the locus of the points which make, with every two opposite sides of the quadrilateral, two triangles (a) the sum of which equals half the area of the quadrilateral for the portion of this line inside the quadrilateral; (b) the difference of which equals half the area of the quadrilateral for that portion of the line outside the quadrilateral.

67. From B, C the angular points of a triangle right-angled at A are drawn straight lines BF, CE respectively parallel to AC, AB and proportional to AB, AC. Find the locus of the intersection of BE and CF.

63. Given the inscribed and circumscribed circles of a triangle, prove that the centers of the escribed circles in every position of the triangle lie on a circle. (Let ABC be one position of the $\triangle$, O_1, O_2, O_3 the centers of the escribed circles, I, O the centers of the inscribed and circumscribed circles. I is the orthocenter of $\triangle O_1O_2O_3$ and $\odot ABC$ is its nine-point circle. If IO is produced to S so that $OS = OI$, S is the center of $\bigcirc O_1O_2O_3$ and I, O are fixed points. Hence S is fixed, and $SO_1 = 2\,AO = \text{constant}$.)

69. ABC is a triangle inscribed in a circle. Show that the tangents to the circle at A, B, C meet the opposite sides produced in three collinear points.

70. The non-parallel sides of a trapezoid are produced to meet in P. Prove that the line joining P to the intersection of the diagonals of the trapezoid bisects the parallel sides.

71. (a) Prove that the lines joining the vertices of a triangle to the points of contact of the inscribed circle are concurrent; (b) show that the lines joining the vertices to the points of contact of the three escribed circles are also concurrent.

72. AD, BE, CF are any three concurrent lines drawn inside the triangle ABC to meet the opposite sides in D, E, F. DE, EF, FD or produced meet AB, BC, CA in G, H, J. Show that GHJ is a straight line.

73. P, Q are the centers of two circles intersecting in A, B. Through A two perpendicular straight lines DAE, CAF are drawn meeting circle P in C, D and circle Q in E, F and cutting PQ in G, H. Show that $GD : GE = CH : HF$.

74. O is any point inside the triangle ABC. If the bisectors of the angles BOC, COA, AOB meet the sides BC, CA, AB in D, E, F respectively, prove that AD, BE, CF are concurrent.

75. $ABCD$ is a trapezoid in which AB, CD are the parallel sides. AD, BC when produced meet in E, and AC, BD intersect in F. From E a straight line GEH is drawn parallel to AB meeting AC, BD produced in G, H. Show that AH, BG, EF are concurrent.

76. $ABCD$, $EBFG$ are two parallelograms having the same angle B. Show that if E lies on AB, then AF, CE, GD are concurrent or parallel.

77. $ABCD$ is a quadrilateral. If AB, CD meet in E and AD, BC in F and if $ED'B'$ cut AD, BC in D', B' and $FA'C'$ cut AB, CD in A', C' respectively, prove that $AA' \cdot BB' \cdot CC' \cdot DD' = A'B \cdot B'C \cdot C'D \cdot D'A$.

78. $ABCD$ is a square and EF, GH are two lines drawn parallel to AB, BC, meeting BC, AD in E, F and AB, CD in G, H. Show that BF, DG, CM are concurrent, M being the point of intersection of EF, GH.

79. On the sides of a triangle ABC, equilateral triangles BCD, CAE, ABF are constructed outside the triangle ABC. Prove that AD, BE, CF are concurrent.

80. The escribed circle opposite to B touches BC at X and CA at Y; also the inscribed circle touches AB at Z. If X, Y, Z are collinear, show that A is a right angle.

81. AP, BQ, CR are three concurrent straight lines drawn from the vertices of a triangle ABC to the opposite sides. If the circle circumscribing the points P, Q, R cuts the sides BC, CA, AB again in X, Y, Z, show that AX, BY, CZ meet in one point.

82. MNL is a right-angled triangle at M. A square $MNPQ$ is described outside the triangle, and LP cuts the perpendicular MT to NL in R. Show that $1/MR = 1/MT + 1/NL$.

83. AD, BE, CF are the altitudes of the triangle ABC and P, Q, R, X, Y, Z are the middle points of BC, CA, AB, AD, BE, CF. Show that PX, QY, RZ are concurrent.

84. The interior angles A, B and the exterior angle C of a triangle ABC are bisected by lines which meet the opposite sides in D, E, F. Prove that D, E, F are collinear.

85. If the ends of three unequal parallel straight lines be joined, two and two, toward the same parts, the joining lines intersect, two and two, in three collinear points.

86. Through a given point P in the side AB of a triangle ABC, draw a straight line meeting BC, AC in Q, R respectively so that AR may be equal to BQ.

87. The altitudes of a triangle meet the sides BC, CA, AB in D, E, F. G is any point taken on AD and EG, FG produced meet FD, ED in P, Q respectively. Show that EF, PQ, BC if produced will meet in one point.

88. Show that the middle points of the three diagonals of a complete quadrilateral are collinear.

89. C is a moving point on the circumference of a circle with center O. AB is a fixed diameter. If AC is produced to D such that $AC = CD$ and OD is joined cutting BC in F, find the locus of F. (See Problem 5.9.)

90. D, E, F are the middle points of the sides BC, CA, AB of a triangle ABC. On the base BC, two points M, N are taken so that $BM = CN$. If DE, AN intersect in X and ME, NF in Y, show that X, Y, B are collinear.

91. Straight lines drawn from the vertices of a triangle ABC parallel respectively to the sides of another triangle $A'B'C'$ in the same plane meet in a point O. Prove that the straight lines drawn through the vertices of the triangle $A'B'C'$ parallel respectively to the sides of the triangle ABC also meet in a point O', and that the three triangles OBC, OCA, OAB are to each other in the same ratios as the triangles $O'B'C'$, $O'C'A'$, $O'A'B'$.

92. From any triangle ABC another triangle $A'B'C'$ is formed by straight lines drawn through A, B, C parallel to the opposite sides such that A', B', C' are opposite to A, B, C respectively. If any straight line through A' meets AB, AC in F, E respectively, prove that (a) $B'F$, $C'E$ intersect in a point D on BC; (b) the area of the triangle DEF is a mean proportional between the areas of ABC, $A'B'C'$; (c) AD, BE, CF are concurrent; (d) $A'D$, $B'E$, $C'F$ are parallel.

93. ABC is a triangle inscribed in a circle. A transversal cuts the sides BC, CA, AB externally in D, E, F. If the lengths of the tangents to the circle from D, E, F be x, y, z, prove that $x \cdot y \cdot z = DC \cdot EA \cdot FB$.

94. O is a point inside the triangle ABC. OD, OE, OF are drawn each perpendicular to OA, OB, OC respectively to meet BC, CA, AB or produced in D, E, F. Show that DEF is a straight line. (Produce AO, BO, CO to meet the opposite sides in G, H, J. Use Menelaus's Th. 5.13.)

95. The sides BC, CA, AB of a triangle ABC are bisected in D, E, F. A transversal PQR cuts the sides DE, DF, EF of the triangle DEF and AR, BQ, CP produced meet BC, AC, AB respectively in X, Y, Z. Show that X, Y, Z are collinear.

96. A line meets BC, CA, AB of a triangle ABC at X, Y, Z and O is any point. Show that $\sin BOX \cdot \sin COY \cdot \sin AOZ = \sin COX \cdot \sin AOY \cdot \sin BOZ$.

97. If on the four sides AB, BC, CD, DA of a quadrilateral $ABCD$ there be taken four points L, M, N, R such that $AL \cdot BM \cdot CN \cdot DR = LB \cdot MC \cdot ND \cdot RA$, show that LM, NR meet on AC and LR, MN on BD. (Apply Menelaus's theorem to ABC and LM also to CDA and NR. Multiply and divide by the given relation.)

98. ABC, $A'B'C'$ are two triangles and O is a point in their plane. Show that if OA, OB, OC meet $B'C'$, $C'A'$, $A'B'$ in collinear points, then OA', OB', OC' meet BC, CA, AB in collinear points. (Apply the result obtained in Exercise 96.)

99. The escribed circles touch the sides BC, CA, AB to which they are escribed at X, Y, Z. If YZ, BC meet at P, ZX, CA at Q, and XY, AB at R, show that P, Q, R are collinear. ($AZ \cdot BX \cdot CQ = ZB \cdot XC \cdot QA$, or $(s - b)(s - c) \cdot CQ = (s - a)(s - b) \cdot QA$. Then $CQ/QA = (s - a)/(s - c)$; s, a, etc., being the semi-perimeter and sides, and so on. Now multiply.)

100. L, M, N are the centers of the circles escribed to the sides BC, CA, AB, of a triangle. Prove that the perpendiculars from L, M, N to BC, CA, AB respectively are concurrent.

CHAPTER 6

GEOMETRY OF LINES AND RAYS

HARMONIC RANGES AND PENCILS

Definitions and Propositions

Definition. *If the rays PA, PB, PC, PD cut a line in A, B, C, D respectively, then the ratio $(AC/CB)/(AD/DB) = AC \cdot DB/CB \cdot AD$ is called the cross ratio of these four points, and is denoted by (AB, CD). If this cross ratio $= 1$, then AC is divided harmonically in B, D.*

Definition. *If a straight line AB is divided harmonically in C, D; C, D are called harmonic conjugates with respect to A, B. Similarly, A, B are harmonic conjugates with respect to C, D, as explained in the next proposition. $ACBD$ is called a harmonic range. AB is called the harmonic mean between AC, AD and DC the harmonic mean between AD, BD.*

Proposition 6.1. *If a straight line AB is divided harmonically in C, D, then CD is also divided harmonically in B and A (Fig. 162).*

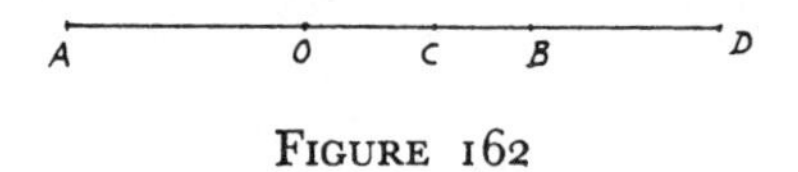

FIGURE 162

Since $AD : BD = AC : BC$ (hypothesis), $\therefore$ *$BD : AD = BC : AC$ and $DB : BC = AD : AC$. Therefore, DC is divided harmonically in B and A.*

Proposition 6.2. *If C, D be harmonic conjugates with respect to A, B and AB be bisected in O, then OB^2 is equal to $OC \cdot OD$, and the converse.*

1. *Since $AC : CB = AD : DB$, $\therefore$ $(AC - CB) : (AC + CB) = (AD - BD) : (AD + BD)$ or $OC : OB = OB : OD$. Hence $OB^2 = OC : OD$.*

2. *Let AB be bisected in O and let $OB^2 = OC \cdot OD$, AB will be divided harmonically in C, D. $\because$ $OB^2 = OC \cdot OD$, $\therefore$ $OC : OB = OB : OD$ and $(OB + OC) : (OB - OC) = (OD + OB) : (OD - OB)$. Therefore, $AC : BC = AD : BD$.*

Proposition 6.3. *The geometric mean between two straight lines is a mean proportional between their arithmetic and harmonic means.*

Let AD, BD be the straight lines. Bisect AB in O and with center O and *radius OA describe a semi-circle APB. Draw tangent DP, and $PC \perp AB$ (Fig. 163). Then $2\,DO = AD + BD$. $\therefore$ OD is the arithmetic mean*

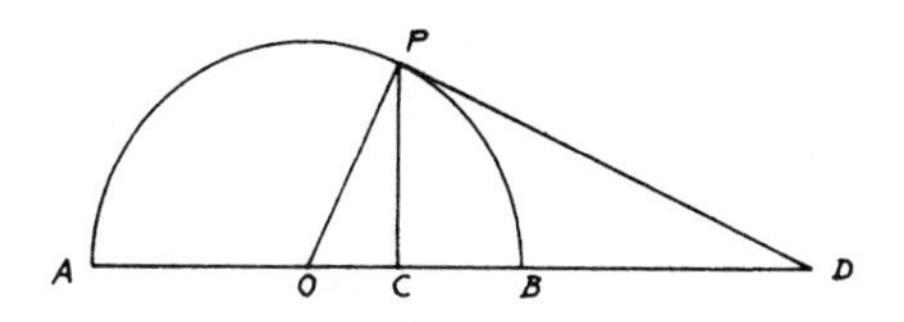

FIGURE 163

between AD, BD. Also $AD \cdot DB = DP^2$. $\therefore$ $AD : DP = DP : BD$. $\therefore$ DP is the geometric mean between AD, BD. Again in the right-angled $\triangle OPD$, $OD \cdot OC = OP^2 = OB^2$. Therefore, $ACBD$ is a harmonic range (Prop. 6.2) and DC is the harmonic mean between AD, BD. And in the right-angled $\triangle OPD$, $OD : PD = PD : DC$.

COROLLARY: *If from a point D in a produced diameter AB of a circle, a tangent DP be drawn to the circle and PC be drawn perpendicular to AB, $(ACBD)$ is a harmonic range.*

Definition. *Any number of straight lines passing through a point P are said to form a pencil; the point P is called the vertex of the pencil and each of the lines is called a ray of the pencil.*

Any straight line cutting the rays of a pencil is called a transversal.

In a pencil of four rays, if each ray passes through a point of a harmonic range, it is called a harmonic pencil. The rays which pass through conjugate points of the range are called conjugate rays. $(P \cdot ABCD)$ denotes a pencil whose vertex is P and whose rays pass through the points A, B, C, D.

Proposition 6.4. *Any straight line drawn parallel to one of the rays of a harmonic pencil is divided into two equal parts by the other three rays, and conversely.*

1. *Let EFG be drawn $\parallel$ PD one of the rays of harmonic pencil $(P \cdot ACBD)$. Through C draw $MCN \parallel EG$ or PD (Fig. 164). $\because$ $\triangle$s BCN, PBD are similar, $\therefore$ $CN : PD = CB : BD$. Similarly, $CM : PD = AC : AD$. But $AC : BC = AD : BD$ (hypothesis). $\therefore$ $AC : AD = BC : BD$. $\therefore$ $CM : PD = CN : PD$. $\therefore$ $CM = CN$* and *$EF = FG$.*

2. *Let EFG drawn $\parallel$ PD, one of the rays of the pencil $(P \cdot ACBD)$ of four rays, be divided into two equal parts by the other three rays PA, PC, PB. $(P \cdot ABCD)$ will be a harmonic pencil. Draw any transversal $ACBD$; through C draw $MCN \parallel EG$ or PD. $\therefore$ $EF = FG$. $\therefore$ $MC = CN$. $\therefore$ $MC : PD = NC : PD$. Also, $MC : PD = AC : AD$ and $NC : PD = CB : BD$. $\therefore$ $AC : AD = BC : BD$ or $AC : BC = AD : BD$.*

3. *Likewise, if EG be divided into two equal parts by three rays PA, PC,*

PB of the harmonic pencil $(P \cdot ACBD)$, *then it is easy to prove that EG will be* $\parallel PD$.

COROLLARY 1. *Any transversal is cut harmonically by the rays of a harmonic pencil.*

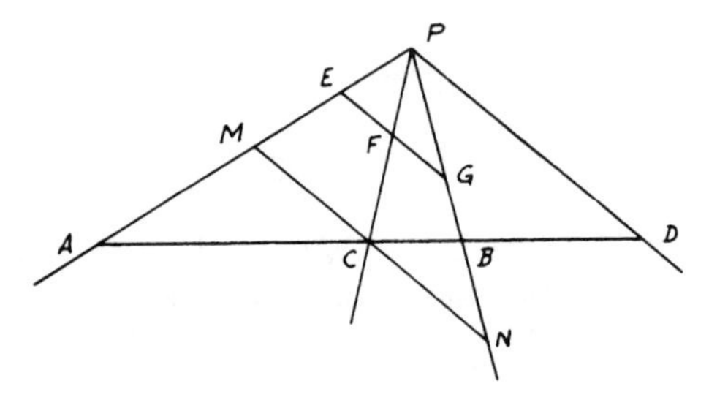

FIGURE 164

COROLLARY 2. *If one of the outside rays DP be produced through P to D′, then* $(P \cdot BCAD')$ *is a harmonic pencil and by producing the other rays in succession four other harmonic pencils may be said to be formed.*

COROLLARY 3. *If three points of a harmonic range are given, the fourth can be found.*

(*i*) *Let the required point be an outside one. Given A, C, B, find D. From any point P draw PA, PC, PB. Draw EFG so as to be bisected in F and* $PD \parallel EG$. *D is the required point.*

(*ii*) *Let the required point be an inside one. Given A, C, D, find B. From any point P draw PA, PC, PD. From any point E in PA draw* $EF \parallel PD$ *and produce EF to G so that* $FG = EF$. *Join PG and produce it to meet AD in B. B is the required point.*

Proposition 6.5. *If one ray of a harmonic pencil bisects the angle between the other pair of rays, the ray conjugate to the first ray is at right angles to it, and conversely.*

1. *Let* $(P \cdot ACBD)$ *be a harmonic pencil and let PC bisect* $\angle APB$. *Through C draw* $ECF \perp PC$ (*Fig.* 165). $\therefore EC = CF$. $\therefore EF$ *is* $\parallel PD$. $\therefore PD$ *is* $\perp PC$.

2. *Conversely, if the conjugate rays PC, PD of the harmonic pencil* $(P \cdot ACBD)$ *be* $\perp$ *to each other, then it is easy to show that they will bisect the* $\angle$*s between the other two rays PA, PB.*

3. *The internal and external bisectors of an angle form with the arms of the angle a harmonic pencil.*

Proposition 6.6. *Any diagonal of a complete quadrilateral is divided harmonically by the other two diagonals.*

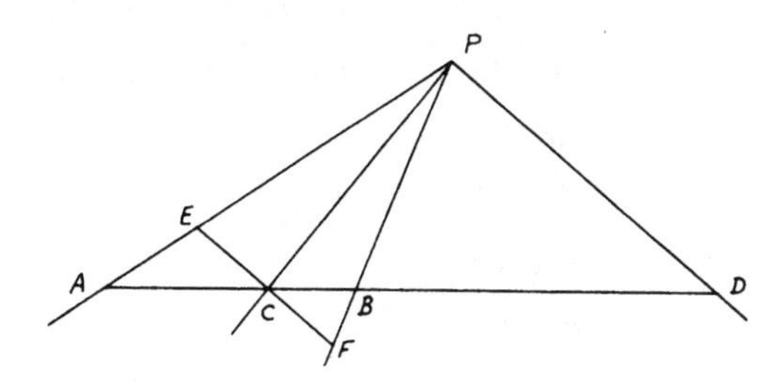

FIGURE 165

Let EF be the third diagonal of the complete quadrilateral ABCDEF and let it be divided by the other diagonals AC, BD in G, H. Through C draw (KCLMN $\parallel$ *AB (Fig.* 166). *In* $\triangle$*s FHB, AHB,* $KL : LN = FB : AB$

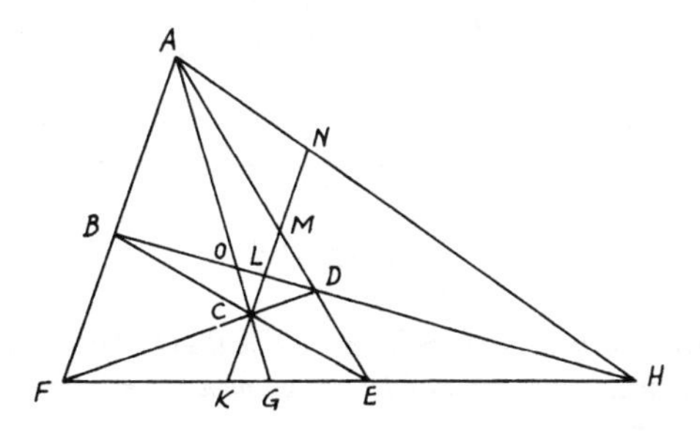

FIGURE 166

and $\therefore$ *in* $\triangle$*s FDB, ADB,* $CL : LM = FB : AB$. $\therefore$ $KC : MN = FB : AB$. *But in* $\triangle$*s FEB, ABE,* $KC : CM = FB : AB$. $\therefore$ $KC : MN = KC : CM$. $\therefore$ $MN = CM$. $\therefore$ $(A \cdot FGEH)$ *is a harmonic pencil (Prop.* 6.4.2). $\therefore$ *BD and FE are divided harmonically in O, H and G, H respectively.*

In the same way, by drawing through O a $\parallel$ *to AD meeting AB in Q, it may be proved that the pencil* $(E \cdot AQBF)$ *is harmonic. Hence AC is divided harmonically in O, G.*

Proposition 6.7. *If four rays OA, OB, OC, OD of a pencil* $(O \cdot ABCD)$ *are cut by two transversals ABCD, EFGH respectively, then (BC, AD) = (FG, EH) and this is true for any other transversals (Fig.* 167).

$$(BC, AD) = \frac{BA}{BD}\,\frac{DC}{AC} = \frac{\triangle AOB}{\triangle BOD}\,\frac{\triangle COD}{\triangle AOC}$$

$$= \frac{AO \cdot OB \sin AOB}{OD \cdot OB \sin BOD} \frac{OD \cdot OC \sin COD}{AO \cdot OC \sin AOC}$$

$$= \frac{\sin AOB}{\sin BOD} \frac{\sin COD}{\sin AOC}$$

$$= \frac{EF}{FH} \frac{GH}{GE}$$

$$= (FG, EH) = (F'G', E'H').$$

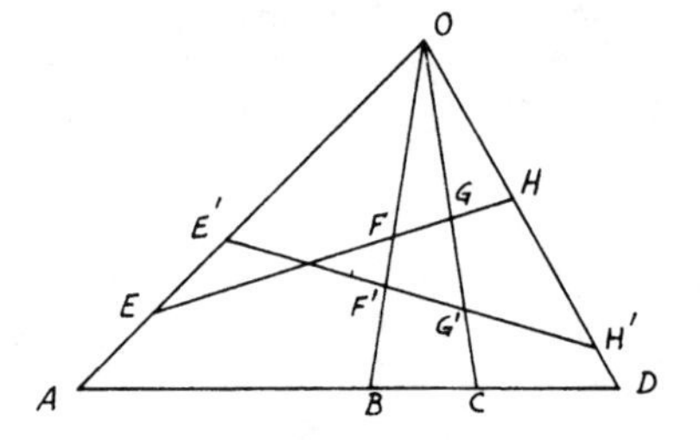

FIGURE 167

Since this cross ratio depends only on the pencil $(O \cdot ABCD)$ *and not on any particular transversal, we may denote the common cross ratio as* $O(AC, BD)$.

Solved Problems

6.1. *D, E are two points on the straight line AX and F any point on another line AY. FD, FE cut a third line AZ in B, C. If M is any point on AZ and DM, EM meet AY in P, Q, show that PC, QB meet on AX.*

CONSTRUCTION: Produce DC, EB to meet AY in R, S. Let PC, QB meet AX in N, N' (*Fig.* 168).

Proof:

$$\begin{aligned} (AN, DE) &= C(AN, DE) \\ &= (AP, RF) \qquad \text{(from Prop. 6.7)} \\ &= D(AP, RF) = (AM, CB) \\ &= E(AM, CB) = (AQ, FS) = B(AQ, FS) \\ &= (AN', DE). \end{aligned}$$

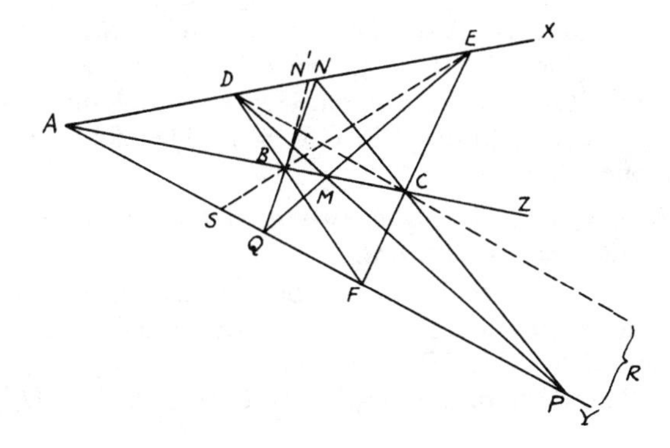

FIGURE 168

Hence N' should coincide with N. Therefore, PC, QB meet on AX. This is one of Pascal's theorems.

6.2. *In a triangle ABC, three concurrent lines AP, BQ, CR meet in S and the opposite sides in P, Q, R respectively. PU meets QR in X, QU meets RP in Y, RU meets PQ in Z. Show that AX, BY, CZ are concurrent.*

CONSTRUCTION: Produce AX, BY, CZ to meet BC, CA, AB in E, F, G respectively (*Fig.* 169).

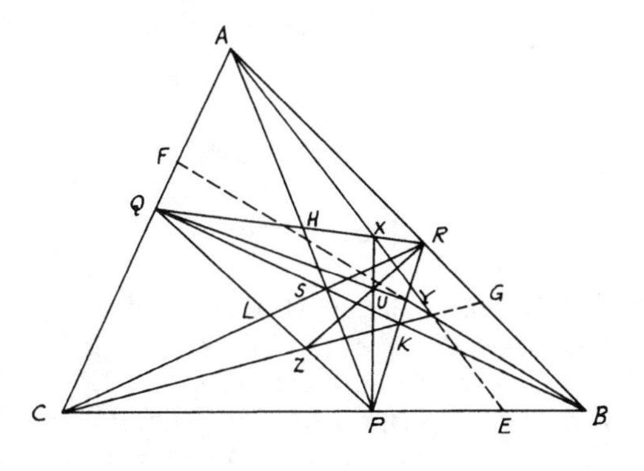

FIGURE 169

Proof: Let AP, BQ, CR meet QR, PR, PQ in H, K, L respectively. AB, AE, AP, AC are rays cut by the two transversals RQ, BC. $\therefore$

$(XH, RQ) = (EP, BC)$. $\therefore$ $(HQ/RH)(RX/XQ) = (PC/PB)(BE/CE)$ (Prop. 6.7). Similarly, $(YK, RP) = (FQ, AC)$. $\therefore$ $(RK/KP)(PY/YR) = (AQ/QC)(CF/FA)$. Also, $(ZL, PQ) = (GR, BA)$. $\therefore$ $(PL/LQ)(QZ/ZP) = (BR/RA)(AG/GB)$, since $(HQ/RH)(RK/KP)(PL/LQ) = 1$ (Ceva's Th. 5.11), and so on. Hence multiplying and simplifying give $(BE/CE)(CF/FA)(AG/GB) = 1$. Therefore AX, BY, CZ meet in one point.

6.3. *The straight lines joining the excenters of a triangle to the middle points of the opposite sides are concurrent.*

CONSTRUCTION: Let O_1, O_2, O_3 be the three excenters of $\triangle ABC$ opposite to the vertices A, B, C respectively and A', B', C' be the mid-points of BC, CA, AB. Join the excenters and produce the lines O_1A', O_2B', O_3C' to meet O_2O_3, O_1O_3, O_1O_2 in R, L, N (*Fig.* 170).

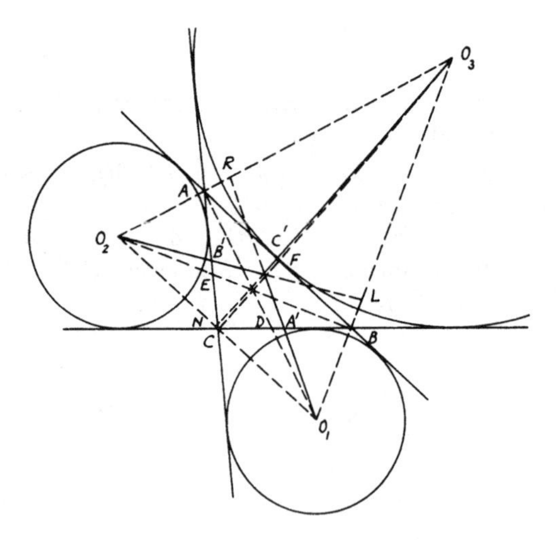

FIGURE 170

Proof: O_1O_2 passes through C, O_2O_3 through A, and O_3O_1 through B. O_1A, O_2B, O_3B are also the altitudes of $\triangle O_1O_2O_3$; hence ABC is the pedal of $\triangle O_1O_2O_3$. According to Prop. 6.7, $(A'D, BC) = (RA, O_3O_2)$ and $(B'E, AC) = (LB, O_3O_1)$. Also $(C'F, AB) = (NC, O_2O_1)$. But, since A', B', C' are the mid-points of the sides, $DC/BD = (AO_2/AO_3)(RO_3/RO_2)$ and $AE/CE = (O_3B/BO_1)(LO_1/LO_3)$ and $BF/AF = (O_1C/CO_2)(O_2N/O_1N)$. Since $(DC/BD)(AE/CE)(BF/AF) = 1$ and $(AO_2/AO_3)(O_3B/BO_1)(O_1C/CO_2) = 1$, $\therefore$ $(RO_3/RO_2)(LO_1/LO_3)(O_2N/O_1N) = 1$. $\therefore$ O_1A', O_2B', O_3C' are concurrent.

6.4. *ABC is a triangle and AD, BE, CF are the altitudes. If O_1, O_2, O_3 are the excenters opposite to A, B, C respectively, show that O_1D, O_2E, O_3F are concurrent.*

CONSTRUCTION: Join O_1O_2, O_2O_3, O_3O_1 passing through C, A, B respectively. Let AO_1, BO_2, CO_3 and O_1D, O_2E, O_3F cut BC, CA, AB, O_2O_3, O_3O_1, O_1O_2 in G, H, J, L, M, N respectively (*Fig.* 171).

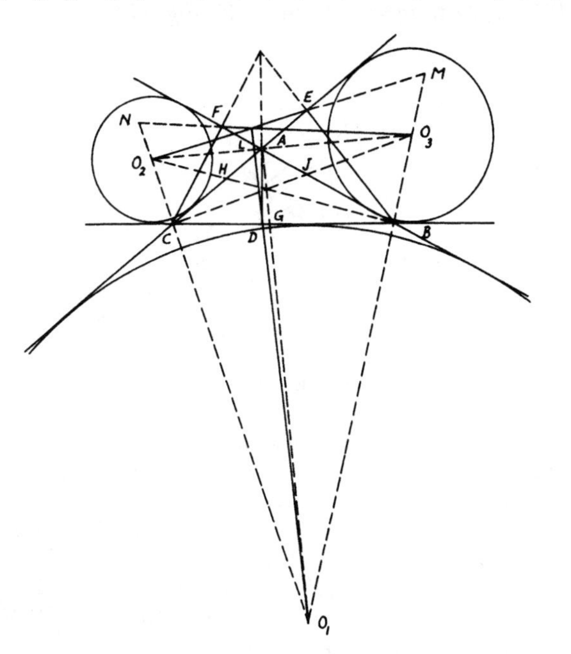

FIGURE 171

Proof: ABC is the pedal $\triangle$ of $\triangle O_1O_2O_3$. $(GD, BC) = (AL, O_3O_2)$ and $(EH, AC) = (MB, O_3O_1)$; also, $(FJ, AB) = (NC, O_2O_1)$. Multiplying the ratios, we get

$$\frac{DC}{BD}\frac{BG}{GC}\frac{EA}{EC}\frac{CH}{AH}\frac{FB}{FA}\frac{AJ}{JB} = \frac{LO_2}{LO_3}\frac{O_3A}{AO_2}\frac{MO_3}{MO_1}\frac{BO_1}{BO_3}\frac{O_1N}{O_2N}\frac{O_2C}{CO_1}$$

But

$$\frac{DC}{BD}\frac{EA}{EC}\frac{FB}{FA} = 1, \qquad \frac{O_3A}{AO_2}\frac{BO_1}{BO_3}\frac{O_2C}{CO_1} = 1, \qquad \frac{BG}{GC}\frac{CH}{AH}\frac{AJ}{JB} = 1.$$

Hence $(LO_2/LO_3)(MO_3/MO_1)(O_1N/O_2N) = 1$. Therefore, O_1D, O_2E, O_3F are concurrent.

6.5. *From a point M inside or outside a triangle ABC, a transversal is drawn to cut the sides BC, CA, AB in D, E, F respectively. If MA, MB, MC, EF, DF, ED are bisected in A', B', C', D', E', F' respectively, show that $A'D'$, $B'E'$, $C'F'$ meet in one point.*

CONSTRUCTION: Let DEF cut the sides of $\triangle A'B'C'$ in K, T, S. Produce $F'C'$, $C'M$ to meet $A'B'$ in R, Y, $A'D'$, $A'M$ to meet $B'C'$ in L, X and $B'E'$, $B'M$ to meet $A'C'$ in N, Z (*Fig.* 172).

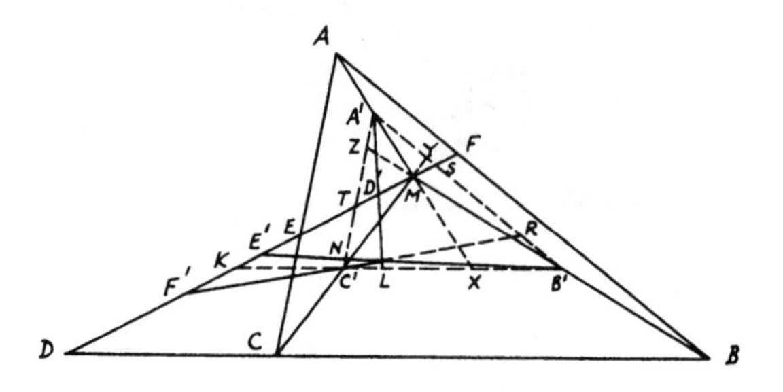

FIGURE 172

Proof: Since A', B', C' are the mid-points of AM, BM, CM, $\therefore$ S, T, K are the mid-points of MF, ME, MD. But $(MD', ST) = (XL, B'C')$ or $(MT/MS)(D'S/D'T) = (XC'/B'X)(LB'/LC')$. $\because$ $D'T = \frac{1}{2}MF = MS$, $\therefore$ $MT = D'S$. $\therefore$ $MT^2/MS^2 = (XC'/B'X)(LB'/LC')$. Similarly, $MK^2/MT^2 = (B'Y/A'Y)(A'R/RB')$ and $MS^2/MK^2 = (A'Z/ZC')(C'N/NA')$. But since $(XC'/B'X)(B'Y/A'Y)(A'Z/ZC') = 1$ (Th. 5.11), $\therefore$ multiplying yields $(LB'/LC')(A'R/RB')(C'N/NA') = 1$. Hence $A'D'$, $B'E'$, $C'F'$ are concurrent.

ISOGONAL AND SYMMEDIAN LINES—BROCARD POINTS

Definitions and Propositions

Definition. *Two straight lines AP, AP' are said to be isogonal or conjugate lines of the straight lines AB, AC if the angles PAP' and BAC have the same bisectors.*

Proposition 6.8. *If P and P' are any points on the isogonal lines AP, AP' of the lines AB, AC and if perpendiculars PN, $P'N'$ are drawn to AB and perpendiculars PM, $P'M'$ to AC, then $PN \cdot P'N' = PM \cdot P'M'$ and, conversely, if $PN \cdot P'N' = PM \cdot P'M'$, then AP, AP' are isogonal lines of AB, AC (Fig. 173).*

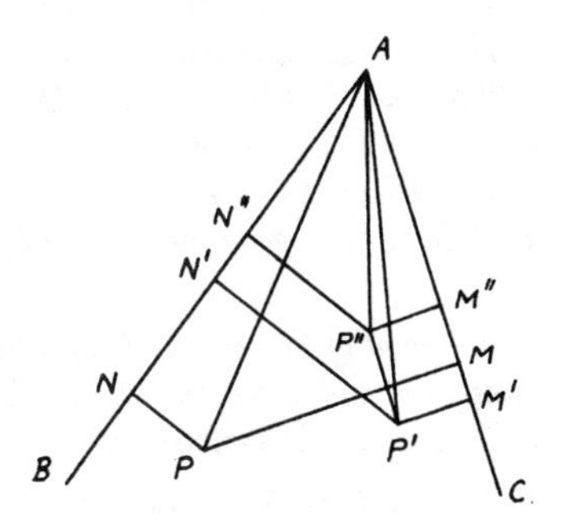

FIGURE 173

Since the angles PAB, $P'AC$ are equal, the $\triangle$s PAN, $P'AM'$ are similar. Hence $AP/PN = AP'/P'M'$ or $AP/AP' = PN/P'M'$. So, from $\angle PAC = \angle P'AB$, $AP/AP' = PM/P'N'$. Hence $PN/P'M' = PM/P'N'$; i.e., $PN \cdot P'N' = PM \cdot P'M'$. Conversely, if $PN \cdot P'N' = PM \cdot P'M'$, AP and AP' are isogonal lines. For, if not, let a parallel to AC through P' cut the isogonal line of AP at P''. Then, by the first part, $PN \cdot P''N'' = PM \cdot P''M'' = PM \cdot P'M'$ (since $P''M'' = P'M'$) $= PN \cdot P'N'$ (by hypothesis). Hence $P'N' = P''N''$ or $P'P''$ is also parallel to AB. Hence P'' coincides with P'. Hence AP and AP' are isogonal lines.

Proposition 6.9. *If AP, BP, CP meet in a point P, the isogonal lines of AP with respect to AB, AC, of BP with respect to BC, BA and of CP with respect to CA, CB meet in a point P' such that the product of the perpendiculars from P and P' on each of the sides is the same* (*Fig.* 174).

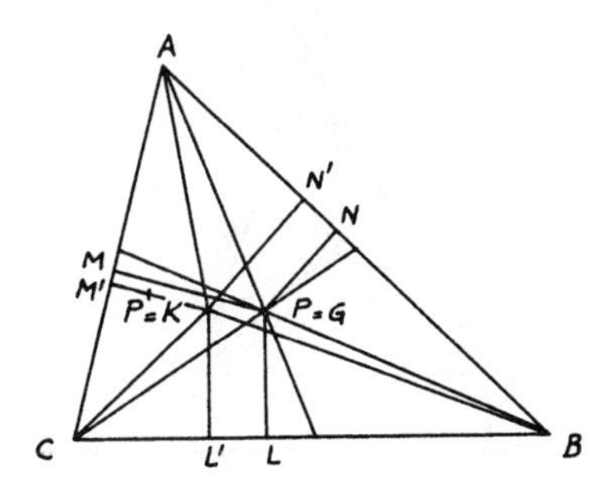

FIGURE 174

Let the isogonals of AP and BP meet at P′. Draw the $\perp$*s PL, P′L′ on BC, and PM, P′M′ on CA and PN, P′N′ on AB. It is easily shown that CP′ is the isogonal of CP. Since AP, AP′ are isogonal,* $PN \cdot P'N' = PM \cdot P'M'$. *So from BP, BP′,* $PN \cdot P'N' = PL \cdot P'L'$. *Hence* $PM \cdot P'M' = PL \cdot P'L'$. *Therefore, CP′ is the isogonal of CP. Hence the isogonals of AP, BP, CP meet in a point P′ such that* $PL \cdot P'L' = PM \cdot P'M' = PN \cdot P'N'$.

Such points P, P′ are said to be isogonal or conjugate points with respect to the triangle ABC.

Definition. *The isogonal point of the centroid G of a triangle ABC is called the symmedian or Lemoine point and is denoted by K. The lines AK, BK, CK isogonal to AG, BG, CG are called the symmedians.*

Proposition 6.10. *Show that the perpendiculars from K on the sides of ABC are proportional to the sides.*

Let the $\perp$*s from G on the sides BC, CA, AB denoted by a, b, c be* a', b', c'. *Then since* $\triangle BGC = \triangle CGA = \triangle AGB$, $\therefore\ aa' = bb' = cc'$. *But according to Proposition* 6.9, *if* $\perp$*s from K on a, b, c be p, q, r, then* $a'p = b'q = c'r$. *Hence* $p : q : r = a : b : c$.

Obvious cases of isogonal points with respect to a triangle are the circumcenter and orthocenter. Also, there are four points, and only four, each of which is isogonal to itself with respect to ABC and these are the incenter and the three excenters.

Proposition 6.11. *One and only one point* Ω *can be found such that* $\angle BA\Omega = \angle AC\Omega = \angle CB\Omega$. *Also one and only one point* Ω' *can be found such that* $\angle AB\Omega' = \angle BC\Omega' = \angle CA\Omega'$, *where* Ω, Ω' *are called the Brocard points and are isogonal with respect to ABC.* $\angle BA\Omega = \angle AB\Omega'$ *is called the Brocard angle and is denoted by* ω.

If a point Ω *exists such that* $\angle BA\Omega = \angle AC\Omega = \angle CB\Omega$, *then the*

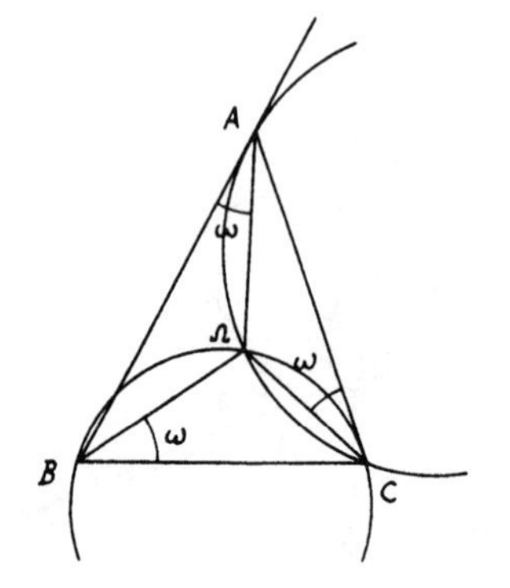

FIGURE 175

circle $A\Omega C$ must touch BA at A, since $\angle BA\Omega = \angle AC\Omega$. So the circle $C\Omega B$ must touch AC at C (Fig. 175). Hence Ω is determined as the second intersection of the circles (1) through C and touching BA at A and (2) through B and touching AC at C. Hence two points Ω cannot exist.

Again the second intersection Ω of the above circles is such that $\angle BA\Omega = \angle AC\Omega$ (from the first circle) $= \angle CB\Omega$ (from the second circle). Hence one such point exists. Similarly for Ω'.

Also Ω and Ω' are isogonal points. For, if not, let Ω'' be isogonal to Ω. Then $\angle BA\Omega = \angle CA\Omega''$, $\angle AC\Omega = \angle BC\Omega''$, $\angle CB\Omega = \angle AB\Omega''$. Hence, since $\angle BA\Omega = \angle AC\Omega = \angle CB\Omega$, then $\angle CA\Omega'' = \angle BC\Omega'' = \angle AB\Omega''$. Hence Ω'' has the property of Ω' and therefore coincides with it, since there is only one point Ω'.

Solved Problems

6.6. *Three lines are drawn through the vertices of a triangle meeting the opposite sides in collinear points (Fig. 176). Show that their isogonal conjugates with respect to the sides also meet the opposite sides in collinear points.*

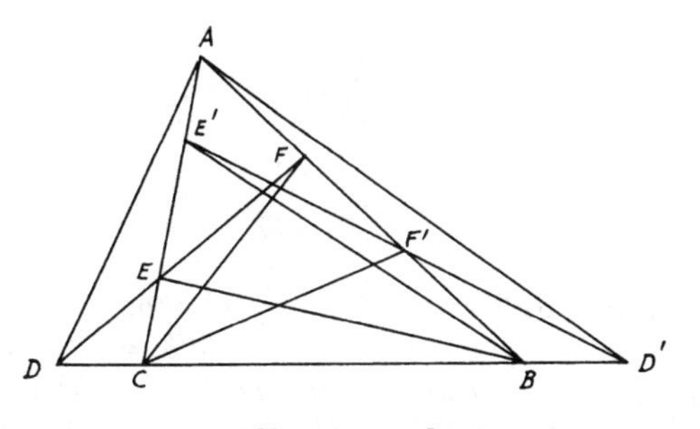

FIGURE 176

Proof: Let D, E, F be the collinear points and AD', BE', CF' the isogonal conjugates. Then $\sin ACF \cdot \sin BAD \cdot \sin CBE = \sin FCB \cdot \sin DAC \cdot \sin EBA$. But $\angle ACF = \angle F'CB$, and so on. Hence $\sin F'CB \cdot \sin D'AC \cdot \sin E'BA = \sin ACF' \cdot \sin BAD' \cdot \sin CBE'$. Therefore, D', E', F' are collinear (see Problem 5.30).

6.7. *The sum of the squares of the perpendiculars from a point to the sides of a triangle has its least value when the point is the symmedian point; evaluate its magnitude.*

Proof: Let the sides BC, CA, AB be a, b, c and the $\perp$s from the symmedian point K be p, q, r. Then $pa + qb + rc$ is constant, being twice the area of the triangle. But $(p^2 + q^2 + r^2)(a^2 + b^2 + c^2) = (pa + qb + rc)^2 + (pb - qa)^2 + (qc - rb^2) + (ra - pc^2)$. Hence $(p^2 + q^2 + r^2)$ is minimum when $pb = qa$, $qc = rb$, $ra = pc$, i.e., when $p/a = q/b = r/c$ or $p : q : r = a : b : c$, i.e., at the symmedian point K (Prop. 6.10). In this case, the minimum value of the sum $(p^2 + q^2 + r^2) = (pa + qb + rc)^2/(a^2 + b^2 + c^2) = 4(\triangle ABC)^2/(a^2 + b^2 + c^2)$.

6.8. *If the symmedians AP, BQ, CR of a triangle ABC meet the opposite sides BC, CA, AB in P, Q, R, show by applying Ceva's theorem that they are concurrent at the symmedian or Lemoine point K (Fig.* 177).

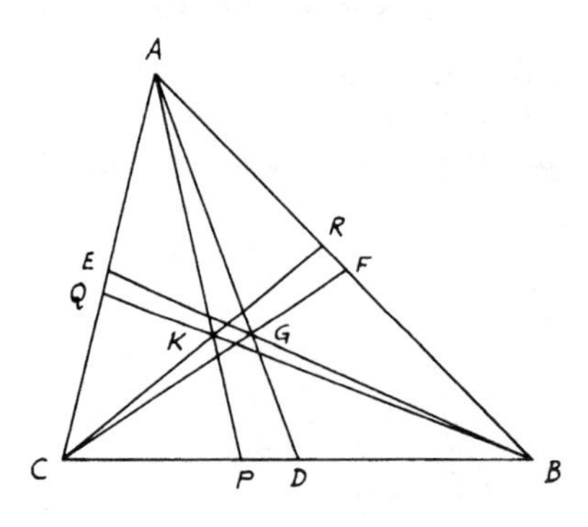

FIGURE 177

Proof: According to Prop. 6.10, if p, q, r be the $\perp$s from K on BC, CA, AB, $\therefore\ p : q : r = a : b : c$ or $p/a = q/b = r/c = k$. $BP/PC = \triangle BKA/\triangle CKA = r \cdot c/q \cdot b = kc \cdot c/kb \cdot b = c^2/b^2$ and similarly $CQ/QA = a^2/c^2$ and $AR/RB = b^2/a^2$. Hence multiplying yields $(BP/PC)(CQ/QA)(AR/RB) = 1$. Therefore, AP, BQ, CR meet in one point which is the symmedian or Lemoine point K (Ceva's Th. 5.11).

6.9. *The external sides of squares described outwardly on the sides of the triangle ABC meet at A′, B′, C′. Show that AA′, BB′, CC′ are the symmedians of ABC.*

CONSTRUCTION: Let AA', BB' meet in K. Draw KL, KM, KN $\perp$s $B'C'$, $C'A'$, $A'B'$ (*Fig.* 178).

Proof: From Prop. 6.10, if K is the symmedian point of $\triangle ABC$,. $\therefore$ $p : q : r = a : b : c$, where a, b, c denote BC, CA, AB. Hence $(p + a) : (q + b) : (r + c) = a : b : c$. Now in $\triangle A'B'C'$, $KL : KN = BE : BH = a : c$. $\therefore\ (p + a) : (r + c) = a : c$.

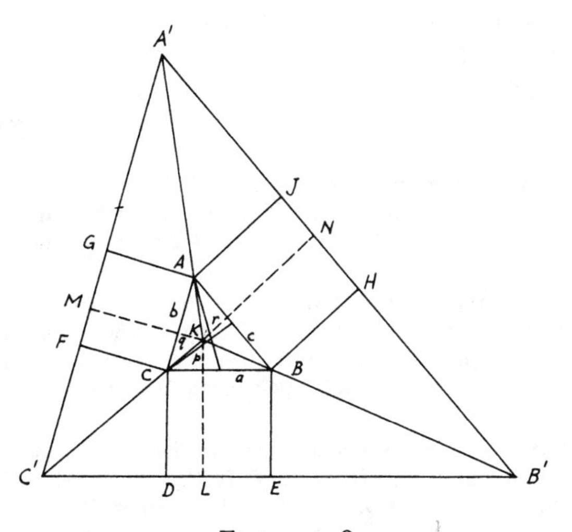

FIGURE 178

Hence BB' is the locus of points the distances of which from BC, AB are in the ratio $a : c$. Similarly, AA', CC' are the loci of the points the distances of which from AC, AB and BC, CA respectively have the ratios $b : c$ and $a : b$.

$\therefore$ $(p + a) : (q + b) : (r + c) = a : b : c$. Hence AA', BB', CC' meet at the symmedian point K.

6.10. *In Prop.* 6.11, *show that if* Ω, Ω' *are the Brocard points of the triangle* ABC, *then* $A\Omega \cdot B\Omega \cdot C\Omega = A\Omega' \cdot B\Omega' \cdot C\Omega'$.

Proof: In *Fig.* 175, $\angle A\Omega C = [180 - \omega - (A - \omega)] = 180° - A$. Now in $\triangle A\Omega C$, $A\Omega/\sin \omega = AC/\sin A\Omega C$ or $A\Omega = AC \sin \omega/\sin A$, and so on; and $A\Omega' = AB \sin \omega/\sin A$, and so on. Hence $A\Omega \cdot B\Omega \cdot C\Omega = AC \cdot AB \cdot BC \cdot \sin^3 \omega/\sin A \cdot \sin B \cdot \sin C = A\Omega' \cdot B\Omega' \cdot C\Omega'$.

Miscellaneous Exercises

1. (BC, XX'), (CA, YY'), (AB, ZZ') are harmonic ranges. Show that if AX, BY, CZ are concurrent, then X', Y', Z' are collinear.
2. AS, BS, CS meet BC, CA, AB at P, Q, R. QR, RP, PQ meet BC, CA, AB at X, Y, Z. Show that (BC, PX), (CA, QY), (AB, RZ) are harmonic ranges and that X, Y, Z are collinear, also that AX, CZ, BQ are concurrent. (From the quadrilateral $ARSQA$, (BC, PX) is harmonic. Hence $BP : PC = BX : CX$. Now use Exercise 1.

3. If (AB, CD) is a harmonic range and P any point collinear with AB, show that $2\, PB/AB = PC/AC + PD/AD$.

4. If in Exercise 3, U and V bisect AB and CD, then $PA \cdot PB + PC \cdot PD = 2\, PU \cdot PV$.

5. If (AB, CD) and $(AB', C'D')$ are harmonic ranges, show that BB', CD' and $C'D$ meet in a point.

6. TA, TB are drawn touching a circle in A, B. From any point C in AB produced, the straight line $CDEF$ is drawn touching the circle in E and cutting TA, TB in D, F. Prove that CE is cut harmonically in D and F.

7. ABC is a triangle inscribed in a circle. The tangent at A meets BC produced in D. Prove that the second tangent from D is cut harmonically by the sides AB, AC and its point of contact with the circle. [Let DE be the second tangent. Then $(A \cdot DCEB)$ is a harmonic pencil (Prop. 6.6), etc.]

8. $ABCD$ is a complete quadrilateral whose diagonals intersect in O. BA, CD meet in E and BC, AD meet in F. If AC, BD intersect EF in O', O'', show that each two of O, O', O'' are harmonic points with respect to two corners of $ABCD$ and that if a parallel is drawn through O to $O'O''$ to meet AB, DC, AD, BC in G, H, I, J, then $OG = OH$ and $OI = OJ$, and similarly with O' and O''.

9. The three triangles ABC, $A'B'C'$, $A''B''C''$ are such that $(A'A'', BC)$, $(B'B'', CA)$, and $(C'C'', AB)$ are harmonic. Show that BC, $B'C'$, $B''C''$ are concurrent and so are CA, $C'A'$, $C''A''$ and AB, $A'B'$, $A''B''$.

10. Through a given point O draw a line cutting the sides BC, CA, AB of a triangle in points A', B', C' such that $(OA', B'C')$ is harmonic.

11. A, B, C, D are four collinear points. Find the locus of P when the circles ABP and CDP touch at P.

12. With any point on a given circle as center a second circle is described cutting the first. Prove that any diameter of the second circle is divided harmonically by the first circle and the common chord or chord produced.

13. The base BC of a triangle ABC is bisected in D and a point O is taken in AD or AD produced. Show how to draw a straight line through D terminated by the sides of the triangle such that the segments into which it is divided at D may subtend equal angles at O. (Draw $OG \perp AD$, $AG \parallel BC$. Let GD meet AC, AB in E, F. EDF is the required line. Use Prop. 6.4.2.)

14. Prove that if a straight line $PQRS$ be drawn intersecting the sides of a square in P, Q, R, S so that $PQ \cdot RS = PS \cdot QR$, $PQRS$ will touch the inscribed circle of the square.

15. Find the locus of a point from which the tangents to two given circles have a given ratio to each other.

16. Prove that the circles on the diagonals of a complete quadrilateral as diameters cut orthogonally the circle circumscribing the triangle formed by the diagonals. (Use Prop. 6.6.)

17. In Prop. 6.9, show that in the figure the points L, L', M, M', N, N' are concyclic.

18. If AB is a diameter of a circle and P a point on the circumference, find the position of P when $l \cdot PA + m \cdot PB$ is greatest, l and m being given.

19. If AP, AQ are two isogonals of AB, AC which cut BC at P, Q show that $AP^2 : AQ^2 = BP \cdot PC : BQ \cdot QC$.

20. ABC is a triangle and AP, AQ are isogonal conjugate lines with respect to AB and AC, P and Q being on the circle ABC. Show that the isogonal conjugate point of P is the point at infinity on AQ. (Let P' be the point at infinity on AQ. It is sufficient to prove that $\angle ABP' = \angle CBP$. Now $BP' \parallel AQ$. Hence $ABP' = BAQ = CAP = CBP$.)

21. If P and P' are isogonal points of ABC, show that $AP' : BP' : CP' = AP \cdot PL : BP \cdot PM : CP \cdot PN$, where PL, PM, PN are the perpendiculars on the sides BC, CA, AB.

22. In a triangle right-angled at A, prove that K bisects the altitude AD, K being the symmedian point.

23. Given the side BC and the angle CBA, show that the locus of one Brocard point is a circle. (In *Fig.* 175, suppose BC is given and also the $\angle BCA$. Then the locus of Ω is the circle touching CA at C and passing through B. So in the given case, the locus of Ω' is the circle touching BA at B and passing through C.)

24. If a parallel to BC through A cuts at R the circle which passes through C and touches BA at A, prove that BR cuts this circle again in Ω.

25. $A\Omega, B\Omega, C\Omega$ meet the circle ABC again at P, Q, R. Show that the triangles ABC, PQR are congruent and have a common Brocard point.

26. If PQ is the tangent from P to a given circle and if any other line through P meets the circle in R and R', show that the triangles PQR, PQR' have the same Brocard angle and have one Brocard point in common.

CHAPTER 7

GEOMETRY OF THE CIRCLE

SIMSON LINE

Definitions and Propositions

Proposition 7.1. *The projections of any point which lies on the circumcircle of a triangle on the sides of the triangle are collinear and conversely, if the projections of a point on to the sides of a triangle are collinear, this point lies on the circumcircle of the triangle. This line is called the pedal or Simson line of the point with respect to the triangle and is equidistant from the given point and the orthocenter of the triangle. Let L, M, N be the projections of P on BC, CA, AB. Let also AK, G be an altitude and the orthocenter of* $\triangle ABC$. *Produce AK to meet the* ⊙ *in H. Join PH cutting BC, LN in F, J. Join FG, PA, PC. Let also PG cut LN in Q (Fig. 179). Now, suppose that P lies on the*

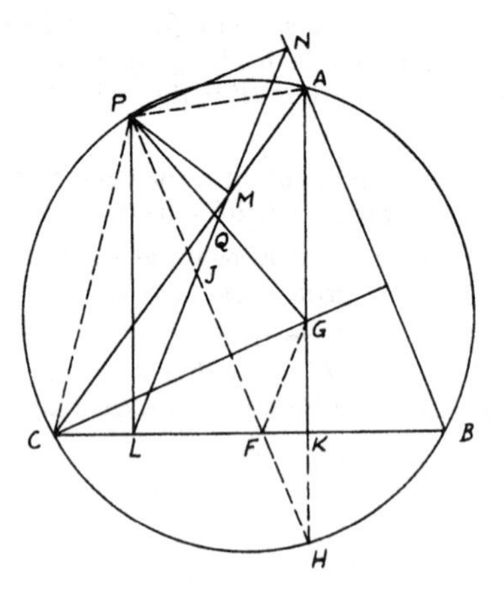

FIGURE 179

circumcircle of ABC. Then, L, M, N will be collinear, if PML + PMN = 2 right angles, which is true, since $\angle PMN = \angle PAN$ *(PNAM is cyclic),* $\angle PAN = \angle PCL$ *(PABC is cyclic) and* $\angle PML + \angle PCL = 2$ *right angles (PMLC is cyclic). Conversely, if LMN is a straight line, it is simple to show that PABC is cyclic. Again,* $\angle PLJ = \angle PCA = \angle PHA$

= ∠JPL (since AH, PL are ⊥s BC). ∴ JL = JP = JF. Since it can be proved that GK = KH and GK is ⊥BC, ∴ ∠GFK = ∠HFK. But ∠HFK = ∠JFL = ∠JLF. ∴ ∠GFK = ∠JLF. ∴ FG is ‖ *LMN. ∴ Q is the mid-point of PG. Hence the ⊥s from P, G on LMN are equal.*

Solved Problems

7.1. *Find a point such that the feet of the four perpendiculars from it to the sides of a given quadrilateral may be collinear.*

ANALYSIS: Let *ABCD* be a quadrilateral. Produce *BA*, *CD* to meet in *E* and *BC*, *AD* in *F*. Now, the two △s *ECB*, *FAB* combine the four sides of the quadrilateral *ABCD* and both △s have the sides *AB*, *BC* in common (*Fig.* 180). Hence the circumcircles on △s *ECB*, *FAB*

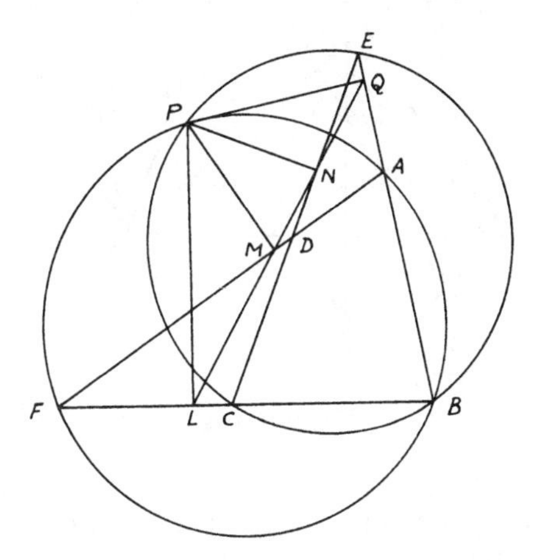

FIGURE 180

intersect in *B*, *P* such that the Simson lines of *P* with respect to △s *ECB*, *FAB*, which are *LNQ*, *LMQ*, coincide, since both have two points in common.

SYNTHESIS: It is obvious that if the two circumcircles *ECB*, *FAB* intersect in *P*, then *P* will be the required point.

7.2. *If p and r are the pedal lines of P and R with respect to the same triangle, then the angle between p and r is equal to the angle subtended by PR at any point on the circumcircle of the triangle. For in Fig.* 180, *∠MLB = right angle − ∠PLM = right angle − ∠PCA.* Hence *∠pr = ∠MLB − ∠M'L'B = ∠RCA − ∠PCA = ∠RCP.*

7.3. *If the perpendiculars PL, PM, PN from any point P on the circumcircle* *of a triangle ABC to the sides BC, CA, AB are produced to meet the circle in E, F, G, show that AE, BF, CG are parallel to the pedal line LMN. Similarly, if P', another point on the circle, lies on the perpendicular from B to AC, then the tangent at B, AE', CG' will be parallel to the pedal line of P' with respect to ABC, where E', G' are the intersections of the perpendiculars $P'L'$, $P'N'$ to BC, BA with the circle.*

CONSTRUCTION: Join PA, PC. $L'M'N'$ is the pedal line of P' with respect to ABC (*Fig.* 181).

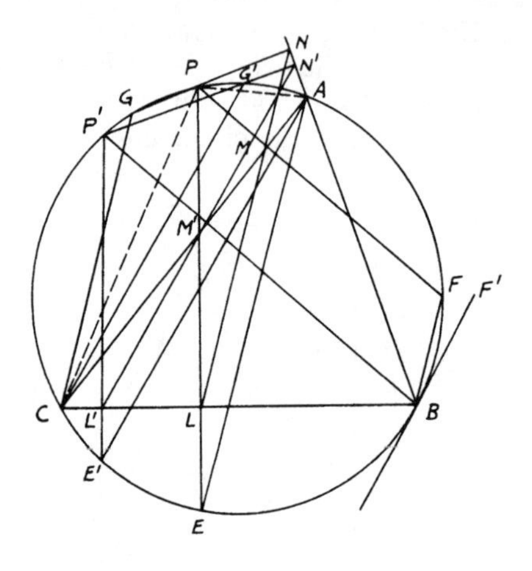

FIGURE 181

Proof: $\because$ $PMAN$ is cyclic, $\therefore$ $\angle NPA = \angle NMA = \angle GCA$. $\therefore$ CG is $\parallel$ MN. Also, $\angle MPA = \angle MNA = \angle FBA$. $\therefore$ BF is $\parallel$ MN. Again, since $PMLC$ is cyclic, $\therefore$ $\angle PLM = \angle PCM = \angle PEA$. $\therefore$ AE is $\parallel$ LMN. Hence AE, BF, CG are $\parallel$ LMN. Similarly, AE' is $\parallel$ $L'M'$. But $\angle L'CM' = \angle L'P'M' = \angle E'P'B = \angle E'AB$. $\because$ $\angle F'BA = \angle BCA$ or $\angle L'CM'$ (since BF' is tangent to $\odot$ at B), $\therefore$ $\angle E'AB = \angle F'BA$. $\therefore$ BF' is $\parallel$ $L'M'$. Similar to AE', CG' is $\parallel$ $L'M'N'$.

7.4. *From a point P on the circumcircle of the triangle ABC, perpendiculars PL, PM, PN are drawn to BC, CA, AB. Find the position of P in order that $ML = MN$.*

ANALYSIS: Assume that LMN is the required pedal line of P with respect to $\triangle ABC$. Join PA, PC (*Fig.* 182). Now, $PCLM$ is cyclic. $\therefore$ $PM = PC \sin PCM = PC \sin PLM$ (since $\angle PMC =$ right angle). $\therefore$ $PM/\sin PLM = PC = ML/\sin MPL = ML/\sin C$ (law of sines for $\triangle PLM$). Hence $ML = PC \sin C$. Similarly, $MN = AP \sin A$. If $LM = MN$ as required, $\therefore$ $AP/PC = \sin C/\sin A = AB/BC$ (law of sines for $\triangle ABC$ and can be proved also by dropping $BE \perp AC$). Therefore, AP/PC = construction ratio of AB/BC, and P can be determined by the intersection of the Apollonius $\odot$ which divides AC internally and externally into the fixed ratio of $AB : BC$ at D, F, with $\odot$ ABC.

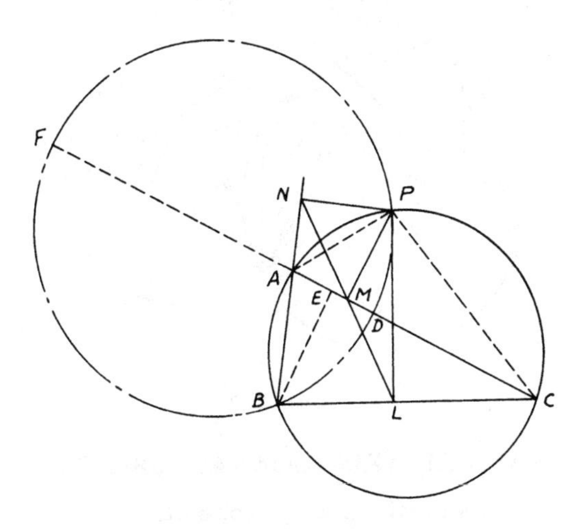

FIGURE 182

SYNTHESIS: Draw the Apollonius $\odot$ on DF as diameter cutting $\odot$ ABC in B, P. $\therefore$ $AP/PC = AD/DC = AB/BC = \sin C/\sin A$. $\therefore$ $PC \sin C = AP \sin A$ or $ML = MN$. Likewise, if it is required that $ML = k \cdot MN$, where k is a fixed integer, then draw the Apollonius $\odot$ that divides AC in the same way into the ratio $AB : k \cdot BC$.

7.5. *The orthocenters of the four triangles formed by four lines are collinear.*

CONSTRUCTION: Let the four given lines be ABE, DCE meeting in E and BCF, ADF meeting in F. Let also O_1, O_2, O_3, O_4 be the orthocenters of the $\triangle$s BCE, ABF, ADE, DCF. Draw the two circumcircles BCE, DCF to intersect in P. From P draw PL, PM, PN, PQ $\perp$s ABE, DCE, BCF, ADF and join O_1P, O_2P, O_3P, O_4P (*Fig.* 183).

Proof: Since P is the point of intersection of the circumcircles BCE, DCF, $\therefore$ $LMNQ$ is one straight line, which is the Simson line of P with respect to both $\triangle$s BCE and DCF (according to Prop. 7.1). Again, it is easy to prove that $ADPE$, $ABPF$ are cyclic quadrilaterals. Hence in the same way as of Prop. 7.1, $LMNQ$ can be proved to be the Simson line of both $\triangle$s ADE, ABF. But, since O_1P, O_2P, O_3P, O_4P are bisected by $LMNQ$ in the four $\triangle$s BCE, ABF, ADE, DFC

(Prop. 7.1) and $LMNQ$ is a straight line, $\therefore O_1, O_2, O_3, O_4$ are collinear.

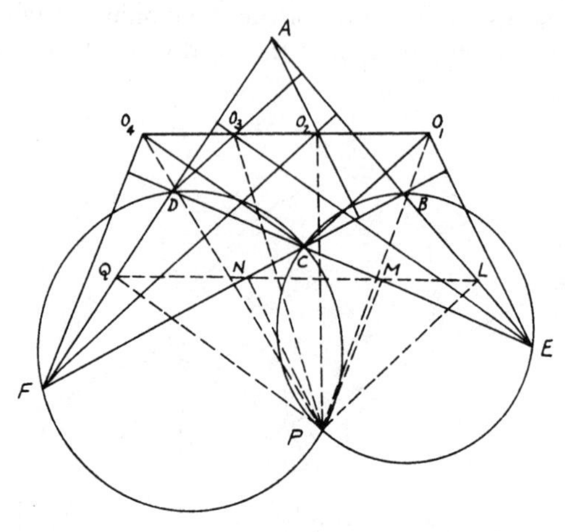

FIGURE 183

RADICAL AXIS—COAXAL CIRCLES

Definitions and Propositions

Proposition 7.2. *Find the locus of a point, the tangents from which to two given circles are equal. (i) If the circles intersect, it is clear that the extension of their common chord is the required locus. (ii) If they touch one another, the common tangent at their point of contact is the required locus. (iii) If they do not meet, let C, C′ be the centers of the ⊙s and suppose the circle C is the greater. Let P be any point on the locus. Draw the tangents PT, PT′ and PO ⊥ CC′. Join PC, PC′, CT, C′T′ (Fig. 184). Hence* $OC^2 - OC'^2$

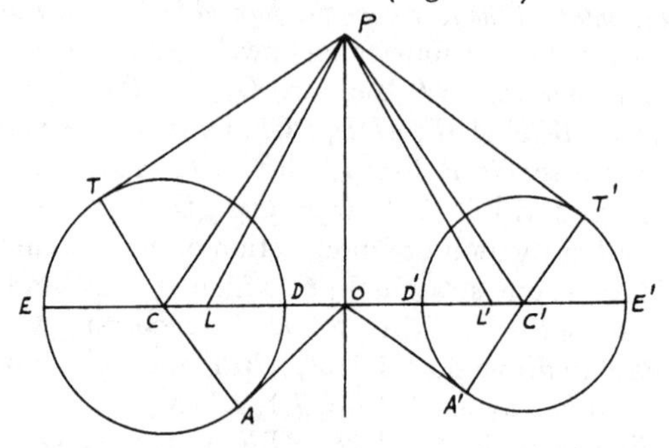

FIGURE 184

$= CP^2 - C'P^2 = CT^2 - C'T'^2$. $\because PT = PT'$, $\therefore$ *the* $\perp$ *PO divides CC' into segments such that the difference of the squares on them = the difference of the squares on the corresponding radii.* $\therefore$ *O is a fixed point and PO therefore a fixed straight line and hence the required locus.*

Definition. *The line PO is called the radical axis of the two circles. From O draw the tangents AO, OA'. These are equal. Take OL, OL' each equal to OA or OA' and join PL, PL'. Hence* $PT^2 = PC^2 - CT^2 = PO^2 + OC^2 - CA^2 = PO^2 + OA^2 = PO^2 + OL^2 = PL^2$. $\therefore PT = PL$. *Similarly,* $PT' = PL'$. $\therefore PT = PL = PL' = PT'$.

COROLLARY: The circle described with any point on the radical axis of two circles as center, and a tangent to one of the circles from this point as radius, will cut both circles orthogonally and if the circles do not meet will pass through two fixed points *L*, *L'* called the *limiting points.*

Proposition 7.3. *If there be three circles whose centers are not in the same straight line, their radical axes taken two and two are concurrent. Let A, B, C be the centers of the* $\bigcirc$*s,* r_1, r_2, r_3 *their radii. Let the radical axes of A, B; B, C; C, A cut AB, BC, CA in E, F, G respectively* (*Fig.* 185). *If EM, EN*

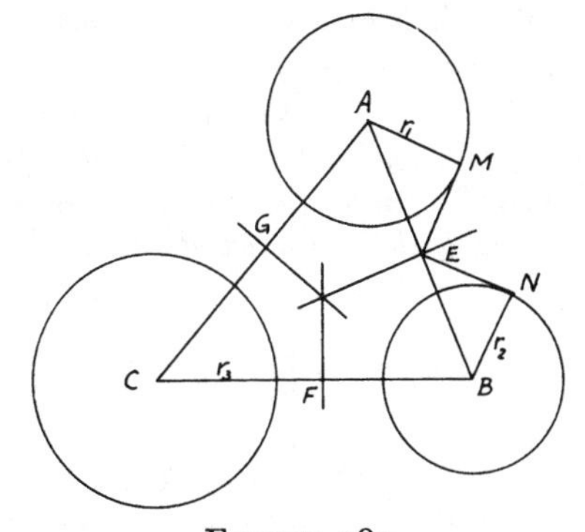

FIGURE 185

are the tangents from E to A, B, then EM = EN and hence $AE^2 - EB^2 = r_1^2 - r_2^2$ (*Prop.* 7.2), $BF^2 - FC^2 = r_2^2 - r_3^2$, $CG^2 - GA^2 = r_3^2 - r_1^2$. *Adding gives* $(AE^2 - EB^2) + (BF^2 - FC^2) + (CG^2 - GA^2) = 0$. *Therefore the* $\perp$*s from E, F, G to AB, BC, CA respectively are concurrent* (*Problem* 2.20).

COROLLARY 1. If three circles intersect, two and two, their common chords are concurrent and if three circles touch each other, two and two, the three common tangents at their points of contact are concurrent.

Definition. *The point of concurrence of the radical axes of three given circles taken two and two is called the radical center of the circles.*

COROLLARY 2. The circle described with the radical center of three circles as center and a tangent from that point to any one of them as radius cuts the three circles orthogonally.

COROLLARY 3. If a variable circle be described through two given points to cut a given circle, the common chord passes through a fixed point on the line which joins the given points.

Proposition 7.4. *Find the radical axis of two given circles. (i) If the circles intersect, the common chord extended is their radical axis. (ii) If the circles touch, the common tangent at their point of contact is their radical axis. (iii) If the circles do not intersect, describe a circle cutting each of the given circles, with any point not in the line of the centers of the circles as center. Draw the common chords. These produced intersect in the radical center of the three circles. Therefore, the perpendicular drawn from this point to the line joining the centers of the given circles is their radical axis. Note the relationship of this proposition to Proposition* 7.2.

Definition. *If any number of circles, taken two and two, have the same radical axis, they are called coaxal. These coaxal circles reduce to the points* L, L' (*Prop.* 7.2) *called the limiting points of the family of the circles.*

Proposition 7.5. *If one coaxial system of circles has its centers on the common chord of another system of intersecting circles, then the two common points of intersection of the second system are the limiting points of the first.*

It can be shown that if a system of circles has its centers collinear and has also a common orthogonal circle, it is a coaxal system. Now in Fig. 186, *one*

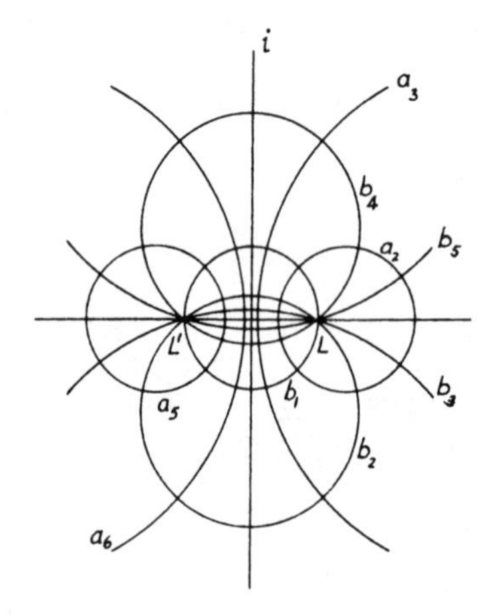

FIGURE 186

system consists of L (a point circle) and the circles a_2, a_3, . . ., which increase in size until eventually the radical axis i is reached. On the other side we get L′, a_5, a_6, . . ., also approaching the radical axis as a limit. The orthogonal system of b_1 on LL′ as diameter (which is the smallest circle), b_2, b_3, . . . and b_4, b_5, . . ., both series approaching the line LL′ as a limit.

Solved Problems

7.6. *If PN is the perpendicular from any point P to the radical axis of two circles the centers of which are A, B, and PQ, PR are the tangents from P to the circles, then $PQ^2 - PR^2 = 2\,PN \cdot AB$.*

CONSTRUCTION: Draw $PM \perp AB$, and join PA, AQ, PB, BR (*Fig.* 187).

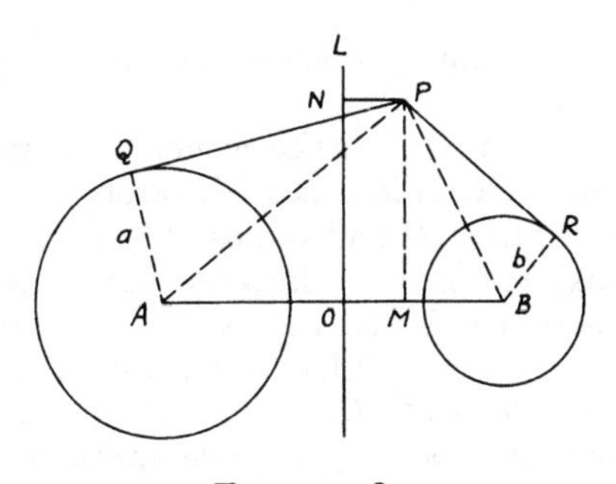

FIGURE 187

Proof: Let a, b be the radii of ⊙s A, B.

$$\begin{aligned} PQ^2 - PR^2 &= (PA^2 - a^2) - (PB^2 - b^2) \\ &= (PA^2 - PB^2) - (a^2 - b^2) \\ &= AM^2 - MB^2 + OB^2 - AO^2 \qquad (\text{since } OA^2 - a^2 = OB^2 - b^2) \\ &= (AM - MB)(AM + MB) + (OB - AO)(OB + AO) \\ &= (AM - MB + OB - AO)AB \\ &= (AM - AO + OB - MB)AB \\ &= 2\,OM \cdot AB = 2\,PN \cdot AB. \end{aligned}$$

If P lies on one of the circles, for instance B, then $PR = 0$ and $PQ^2 = 2\,PN \cdot AB$.

7.7. *Given two circles, construct a system of circles coaxal with them.*

CONSTRUCTION: (i) If the given ⊙s intersect, any ⊙ through their points of intersection will be coaxal with them.

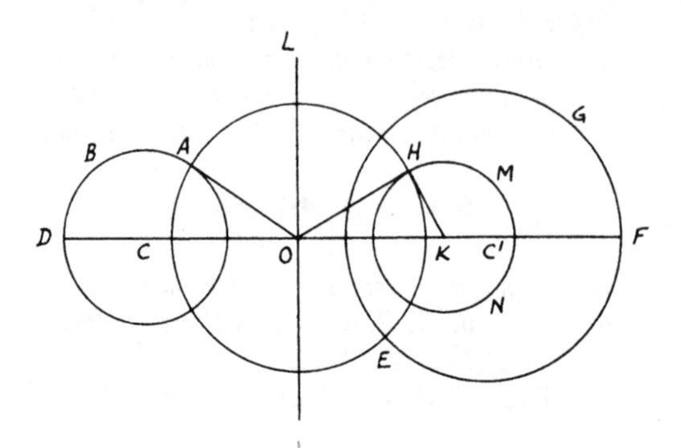

FIGURE 188

(ii) If the ⊙s touch, any ⊙ touching them at their point of contact will be coaxal with them.

(iii) If the given ⊙s ABD, EFG do not meet, take C, C' their centers, and draw their radical axis OL (Prop. 7.4). From O draw tangent OA. With center O and radius OA describe ⊙AEH; from any point H on its circumference draw tangent HK meeting CC' in K (*Fig.* 188). With center K and radius KH describe ⊙HMN. OH touches ⊙HMN and $OH = OA$. $\therefore$ OL is radical axis of ⊙s ABD, HMN. Hence ⊙s ABD, EFG, HMN are coaxal. Similarly, by drawing tangents from other points on the circumcircle of ⊙AEH, any number of ⊙s can be described coaxal with ABD, EFG.

7.8. *Construct a circle which will pass through two given points and touch a given circle.*

CONSTRUCTION: Let A, B be the given points and CDE the given ⊙. Describe a ⊙ passing through A, B and cutting a given ⊙ in C,

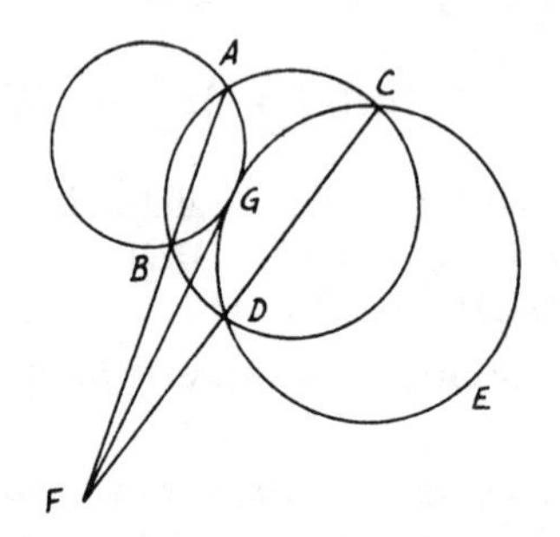

FIGURE 189

D. Let AB, CD meet in F. F is the radical center of ⊙s CDE, ABC, and the required ⊙. Draw tangent FG to ⊙CDE. FG is the radical axis of ⊙CDE and the required ⊙ (*Fig.* 189). Hence ⊙ through A, B, G is the required ⊙. Describe this ⊙. Since $FA \cdot FB = FC \cdot FD = FG^2$, $\therefore$ FG touches ⊙ABG. $\therefore$ ⊙ABG touches ⊙CDE. If the other tangent from F to ⊙CDE meet that ⊙ in G', the ⊙ through A, B, G' will also touch given ⊙CDE.

7.9. *AD, BE, CF are perpendiculars from A, B, C to the opposite sides of the triangle ABC. EF, FD, DE meet BC, CA, AB respectively in L, M, N. Show that L, M, N are collinear and that the straight line through them is perpendicular to the line joining the orthocenter and the circumcenter.*

CONSTRUCTION: Draw the nine-point circle of $\triangle ABC$ with center S the mid-point of OG, O, G being the circumcenter and orthocenter (*Fig.* 190).

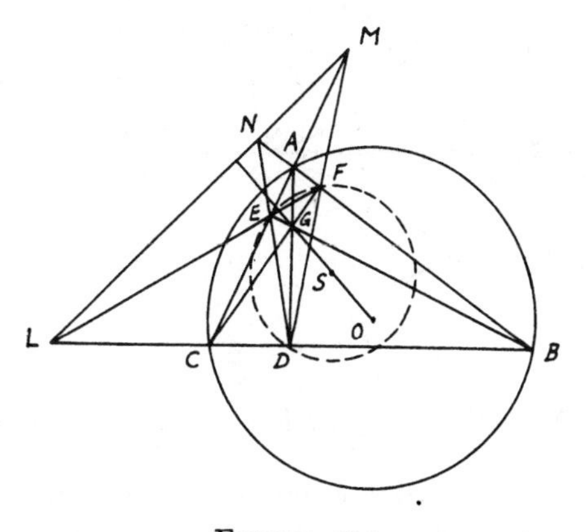

FIGURE 190

Proof: Since the nine-point circle passes through D, E, F, then ⊙DEF is the nine-point circle of $\triangle ABC$. But $BDEA$ is cyclic (since $\angle ADB = \angle AEB =$ right angle). $\therefore$ $NA \cdot NB = NE \cdot ND$. Hence tangent from N to ⊙ABC = tangent from N to ⊙DEF. Therefore, N lies on the radical axis of these two ⊙s. Similarly, L, M lie also on the radical axis of these two ⊙s. Hence LMN is one straight line and is the radical axis of these two ⊙s. $\therefore$ The line of the centers OSG is $\perp$ the radical axis of the two ⊙s LMN.

7.10. *ABCD is a quadrilateral in which the angles B, D are equal. Two points E, F are taken on BC, CD respectively, such that $\triangle AED$, $\triangle AFB$ are equal. Show that the radical axis of the two circles described on BF, DE as diameters passes through A.*

CONSTRUCTION: Draw EP, $FQ \parallel AD$, AB respectively meeting AB, AD produced in P, Q. Join BD, PQ, DP, BQ. Let ⊙s on BF, DE as diameters cut AB, AD in G, H respectively and join EH, FG (*Fig.* 191).

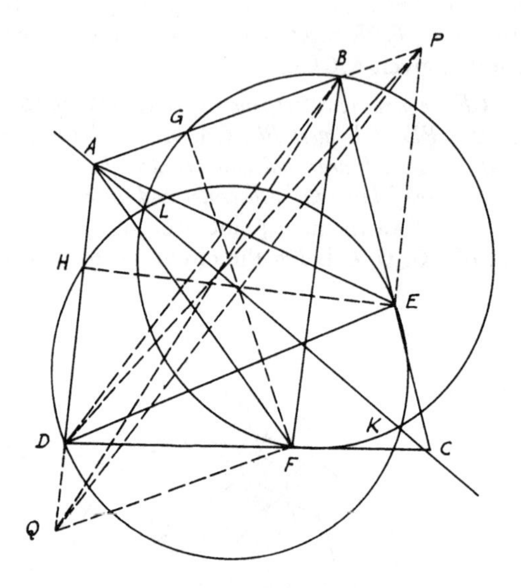

FIGURE 191

Proof: $\because$ $EP \parallel AD$, $\therefore$ $\triangle AED = \triangle APD$. Also, $FQ \parallel AB$. $\therefore$ $\triangle AFB = \triangle AQB$. Since $\triangle AED = \triangle AFB$, $\therefore$ $\triangle APD = \triangle AQB$. Eliminating common $\triangle ADB$ yields $\triangle BPD = \triangle DQB$. $\therefore$ $DB \parallel PQ$. $\therefore$ $AB : BP = AD : DQ$ (1). But $\triangle$s BEP, DFQ are similar, $\therefore$ $BP : BE = DQ : DF$ (2). Multiplying (1) and (2) gives $AB : BE = AD : DF$. But $\angle B = \angle D$. $\therefore$ $\triangle$s ABE, ADF are similar. $\therefore$ $\angle BAE = \angle DAF$. Adding $\angle EAF$ to both, $\therefore$ $\angle BAF = \angle DAE$. Since $\angle BGF = \angle DHE =$ right angle, $\therefore$ $\triangle$s AGF, AHE are similar. $\therefore$ $GF : AG = HE : AH$. $\because$ $\triangle AED = \triangle AFB$, $\therefore$ $AB : GF = AD \cdot HE$. $\therefore$ $AG \cdot AB = AH \cdot AD$,

Hence the tangent from A to ⊙BFG = tangent from A to ⊙DEH. Therefore, A lies on the radical axis of these two ⊙s, which, in this case, is their common chord LK.

POLES AND POLARS

Definitions and Propositions

Definition. *If the radius CA of a circle, C being the center, be divided*

internally and externally in P and D so that the rectangle $CP \cdot CD$ *is equal to the square on CA, the straight line drawn through either P or D perpendicular to CA is called the polar of the other point with respect to the circle and that other point is called the pole of the perpendicular. P and D are sometimes called conjugate poles. Hence when the pole is inside the circle, the polar does not meet the circle and when the pole is on the circumference of the circle its polar is the tangent drawn at the pole.*

Proposition 7.6. *If a line cuts a circle, the point of intersection of the tangents drawn at the points where the line cuts the circle is the pole of the line.*

Let TT' *cut* $\odot ATT'$ *of center C. Draw tangents TP,* $T'P$. *Join CT, CP cutting* TT' *in D (Fig. 192). Now,* $\angle CTP$ *is right and* $TT' \perp CP$.

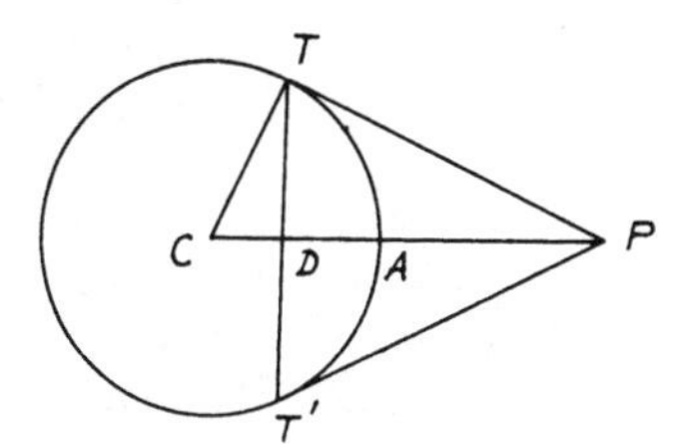

FIGURE 192

$\therefore CP \cdot CD = CT^2$. $\therefore$ *P is the pole of* TT'. *Hence when the pole is outside the circle, its polar cuts the circle and coincides with the chord of contact of tangents to the circle drawn from the pole.*

Proposition 7.7. *If from every point in a given straight line pairs of tangents be drawn to a given circle, all the chords of contact intersect in the pole of the given line.*

Let MN be the given line, P its pole with respect to the $\odot$ *with center C. From any point M in MN draw tangents MA, MB and draw* $PE \perp CM$ *(Fig. 193).* $\because$ $\angle$*s PDM, PEM are right,* $\therefore$ *DPEM is cyclic.* $\therefore CE \cdot CM = CP \cdot CD =$ *square on radius.* $\therefore$ *PE is the polar of M. Hence AB passes through P. This proposition may also be enunciated thus: the polars of all points in a straight line intersect in the pole of the line.*

COROLLARY 1. If the polar of *P* passes through *Q*, then the polar of *Q* passes through *P*. Such points as *P*, *Q* are called *conjugate points* with respect to circle.

COROLLARY 2. The polar of the intersection of two straight lines is the line joining their poles.

COROLLARY 3. The poles of all straight lines meeting in a point lie on the polar of that point.

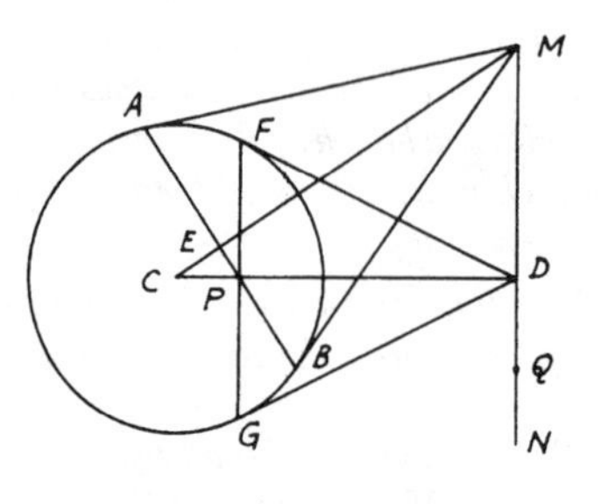

FIGURE 193

Proposition 7.8. *The locus of the intersection of pairs of tangents to a circle drawn at the extremities of a chord which passes through a given point is the polar of the point.*

Let P (Fig. 193) be the given point, C the center of given ⊙. Through P draw any chord AB. Join CP and draw FPG ⊥ CP. Draw also tangents AM, BM, FD, GD. Let CM, AB intersect in E. Join MD. ∵ CP·CD = square on radius = CM·CE, ∴ DPEM is cyclic and ∠PEM is right. ∴ ∠PDM is right.

Hence MD is the polar of P, and this is the locus of M.

This proposition may also be enunciated thus: the locus of the poles of all secants to a given circle which pass through a given point is the polar of the point.

COROLLARY: The pole of the straight line joining two points is the point of intersection of the polars of the points.

Proposition 7.9. *Any straight line drawn through a fixed point and cutting a given circle is divided harmonically by the point, the circle, and the polar of the point, and the converse.*

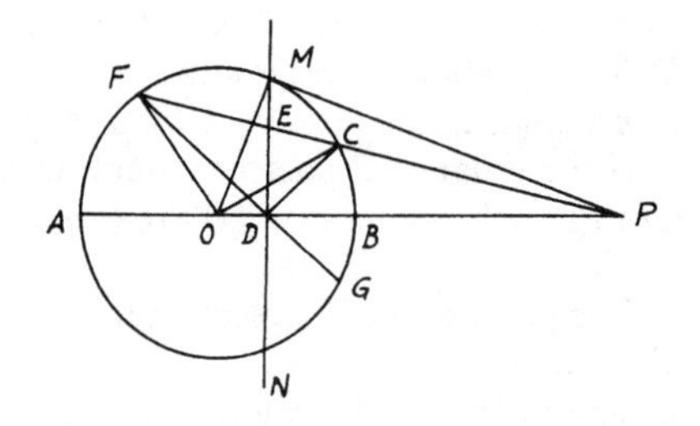

FIGURE 194

(*i*) *Let PCEF cut MN the polar of P in E. Draw PBDOA through the center O. Join OF, OM, OC, DF, DC, PM; produce FD to G* (*Fig.* 194). $\because$ $\angle PMO$ is right, $\therefore$ $PO \cdot PD = PM^2 = PC \cdot PF$. $\therefore$ *CDOF is cyclic.* $\therefore$ $\angle PDC = \angle OFC = \angle OCF = \angle ODF$ *in some segment.* $\therefore$ *DM bisects* $\angle FDC$. $\therefore$ *DP bisects the adjacent* $\angle CDG$. $\therefore$ *PCEF is a harmonic range. A similar proof holds when P lies inside the* $\odot$.

(*ii*) *If the chord CF be divided harmonically in E and P, then the polar of P passes through E. This is easily proved indirectly.*

Solved Problems

7.11. *The distances of two points from the center of a circle are proportional to the distance of each point from the polar of the other.*

CONSTRUCTION: Let DN be the polar of P and EM the polar of Q with respect to a circle whose center is C. Draw $PF \perp CQ$, $PM \perp EM$, $QG \perp PC$, and $QN \perp DN$ (*Fig.* 195).

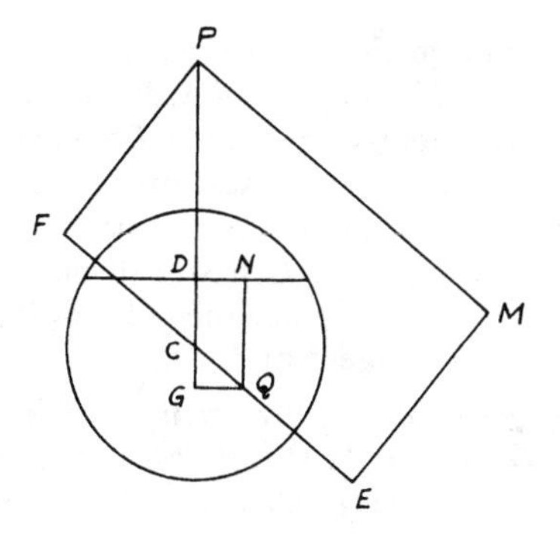

FIGURE 195

Proof: Since QGP, QFP are right angles, $\therefore$ $QGFP$ is cyclic. $\therefore$ $GC \cdot CP = QC \cdot CF$. Also, $CD \cdot CP =$ square on radius $= CQ \cdot CE$. $\therefore$ $GC \cdot CP + CD \cdot CP = QC \cdot CF + QC \cdot CE$ or $CP \cdot GD = QC \cdot EF$. Hence $PC : QG = EF : GD = PM : QN$. This is called *Salmon's theorem.*

7.12. (*i*) *Parallel tangents to a circle at P, Q meet the tangent at the point R in S, T. PQ meets this tangent in U. Show that RV is the polar of U, V being the intersection of PT and QS* (*Fig.* 196).

Proof: Let a line $\parallel$ to PS through V cut PQ at M. $\therefore$ VM is $\perp PQ$. Again the harmonic points of the quadrangle $PSTQ$ are U, V and the point I at infinity on PS. Hence $V(PQ, UI)$ is harmonic. $\therefore$ (PQ, UM) is harmonic. Hence the polar of U passes through M and it is $\perp PQ$. $\therefore$ It is VM which also passes through R.

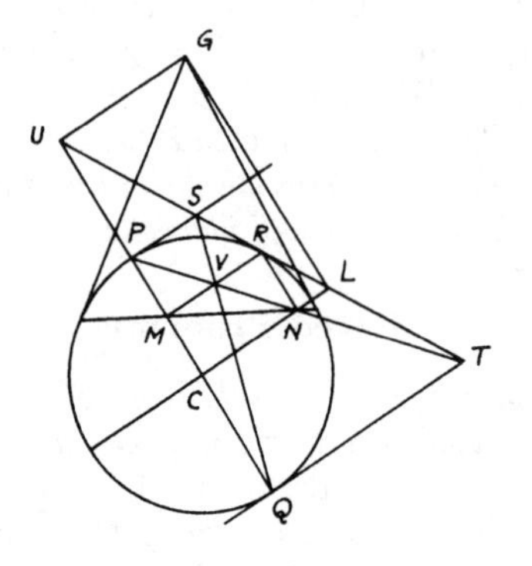

FIGURE 196

(ii) M, N are the projections of a point R on a circle on two perpendicular diameters. G is the pole of MN and U, L are the projections of G on the diameters. Show that UL touches the circle at R.

Proof: Since the polar of G passes through M, the polar of M passes through G and, being $\perp$ the radius CM, is GU, $\therefore$ passes through U. Hence the polar of U passes through M and therefore RM. So the polar of L is RN. Hence the polar of R (on RM and RN) is UL. Hence UL is the tangent at R.

7.13. *ABCD is a quadrilateral inscribed in a circle. AD, BC produced meet in E and AB, DC meet in F. O is the point of intersection of AC, BD. Show that each vertex of the triangle OEF is the pole of the opposite side* (*Fig.* 197).

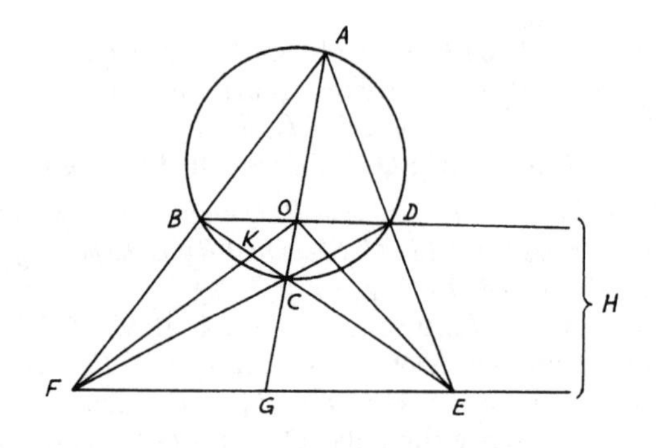

FIGURE 197

Proof: Let BD, FE produced meet in H and let FO, BE meet in K. $\because$ AC is cut harmonically in O, G (Prop. 6.6), $\therefore$ the polar of O passes through G (Prop. 7.9) and $\because$ BD is cut harmonically in O, H, $\therefore$ the polar of O passes through H. $\therefore$ EF is the polar of O. Again, $\because$ the polar of O passes through E, $\therefore$ the polar of E passes through O (Prop. 7.7, Cor. 1). $\because$ $F(BODH)$ is a harmonic pencil, $\therefore$ BC is cut harmonically in K, E. $\therefore$ The polar of E passes through K. $\therefore$ FO is the polar of E and similarly EO is the polar of F.

7.14. *If a quadrilateral be inscribed in a circle and tangents be drawn at its angular points forming a circumscribed quadrilateral, the interior diagonals of the two quadrilaterals are concurrent and their third diagonals coincident* (*Fig.* 198).

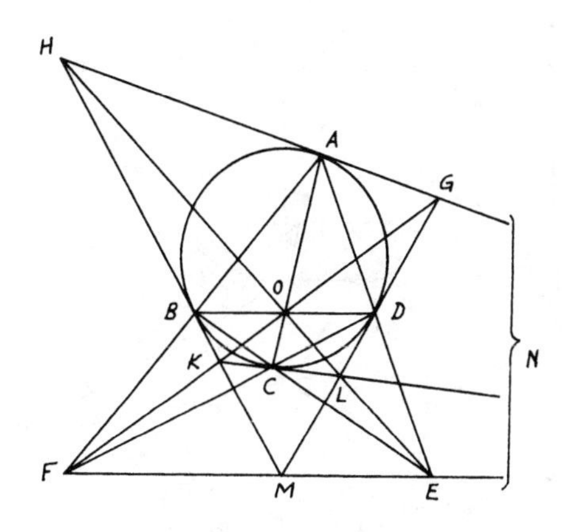

FIGURE 198

Proof: Let the tangents at A, C meet in N. $\because$ G is the pole of AD and K the pole of BC, $\therefore$ E is the pole of GK.

Similarly, F is the pole of HL. $\therefore$ The intersection of GK, HL is the pole of EF. But the intersection of AC, BD is the pole of EF. $\therefore$ The four diagonals AC, BD, GK, HL intersect in O. Again, $\because$ M is the pole of BD and N is the pole of AC, $\therefore$ MN is the polar of O. $\therefore$ EF, MN coincide. It can be deduced also that:

1. E is the pole of GK, but E is also the pole of FO. Therefore, GK passes through F. Similarly, HL passes through E.

2. HL is the polar of F, therefore $H(AOBF)$ is a harmonic pencil. Hence $FMEN$ is a harmonic range; i.e., the extremities of the third diagonal of one quadrilateral are harmonic conjugates to the extremities of the other.

3. Also $O(AHBF)$ is a harmonic pencil; i.e., the diagonals of the two quadrilaterals form a harmonic pencil.

7.15. *Construct a cyclic quadrilateral given the circumscribing circle and the three diagonals.*

Analysis: Let $ABCD$ be the required quadrilateral inscribed in $\odot I$. Let the tangents to the $\odot$ at A, B, C, D meet in H, K, L, G. HK, GL produced meet in M and GH, LK produced meet in N. Also AD, BC meet in E and AB, DC in F. $\therefore$ MN coincides with EF (Problem 7.14). Join IM, IN cutting BD, AC at right angles in Q, P (*Fig.* 199).

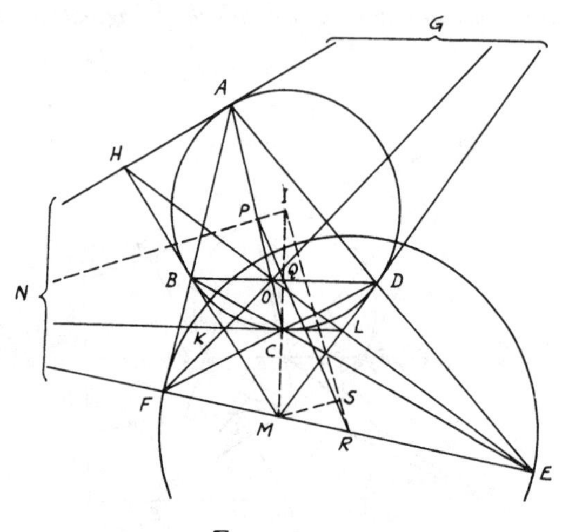

Figure 199

Hence Q, P are the mid-points of BD, AC. Produce PQ to meet EF in R. Since the mid-points of the diagonals of a complete quadrilateral are collinear (Problem 2.16), $\therefore$ R is the mid-point of EF. $\because$ M, N are the poles of BD, AC. $\therefore$ $IP \cdot IN = IQ \cdot IM =$ square on radius of $\odot I$. $\because$ AC, BD are given, $\therefore$ their distances IP, IQ from the center I are known. Therefore, IN, IM are known. But $IR^2 =$ tangent^2 from R to I + radius^2 of I, and since the tangent from R to $\odot I =$ half the third diagonal EF (Problem 3.23), $\therefore$ $IR^2 = RE^2 +$ radius^2 of I and hence is known.

But since the transversal PQR cuts the sides of $\triangle IMN$, $\therefore$ $(RM/RN)(IQ/QM)(NP/IP) = 1$ (Menelaus' Th. 5.13). $\therefore$ $RM/RN = (QM/IQ)(IP/NP)$, and since $QM = IM - IQ =$ known, so also NP is a known quantity. Hence $RM : RN$ is a known ratio. This reduces the problem to drawing the $\triangle IMN$ given IM, IN and the length of the line IR which divides MN into a known ratio $RM : RN$. Draw $MS \parallel IN$. $\therefore MS : IN = RM : RN$. $\therefore MS = IN(RM : RN)$ and is known and $IS : IR = MN : RN$. $\therefore IS = IR(MN : RN) = IR[1 - (RM/RN)]$ and is also known, and since IM is known, the triangle IMS can be constructed knowing the three sides.

SYNTHESIS: Take the known length IR on IS produced and join RM and produce it to meet the $\parallel$ to MS from I in N. On IN, IM, take P, Q with the known lengths IP, IQ. Describe the given $\odot$ with center I and from P, Q draw AC, $BD \perp IN$, IM. $ABCD$ is the required quadrilateral.

SIMILITUDE AND INVERSION

Definitions and Propositions

Definition. *If the straight line joining the centers of two circles be divided internally and externally in the points S', S in the ratio of the radii of the circles, S' is called the internal center and S the external center of similitude of the circles. The circle described on SS' as diameter is called the circle of similitude.*

It follows from the definition that if two circles touch each other, the point of contact is their internal center of similitude if the contact is external, their external center of similitude if the contact is internal (Fig. 200). If O, O' are

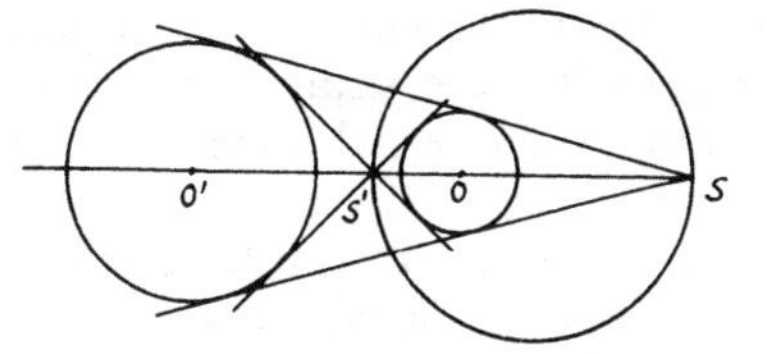

FIGURE 200

the centers of the circles, $SOS'O'$ is a harmonic range.

Proposition 7.10. *The straight line joining the extremities of parallel radii of two circles passes through their external center of similitude, if they are turned in the same direction; through their internal center, if they are turned in opposite directions. And the converse.*

(i) Let O, O' be the centers of the $\odot$s; let OX be $\parallel$ $ZO'X'$ and let XX' and ZX meet OO' in S and S' (Fig. 201). In $\triangle$s $SO'X'$, SOX, $SO'/SO = O'X'/OX$. $\therefore$ S is the external center of similitude of the $\odot$s. In $\triangle$s $S'O'Z$, $S'OX$, $S'O'/S'O = O'Z/OX$. $\therefore$ S' is the internal center of similitude.

(*ii*) *Let $SXYX'Y'$ drawn through S the external center of similitude of the* ⊙*s cut the* ⊙*s in X, Y, X', Y'. Then OX will be $\parallel$ $O'X'$ and $OY \parallel O'Y'$. Since each of the* ∠*s OXY, $O'X'Y'$ is acute, ∴ each of the* ∠*s OXS, $O'X'S$ is obtuse and* ∠OSX *is common to* △*s OSX, $O'SX'$ and $SO : OX = SO' : O'X$. ∴ $\angle SOX = \angle SO'X'$. ∴ OX is $\parallel$ $O'X'$. Similarly, OY is $\parallel$ $O'Y'$. Similarly for the internal center of similitude.*

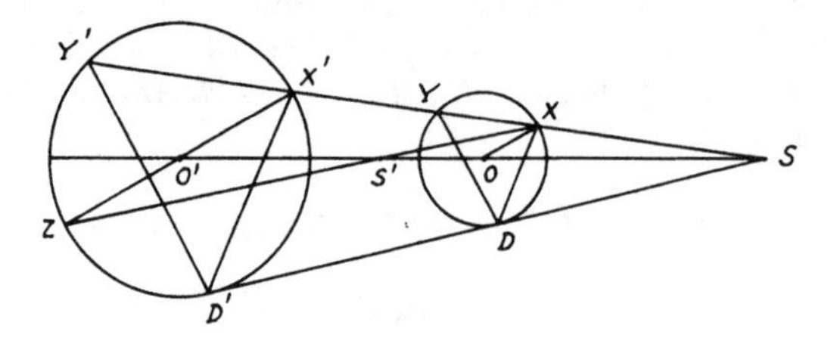

FIGURE 201

COROLLARY 1. From similar triangles $SX : SX' = OX : O'X' = SY : SY'$. Therefore, the ratio of $SX : SX'$ is constant, as also the ratio of $SY : SY'$.

COROLLARY 2. If a variable circle touch two fixed circles both externally or both internally, the straight line joining the points of contact passes through the external center of similitude of the fixed circles.

COROLLARY 3. If a variable touches two fixed circles, one internally, the other externally, the straight line joining the points of contact passes through the internal center of similitude of the fixed circles.

COROLLARY 4. The direct common tangent of two circles passes through their external center of similitude, the transverse common tangent through their internal center of similitude.

Definition. *Given a fixed point O called the center of inversion. If on OP a point P' is taken such that $OP \cdot OP' = k$, where k is a constant called the constant of inversion, then the points P and P' are said to be inverse points with respect to O.*

If the constant k is positive, the two points P, P' lie on the same side of O and if k is negative, they lie on opposite sides of O. If P, P' are on the same side of O so that $OP \cdot OP' = c^2$, then P, P' are inverse points with respect to the circle with center O and radius c. This circle is called the circle of inversion. The inverse of a point on the circle of inversion is the point itself; for in $OP \cdot OP' = c^2$, if $OP = c$, then $OP' = c$ and therefore P' coincides with P. If P describes a curve or a figure of any kind, then P' describes another figure and the two figures are said to be inverse figures.

Notice that an intersection of two curves inverts into an intersection of the two inverse curves. Also, that if OT touches the given curve at T, then OT also touches the inverse curve at T', the inverse of T. For, if P and Q coincide at T, P' and Q' coincide at T'.

Proposition 7.11. *Two curves intersect at the same angle as the inverse curves. For this reason inversion is said to be a conformal transformation.*

Proposition 7.12. *If P, P' and Q, Q' are pairs of inverse points with respect to O, then $P'Q' = k \cdot PQ/OP \cdot OQ$, where k is the constant of inversion.*

For since $OP : OQ = OQ' : OP'$, the $\triangle$s OPQ and $OP'Q'$ are similar. Hence, $P'Q' : PQ = OQ' : OP = OP' \cdot OQ' : OP \cdot OP' = OP' \cdot OQ' : k$. So $PQ = k \cdot P'Q'/OP' \cdot OQ'$. If, however, P, Q are collinear with O, then $P'Q' = OQ' - OP' = k/OQ - k/OP = k(OP - OQ)/OQ \cdot OP = k \cdot PQ/OP \cdot OQ$.

Proposition 7.13. *The inverse of a straight line is a circle through the center of inversion O, and, conversely, the inverse of a circle through O is a straight line.*

(i) Draw $OA \perp$ the line and take any point P upon the line. Let A', P' be the inverses of A, P (Fig. 202). Then $OA \cdot OA' = OP \cdot OP'$. Hence, A, A', P', P are concyclic. Hence $\angle OP'A' = \angle OAP =$ right angle. Therefore, the inverse of the line AP (i.e., the locus of P') is the $\odot$ on OA' as diameter, OA' being $\perp$ the line.

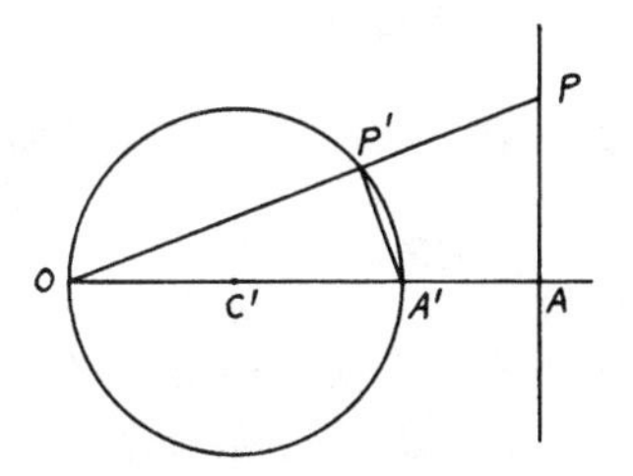

FIGURE 202

(ii) Take the inverse A, P of A', P'; OA' being a diameter of the given $\odot$. Then, as before, $\angle OAP = \angle OP'A' =$ right angle. Hence the inverse with respect to O of the circle on OA' as diameter (i.e., the locus of P) is the straight line $\perp OA'$ through A. Therefore, $OC' = \frac{1}{2} OA' = \frac{1}{2} k/OA$. Hence the center C' and the radius OC' of the inverse to a given line are known.

An exceptional case is when the straight line passes through O, e.g., OP in the above figure. Then P' lies also on the line OP. Hence the inverse of a straight line through O is the straight line itself. Notice that the inverse of a *circle with respect to a point O on it is a straight line. This property can be used mechanically to convert circular to linear motion.*

It is obvious that we can invert any two circles which intersect into straight lines by taking O at one of the intersections. If the circles touch, their inverses would then be parallel straight lines.

Proposition 7.14. *The inverse of a circle with respect to a point O not on it is a circle.*

Let P' be the inverse of P, a point on the given circle whose center is C. Let OP cut this ⊙ again at Q. Draw $P'D' \parallel QC$ cutting OC at D' (Fig. 203). The $OP \cdot OQ$ is constant and $OP \cdot OP'$ is constant. Hence $OP' : OQ$ is constant. But $OD' : OC = OP' : OQ$. Hence OD' is constant and hence D' is fixed. Also, $P'D' : QC = OP' : OQ$. Hence $D'P'$ is fixed in length. Therefore, the locus of P' is a circle; i.e., the inverse of the given ⊙ is a ⊙.

In exactly the same way, it can be shown that the inverse of a sphere with respect to a point O not on it is a sphere.

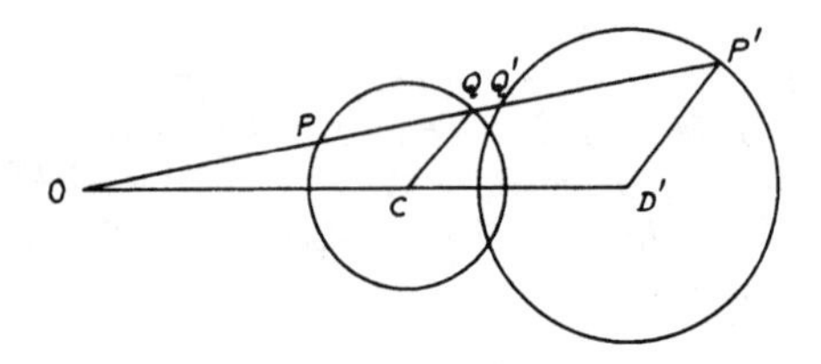

FIGURE 203

Proposition 7.15. *Find the center and radius of the inverse circle.*

In the above figure, it can be seen that $OD'/OC = P'D'/QC = OP'/OQ = OP \cdot OP'/OP \cdot OQ = k/(OC^2 - CP^2)$.

Hence $OD' = k \cdot OC/(OC^2 - CP^2)$ and $P'D' = k \cdot QC/(OC^2 - CP^2)$ or D' is the center of the inverse ⊙ whose radius is $P'D'$.

Proposition 7.16. *Inverse circles and the circle of inversion are coaxal.*

For let the circle of inversion cut the given circle at E and F. Then the inverse of E is E. Hence the inverse circle passes also through E and similarly through F.

As a particular case, a straight line is the radical axis of its inverse circle and the circle of inversion.

Proposition 7.17. *The inverse of a circle with respect to an orthogonal circle is the circle itself.*

For if O is the center and c the radius of the orthogonal circle and OPP' a chord (through O) of the given circle, then $OP \cdot OP' = c^2$, since a radius of one circle touches the other.

Proposition 7.18. *Any two circles are inverse with respect to either center of similitude and with respect to no other point.*

(*i*) *In Fig.* 203, *let* O *be a center of similitude of the circles, then* $OP' : OQ = r' : r$. *Also* $OP \cdot OQ$ *is constant. Hence,* $OP \cdot OP'$ *is constant. Hence the circles are inverse with respect to* O.

(*ii*) *Let the circles be inverse with respect to* O. *Then it is proved in Prop.* 7.14 *that* $OC : OD' = r : r'$. *Hence,* O *is the center of similitude. Notice that the radius of inversion is* OE, E *being one of the intersections of the given circles.*

If in the limiting case one circle becomes a straight line, then we can still speak of the centers of similitude of this circle and this line. Hence, the centers of similitude of a circle and a line are the ends of the diameter perpendicular to the line, for these are the only possible centers of inversion.

Proposition 7.19. *If a circle touches two given circles the points of contact are inverse points with respect to the external center of similitude.*

For let the line joining the points of contact P, Q *cut the line of centers of the given circles* a, b *at* O. *Invert with respect to* O, *taking* P, Q *as inverse points. Then the touching circle* c *inverts into itself. Also* a *which touches* c *at* P *inverts into a circle touching* c *at* Q *and having, by symmetry, its center on the line of centers, i.e., into* b. *Hence* P *and* Q *are inverse points on the circles.*

Solved Problems

7.16. *If through the external center of similitude* S *of two circles any straight line* $SXYX'Y'$ *cutting the first circle in* X, Y *and the other in* X', Y' *be described, and also the common tangent* SDD', *then each of the rectangles* SX, SY' *and* SY, SX' *is constant and equal to the rectangle* SD, SD'.

CONSTRUCTION: Join XD, YD, $X'D'$, $Y'D'$ (*Fig.* 204).

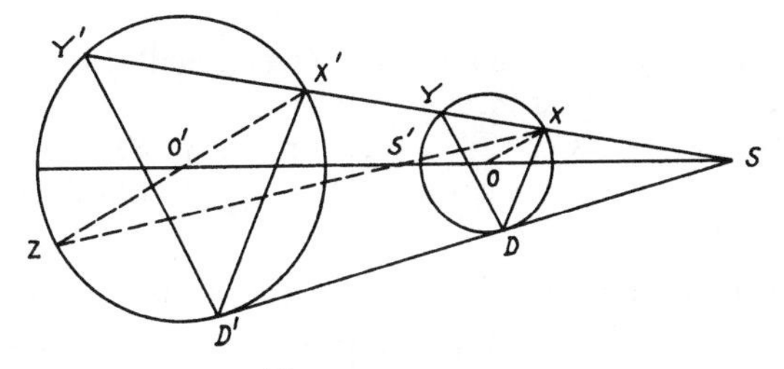

FIGURE 204

Proof: $\because SX : SX' = OD : O'D' = SD : SD'$, $\therefore XD$ is $\parallel X'D'$. Similarly, YD is $\parallel Y'D'$. $\therefore \angle SDX = \angle SD'X' = \angle D'Y'X'$. $\therefore DD'Y'X$ is cyclic. $\therefore SX \cdot SY' = SD \cdot SD'$. Similarly, $DD'X'Y$ is cyclic. $\therefore$ $SY \cdot SX' = SD \cdot SD'$. Similar results can be proved in a similar way for the internal center of similitude.

COROLLARY: XD, $Y'D'$ intersect in the radical axis of the two circles.

For since a circle goes through $XDD'Y'$, therefore XD is the radical axis of this circle and the circle O and $Y'D'$ is the radical axis of this circle and the circle O'. Therefore, the point of intersection of XD, $Y'D'$ is the radical center of the three circles and hence a point in the radical axis of the circles O, O'.

7.17. *If two circles touch two others, so that the contacts of the two circles with the two others are both internally or externally or even one internally and the other externally, the radical axis of each pair passes through the center of similitude of the other pair (Fig. 205).*

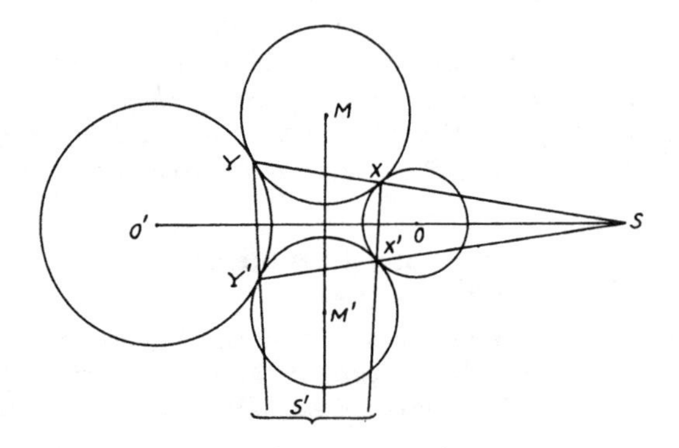

FIGURE 205

Proof: Let ⊙s M, M' touch ⊙s O, O' externally in X, Y and X', Y'. YX and $Y'X'$ both pass through S the external center of similitude of O, O' (Prop. 7.10, Cor. 2). $\therefore$ $SX \cdot SY = SX' \cdot SY'$. Hence tangents from S to M and M' are equal. $\therefore$ S is on the radical axis of M, M'. Similarly with S'.

7.18. *A straight line drawn through a center of similitude S of two circles meets them in P, Q and P', Q' respectively. If PQ' is divided in R in the ratio of the radii of the circles, prove that the locus of R is a circle.*

CONSTRUCTION: Let $PQP'Q'$ be drawn from the external center of similitude of the two ⊙s O, O' and let S' be the internal center of similitude. Then the locus of R is the ⊙ on SS' as diameter (*Fig.* 206).

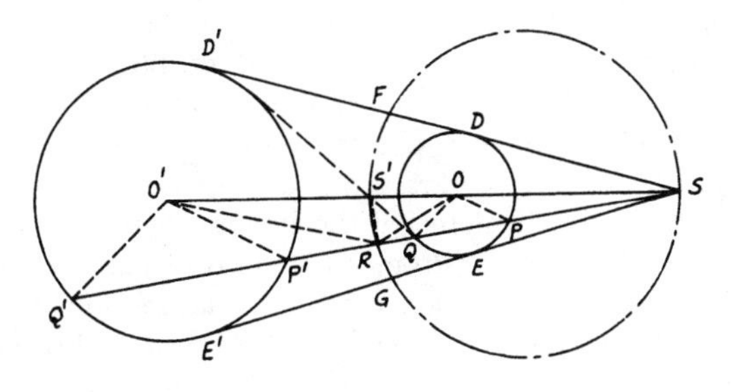

FIGURE 206

Proof: Join OP, OQ, OR, $O'P'$, $O'Q'$, $O'R$, $S'R$ and draw SDD', SEE' common external tangents of the ⊙s. ∵ OP is $\parallel O'P'$, ∴ $\angle OPR = \angle O'P'Q' = \angle O'Q'P'$. ∵ $PR : RQ' = OP : O'Q'$ (hypothesis), ∴ $PR : OP = RQ' : O'Q'$, and since $\angle OPR = \angle O'Q'P'$ or $\angle O'Q'R$, ∴ △s OPR, $O'Q'R$ are similar. ∴ $\angle ORP = \angle O'RQ'$ and $OR : O'R = OP : O'Q'$. ∵ S' is the internal center of similitude of ⊙s, ∴ $OS' : O'S' = OP : O'Q' = OR : O'R$. ∴ $\angle ORS' = \angle O'RS'$, i.e., RS' bisects $\angle ORO'$. Hence $\angle ORP + \angle ORS'$ = right angle; i.e., $\angle SRS'$ is right. Since SS' is fixed for the two ⊙s, then the locus of R is a ⊙ on SS' as diameter. Notice that this locus ⊙ also cuts SDD', SEE' in F, G in the ratio of the radii.

7.19. *Peaucellier's Cell. OB, OD, AB, AD, BC, CD, EA are rigid, very thin, straight rods, freely jointed at their ends as in the figure. O and E are fixed points. Also,* $OB = OD$, $OE = AE$, *and* $AB = BC = CD = DA$. *Show that as the rods move, C describes a straight line* (*Fig.* 207).

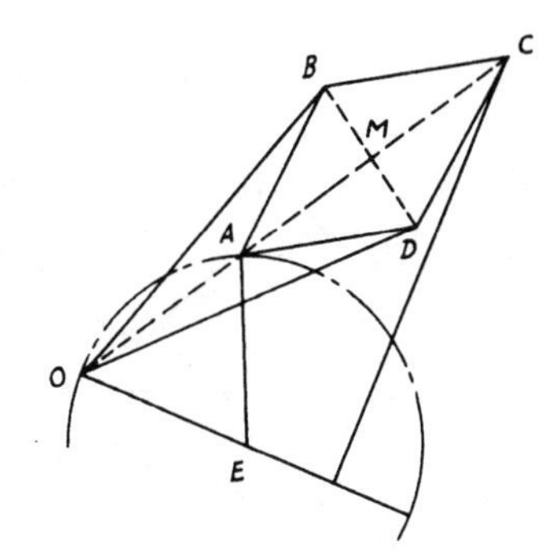

FIGURE 207

Proof: Since $EA = EO$, the locus of A is a ⊙ through O with E as center. Hence it is sufficient to prove that A and C are inverse points with respect to O, i.e., that $OA \cdot OC$ is constant. But $OA \cdot OC = (OM - MA)(OM + MA) = OM^2 - MA^2 = OB^2 - AB^2 =$ constant. Therefore, A, C are inverse points in the inversion whose ⊙ of inversion has O as center and radius equals $\sqrt{(OB^2 - AB^2)}$. Thus if A describes a ⊙, C will describe the inverse of this ⊙ which is a straight line $\perp$ OE from C (Prop. 7.13). This mechanism, known as Peaucellier's cell, is useful in converting circular motion into straight-line motion. If we remove links OE and OA, then as A traverses any curve, C will traverse its inverse curve. This cell, however, is sometimes called an *inverter* and is also used in the design of compound compasses for drawing arcs of circles with large radii.

7.20. *Invert any triangle into a triangle of given shape.*

ANALYSIS: $\because$ $A'B'/B'C' = (k \cdot AB/OA \cdot OB)(OB \cdot OC/k \cdot BC)$ (Prop. 7.12), $\therefore$ $A'B'/B'C' = (AB/BC)(OC/OA)$. $\because$ $A'B/B'C'$ and AB/BC are given from shape of △, $\therefore$ OC/OA is a given ratio. Similarly, OB/OA is also a given ratio. Hence O is either one of intersections of two ⊙s of Apollonius (Th. 5.10) drawn on $M'N'$ (on AC), and MN (on AB) as diameters such that $OC/OA = M'C/M'A = N'C/N'A =$ given and so on (O, O' are two inversion centers).

SYNTHESIS: Draw a ⊙ with AB as chord cutting OA, OB in A', B' such that $OA \cdot OA' = OB \cdot OB' = k$. Draw another ⊙ around $\triangle ACA'$ cutting OC in C' such that $OA \cdot OA' = OC \cdot OC' = k$ (*Fig.* 208). Hence $A'B'C'$ is one of the two inverted △s.

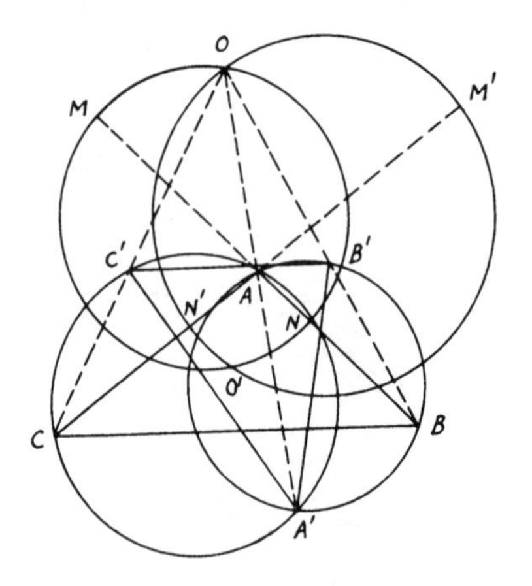

FIGURE 208

Miscellaneous Exercises

1. If PL, PM, PN, the perpendiculars from any point P on the circumference of a circle to the sides BC, CA, AB of an inscribed triangle ABC, be turned through the same angle about P in the same direction and thus cut BC, CA, AB in L', M', N' then L', M', N' are collinear. (They lie on a parallel to the pedal line turned through this angle, for $PL' : PL = PM' : PM = PN' : PN$.)
2. P, Q are diametrically opposite points on a circle circumscribing a triangle. Perpendiculars from P and Q on their pedal lines with respect to the triangle meet at R. Show that R is on the circle.
3. In Problem 7.1, prove that the circles ECB, EDA, FAB, FDC meet in a point (i.e., the point P of Problem 7.1.)
4. If O, A, B, C, D are any concyclic points, show that the projections of O on the pedal lines of O with respect to the triangles BCD, CDA, ABD, ABC are collinear.
5. A, B, C, D are fixed points on a circle on which moves the variable point P. The pedal lines of C and D with respect to ABP meet at Q. Find the locus of Q. (The pedal x of C with respect to ABP passes through a fixed point, i.e., the projection of C on AB, so the pedal y of D passes through a fixed point on AB. Also x, y meet at an angle equal to DAC. Hence the locus of Q is a circle.)
6. A, B, C, D are concyclic points. Show that the angle between the pedal lines of A with respect to BCD and of B with respect to ACD is the angle subtended by A and B at the center of the circle.
7. Given the direction of the pedal line of P with respect to a given triangle, find the position of P.
8. Prove the following construction for the pedal line p of P: Bisect PG, G being the orthocenter, in P' and draw p through P' perpendicular to the line AQ which is the reflection of AP in the internal bisector of BAC.
9. Prove that the pedal lines of three points P, Q, R on the circumference of ABC with respect to ABC form a triangle similar to PQR.
10. Lines drawn through a point P on the circle ABC parallel to BC, CA, AB are turned about P through a given angle in the same direction to cut BC, CA, AB at L, M, N. L', M', N' are formed in the same way from P'. Show that LMN, $L'M'N'$ are perpendicular lines if PP' is a diameter of the circle. (By Exercise 1, LMN, $L'M'N'$ are $\parallel$ to the lines obtained by turning the pedal lines of P, P' through a given angle. Hence the angle between them is right by Problem 7.2.
11. P is a variable point on a given circle. L, M, N are the projections of P on the sides of a fixed inscribed triangle. O_1, O_2, O_3 are the centers of the circles PMN, PNL, PLM. Show that the circle $O_1O_2O_3$ is of fixed size.

12. $ABCD$ is a quadrilateral inscribed in a circle. Show that the pedal lines of A with respect to BCD, of B with respect to CDA, of C with respect to DAB, and of D with respect to ABC meet in one point, and are equal to one another.

13. P is a point on the circumcircle of the triangle ABC and the perpendiculars from P to BC, CA, AB meet the circle again in X, Y, Z. Show that the perpendiculars from A, B, C to the sides of XYZ meet in a point Q on the circle such that the pedal lines of P and Q with respect to ABC and XYZ respectively are coincident.

14. P is any point on the circumscribed circle of a triangle ABC. PL, PM, PN are the perpendiculars from P to BC, CA, AB. If PL, PM, PN are produced to L', M', N' respectively, such that $PL = LL'$, $PM = MM'$ and $PN = NN'$, show that L', M', N' are collinear and pass through the orthocenter of the triangle ABC.

15. Find the locus of a point, given (a) the sum; (b) the difference of the squares on its tangents to two given circles.

16. In the triangle ABC, AD, BE, CF are the altitudes on BC, CA, AB. BC, EF meet at P, CA, FD at Q and AB, DE at R. Show that P, Q, R lie on the radical axis of the circumcircle and the nine-point circle.

17. If A', B', C' bisect BC, CA, AB, show that each bisector of the angles of the triangle $A'B'C'$ is a radical axis of two of the circles touching the sides of ABC. (For a bisector of A' is perpendicular to that bisector of A which passes through the centers of two of the circles and A' has equal tangents to these circles.)

18. A variable circle passes through two fixed points and cuts a fixed circle in the points P and Q. Show that PQ passes through a fixed point.

19. Pairs of points are taken on the sides of a triangle such that each two pairs are concyclic. Show that all six points lie on the same circle. (Otherwise the radical axes form a triangle.)

20. Prove that six radical axes of four circles, taken in pairs, form the six sides of a complete quadrangle. (For they meet, three by three, in four points.)

21. In the triangle ABC, prove that the orthocenter is the radical center of the circles on BC, CA, AB as diameters.

22. The radical axes of a given circle and the circles of a coaxal system are concurrent. (Compare with Exercise 27.)

23. If A, B, C are the centers and a, b, c the radii of three coaxal circles, then $a^2 \cdot BC + b^2 \cdot CA + c^2 \cdot AB = BC \cdot CA \cdot AB$.

24. Through one of the limiting points of a system of coaxal circles is drawn a chord PQ of a fixed circle of the system. Show that the product of the perpendiculars from P and Q on the radical axis is the same for all such chords. (Use Problem 7.6.)

25. If a line cuts one circle at P, P' and another at Q, Q', show that PQ and $P'Q'$ subtend, at either limiting point of their coaxal family, angles which are equal or supplementary. (For, if the line cuts the radical axis at X, XL touches the circles PLP' and QLQ', since $XL^2 = XP \cdot XP' = XQ \cdot XQ'$. Hence in the case in which PP' and QQ' are on the same side of the radical axis, $\angle XLQ = LQ'P$ and $\angle XLP = LP'P$. Hence, $\angle QLP = Q'LP'$. So also for the other case.)

26. If PQ is a common tangent of two circles, then PQ subtends a right angle at L and L', their limiting points. (For $XP = XQ = XL = XL'$.)

27. If two lines AP and AP' are divided at Q, R, . . . and Q', R', . . . so that $PQ : QR : \ldots = P'Q' : Q'R' : \ldots$, then the circles APP', ARR', . . . are coaxal.

28. Three circles 1, 2, 3 are such that the radical axes of 1, 2 and 2, 3 pass respectively through the centers of 3 and 1. Show that the radical axis of 3, 1 passes through the center of 2.

29. If the radical center of three circles is an internal point, show that a circle can be drawn which is cut by each of the three circles at the ends of a diameter of the circle.

30. If from any point in the radical axis of two circles a line be drawn to each circle, the four points of intersection lie on the circumference of a circle.

31. Prove that the orthocenter of a triangle is the radical center of the three circles described on the sides of the triangle as diameters.

32. Give a circle and a straight line MN. Describe a circle with given radius such that MN shall be the radical axis of it and the given circle. (If MN cuts the given circle, the two circles described with given radius through the points of intersection will be the required circles. If MN touches the circle, the two circles described with given radius touching the given circle at the point of contact will be the required circles.)

33. Given a circle, a straight line MN and a point P. Describe a circle passing through P such that MN shall be the radical axis of it and the given circle.

34. A, B, C, D are four points in order in a straight line. Find a point P in that line such that $PA \cdot PB = PC \cdot PD$. (This is the intersection of the radical axis of any two circles through A, B and C, D with the straight line $ABCD$.)

35. Prove that the radical axis of two circles is farther from the center of the larger circle than from the center of the smaller, but nearer its circumference.

36. Prove that if a circle be described with its center on a fixed circle and passing through a fixed point, the perpendicular from the fixed point on the common chord of the two circles will be of constant length.

37. From A two tangents AB, AB' are drawn to two circles. AB, AB' are bisected in D, D' and DE, $D'E$ drawn perpendicular to the lines joining A and the centers of the circles intersect in E. Show that E is on the radical axis of the circles.

38. A is a fixed point on a given circle of a coaxal system whose limiting points are L and M, and P is any point on the tangent at A. Show that $AP^2 - PL^2 : AP^2 - PM^2 = AL^2 : AM^2$.

39. With given radius, describe a circle orthogonal to two given circles, a, b with centers A, B. (Let the center of the required ⊙ of radius x be X; then X lies on the radical axis of the given circles a, b. Let XP be the radius of x which touches a; then $x^2 = XP^2 = XA^2 - a^2 = XO^2 + OA^2 - a^2$. But $OA^2 - a^2$ is known and x is given. Hence OX and X are known.)

40. Two variable circles pass through the fixed points A and B. Through A is drawn the line PAQ to meet one circle at P and the other at Q. Given that the angles ABP and ABQ are equal, find the locus of the intersection of the tangents at P and Q.

41. ABC is a triangle. I, I_1, I_2, I_3 are the centers of its inscribed and escribed circles. α, β, γ, δ are the radical centers of these four circles, δ being that of the escribed circles. Show that the triangle $\alpha\beta\gamma$ is similar to the triangle $I_1I_2I_3$ and that δ, α, β, γ are the centers of the inscribed and escribed circles of the triangles formed by joining the middle points of the sides of ABC.

42. Construct a circle to pass through two given points A, B so that its chord of intersection with a given circle shall pass through a given point C. (On the given circle take any point P and let the ⊙ABP cut the given circle again at Q. Let PQ, AB meet at R, and let RC cut the given circle at X, Y. Then ABX is the required circle.)

43. If tangents from a variable point to two given circles have a given ratio, the locus of the point is a circle coaxal with the given circles. (Through any position P of the point draw a circle z with center Z coaxal with the given circles x, y (centers X, Y). Drop $PN \perp$ the radical axis of x and y. Draw the tangents PQ and PR to x and y. Then $PQ^2 = 2\,PN \cdot XZ$ and $PR^2 = 2\,PN \cdot YZ$. But $PQ : PR$ is constant. Hence $XZ : YZ$ is constant. Hence Z is fixed. Therefore the coaxal through P is fixed; i.e., the locus of P is a circle coaxal with x and y.)

44. A straight line meets two circles in four points. Show that the tangents at these points intersect in four points, two of which lie on a circle coaxal with the two given circles. (Use Exercise 43.)

45. A, B are points on two circles ACD, BEF. It is required to find on the radical axis of these circles a point P such that if the secants PAC, PBE be drawn, CE will be perpendicular to the radical axis.

46. From a point P outside a given circle two straight lines PAB, PCD are drawn, making equal angles with the diameter through P and cutting

the circle in A, B and C, D respectively. Prove that AD, BC intersect in a fixed point. [Use Prop. 7.9(ii).]

47. Two fixed circles intersect and their common tangents intersect in F. Prove that all circles which touch these given circles are intersected orthogonally by a circle of which F is the center. (Use Problem 7.14.)

48. Prove that the poles of the four rays of a harmonic pencil with respect to a given circle are collinear and form a harmonic range.

49. Prove that the polars of any point on a circle which cuts three circles orthogonally with respect to these circles are concurrent.

50. If A, B, C, D form a harmonic range, their polars a, b, c, d with respect to a given circle form a harmonic pencil. [For if AB is l, a is the line through L, the pole of AB, $\perp OA$. Hence, (a, b, c, d) is superposable to $O(ABCD)$, O being the center of the circle.]

51. A circle touches the sides BC, CA, AB of a triangle at P, Q, R and RQ cuts BC at S. Show that $(BPCS)$ is harmonic. (For the polar of S passes through P and also through A, since S is on the polar of A.)

52. If P, Q are conjugate points with respect to a circle (see Prop. 7.7, Cor. 1), show that the circle on PQ as diameter is orthogonal to the given circle.

53. If P, Q are conjugate points with respect to a circle and t_1 and t_2 are the tangents from P, Q to the circle, show that the circles with centers P, Q and radii t_1, t_2 are orthogonal. (For $PQ^2 = P'P^2 + P'Q^2 = t_1^2 - P'R^2 + P'Q^2 = t_1^2 + QR \cdot QR' = t_1^2 + t_2^2$, where $RP'R'$ is the chord of contact of P.)

54. P, Q are conjugate points of a circle. U is the projection of the center of the circle on PQ. Show that $PU \cdot UQ$ is equal to the square of the tangent to the circle from U.

55. The chord PQ is the polar of R with respect to a circle. S bisects any chord through R. Show that SR bisects the angle PSQ.

56. Two circles cut each other orthogonally. Show that the distances of any point from their centers have the same ratio as the distances of the centers each from the polar of the point with respect to the circle of which the other is the center.

57. The line joining any two of the six centers of similitude of three circles taken in pairs passes through a third center of similitude.

58. The internal and external centers of similitude of the circumcircle and the nine-point circle of a triangle are the centroid and the orthocenter.

59. Given two circles centers A, B and their circle of similitude described on the line joining their centers of similitude as diameter. Prove that if P is any point on the circle of similitude, then $PA : PB$ is constant.

60. The circles of similitude of three circles taken in pairs are coaxal. (Let the centers be A, B, C and radii a, b, c. Then if P is a point on the circles of similitude of the circles a, b and b, c, then $PA : PB = a : b$

and $PB : PC = b : c$ $\therefore$ $PA : PC = a : c$. Hence P is on the other circle of similitude.)

61. PQ, $P'Q'$ are parallel chords of two circles the lengths of which are in the ratio of the radii. Show that PP', QQ' or PQ', $P'Q$ meet at a center of similitude.

62. A common tangent to the circles the centers of which are A and B touches the circles at P and Q. Show that PB and QA meet at a point which bisects the perpendicular from one of the centers of similitude to PQ.

63. If the line joining the centers of two circles cuts the circles at B, C and B', C', show that the squares of the common tangents are equal to $BB' \cdot CC'$ and $BC' \cdot B'C$. [Draw $\perp$ from center of smaller circles to radius of large circles at point of contact of a common tangent of length t. If d, a, b are the distances between the centers and the radii, then $t^2 = d^2 - (a \pm b)^2$.]

64. t_1 and t_2 are the common tangents of two circles. Show that $t_1{}^2 - t_2{}^2 = d_1 d_2$, where d_1 and d_2 are the diameters. (Use Exercise 63.)

65. The circles of similitude of three circles taken in pairs cut orthogonally the circle through the centers of the given circles.

66. AA', BB', CC' are the pairs of opposite vertices of a complete quadrilateral circumscribing a circle c, A', B', C' being collinear. Show that half of the centers of similitude of the four circles ABC, $AB'C'$, $A'BC'$, $A'B'C$, taken in pairs, lie on a line which bisects at right angles the line joining the center of c to the point through which the four circles pass.

67. Prove Problem 7.17 using inversion. (Invert with respect to S taking X, Y as inverse points. $\because$ $SX' \cdot SY' = SX \cdot SY = k$, $\therefore$ X', Y' are also inverse points, and $\odot$s M, M' invert into themselves. Also $\odot O$ inverts into $\odot O'$. $\therefore$ S is center of inversion and center of similitude of circles O, O'.)

68. Two circles intersect at A and touch one line at P, Q and another line at R, S. Show that the circles PAQ and RAS touch at A. (Inverting with respect to A, we get the parallel lines $P'Q'$, $R'S'$.)

69. The inverse C' of the center C of a circle is the inverse point of O with respect to the inverse circle. (See Prop. 7.14.)

70. Two circles are drawn to touch each of two given circles and also one another. Show that the locus of the point of contact of these two circles consists of the given circles when these do not intersect and the two coaxal circles which bisect the angles between the given circles when these intersect.

71. If tangents are drawn at A, B, C to the circumcircle forming the triangle LMN, then the circumcircle of LMN belongs to the above system of coaxals. (Let OL cut BC in A'. Then A' bisects BC and is the inverse of L with respect to the circumcircle.)

72. A circle orthogonal to two given circles cuts them in A', B' and C', D' and the line of centers cuts them in A, B and C, D. Show that the lines AA', BB', CC', DD' are concurrent for two locations of the points A, B, C, D, A', B', C', D'. (Invert the circles into themselves with respect to either intersection of the orthogonal circle and the radical axis.)

73. If three circles c_1, c_2, c_3 are such that c_3 is the inverse of c_1 with respect to c_2 and c_1 of c_2 with respect to c_3, then c_2 is the inverse of c_3 with respect to c_1.

74. A line is drawn through the fixed point O to touch two circles of a given coaxal system at P and Q. Find the locus of R, given that (OR, PQ) is harmonic.

75. Two fixed circles touch internally at A. Show that the locus of the inverse point of A, with respect to a variable circle touching the given circles, is a circle whose radius is the harmonic mean between the radii of the given circles.

CHAPTER 8

SPACE GEOMETRY

Theorems and Corollaries

8.1. *One part of a straight line cannot be in a plane and another part without.*

8.2. *Two straight lines which intersect are in one plane and three straight lines each of which cuts the other two are in the same plane.*

COROLLARY 1. If three straight lines not in one plane intersect, two and two, they intersect in the same point.

COROLLARY 2. Similarly if four straight lines, no three of which are in the same plane, are such that each meets two of the others, they all meet in the same point.

8.3. *If two planes intersect, their common section is a straight line.*

8.4. *If a straight line be perpendicular to each of two intersecting straight lines at their point of intersection, it is perpendicular to the plane in which they lie.*

8.5. *If three straight lines meet in a point and a straight line be perpendicular to each of them at that point, the three lines are in one plane.*

8.6. *Two straight lines which are perpendicular to the same plane are parallel.*

8.7. *If two straight lines are parallel, the straight line joining any point in the one to any point in the other is in the same plane with the parallels.*

8.8. *If two straight lines be parallel and one of them be perpendicular to a plane, the other is perpendicular to that plane.*

8.9. *Straight lines which are parallel to the same straight line, even though not in the same plane with it, are parallel to each other.*

8.10. *If two intersecting straight lines be respectively parallel to two other intersecting straight lines, though not in the same plane with them, the first two and the second two contain equal angles.*

8.11. *It is possible to draw a straight line perpendicular to a given plane from a given point outside the plane.*

8.12. *It is also possible to draw a straight line perpendicular to a given plane, from a given point in it.*

8.13. *Only one perpendicular can be drawn to a given plane from a given point, whether the point be in or outside the plane.*

8.14. *Planes to which the same straight line is perpendicular are parallel to each other.*

8.15. *If two intersecting straight lines in one plane be respectively parallel to two intersecting straight lines in another plane, the planes are parallel.*

8.16. *If two parallel planes be cut by any plane, their common sections with it are parallel.*

8.17. *If two straight lines be cut by parallel planes, they are cut in the same ratio.*

8.18. *If a straight line be perpendicular to a plane, every plane which passes through it is perpendicular to that plane.*

8.19. *If two planes which intersect be each of them perpendicular to a plane, their common section is perpendicular to that plane.*

8.20. *If a solid angle be contained by three plane angles, any two of them are together greater than the third.*

8.21. *Every solid angle is contained by plane angles which are together less than four right angles.*

COROLLARY: There can be only five regular polyhedra.

(i) Three faces at least must meet to form each solid angle of a regular polyhedron.

(ii) The sum of the plane angles forming each of the solid angles is less than four right angles.

Now three angles of a regular hexagon are together equal to four right angles and three angles of any regular polygon of a greater number of sides are together greater than four right angles. Hence the faces of a regular polyhedron must be either equilateral triangles, squares, or regular pentagons.

8.21.1. *If the faces are equilateral triangles, each solid angle of the polyhedron may be formed by: (i) Three equilateral triangles. The solid thus formed is a tetrahedron. (ii) Four equilateral triangles. The solid thus formed is an octahedron. (iii) Five equilateral triangles. The solid thus formed is an icosahedron. The angles of six equilateral triangles are together equal to four right angles and therefore cannot form a solid angle.*

8.21.2. *If the faces are squares, each solid angle will be formed by three squares. The solid thus formed is a cube. The angles of four squares are together equal to four right angles and cannot form a solid angle.*

8.21.3. *Similarly, if the faces are regular pentagons, each solid angle will be formed by three such pentagons. The solid thus formed is a dodecahedron.*

8.22. *The section of a sphere by a plane is a circle.*

8.23. *The curve of intersection of two spheres is a circle.*

8.24. *The properties of the radical plane, coaxal spheres, and the centers of similitude of two spheres follow by natural generalization from the corresponding concepts with circles in a plane.*

8.25. *If the radius of a sphere is r, then the surface area and volume of the sphere are $4\pi r^2$ and $\frac{4}{3}\pi r^3$ respectively.*

8.26. *Only one sphere can be drawn through four points which do not lie in a plane and no three of which lie on a line.*

8.27. *Eight spheres can, in general, be drawn to touch the faces of a tetrahedron.*

8.28. *If the vertices A, B, C, D of a tetrahedron are joined to the centroids A', B' ,C', D' of the opposite faces, the lines AA', BB', CC', DD' meet at a point G called the centroid of the tetrahedron such that $AG = 3\ GA'$, and so on.*

8.29. *If the radius of the base circle of a cone is r and its generator and altitude from the vertex to the base circle are l and h, then the side surface area and volume of cone are πrl and $(\pi/3)r^2h$ respectively.*

8.30. *Any surface area and volume of revolution of a plane figure are equal to the perimeter and area of that figure times the path of the center of area of the figure respectively: Pappus' theorem.*

Solved Problems

8.1. *AD and BC are two perpendiculars from the points A, B on a given plane. If a plane through A is drawn perpendicular to AB to cut the given plane in EF, show that CD is ⊥ EF.*

CONSTRUCTION: Let *DC* and *EF* in the given plane intersect in *G*. Join *AG*, *AE* (*Fig.* 209).

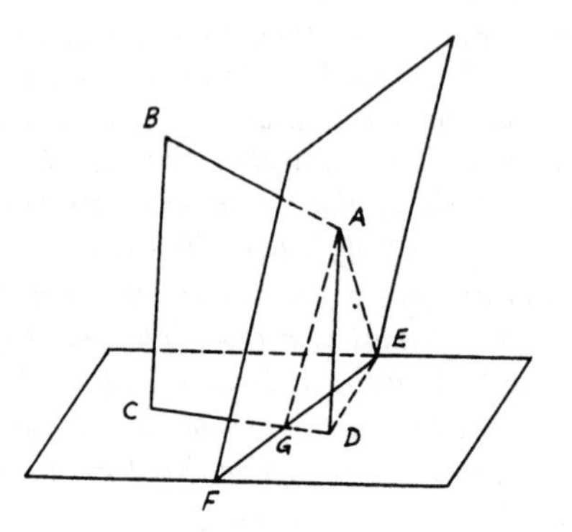

FIGURE 209

Proof: Since AD, BC are $\perp$ given plane, $\therefore$ they are $\parallel$ (Th. 8.6). Hence A, B, C, D lie in one plane (Th. 8.7). $\because$ AB is $\perp$ plane AGE, $\therefore$ plane $ABCD$ is $\perp$ plane AGE (Th. 8.18) or plane AGD is $\perp$ plane AGE. But since AD is $\perp$ plane GDE, $\therefore$ plane AGD is $\perp$ plane GDE (Th. 8.18). Therefore, plane AGD is $\perp$ both planes AGE and GDE. $\therefore$ GE is $\perp$ plane AGD (Th. 8.19). $\therefore$ GE is $\perp$ any line in plane AGD or GE is $\perp$ CD; i.e., CD is $\perp$ EF.

8.2. *CEDF is a plane and A is a point outside it. AC is perpendicular to the plane and CD is perpendicular to EF in the plane. Show that AD is $\perp$ EF.*

CONSTRUCTION: Join AE, CE (*Fig.* 210).

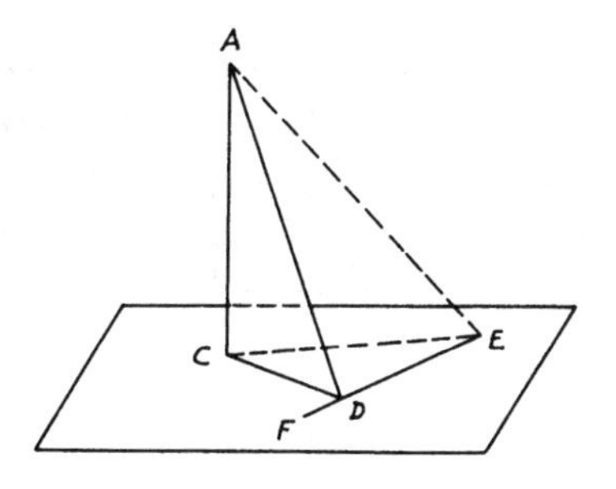

FIGURE 210

Proof: Since AC is $\perp$ plane $CEDF$, $\therefore$ AC is $\perp$ CD and CE (converse, Th. 8.4). Hence $AD^2 = AC^2 + CD^2$ and $AE^2 = AC^2 + CE^2$. $\therefore$ Subtraction gives $AE^2 - AD^2 = CE^2 - CD^2$. But $\because$ CD is $\perp$ EF, $\therefore CE^2 - CD^2 = DE^2$. $\therefore AE^2 - AD^2 = DE^2$. $\therefore$ $\angle ADE$ is right or AD is $\perp$ EF.

8.3. *Draw a straight line perpendicular to each of two straight lines not in the same plane. Prove that this common perpendicular is the shortest distance between the lines.*

CONSTRUCTION: (i) Let AB, CD be the given straight lines. Through any point A in AB draw $AE \parallel CD$. Draw $DF \perp$ plane ABE and $FG \parallel AE$ meeting AB in G. Make DC equal to FG and join CG (*Fig.* 211).

Proof: $\because$ CD, FG are each $\parallel$ AE, $\therefore$ CD is $\parallel$ FG and $CD = FG$ (construct). $\therefore$ CG is $\parallel$ DF and DF is $\perp$ plane ABE. $\therefore$ CG is $\perp$ plane ABE. $\therefore$ CG is $\perp$ AB and GF and GF is $\parallel$ CD. $\therefore$ CG is $\perp$ both AD and CD.

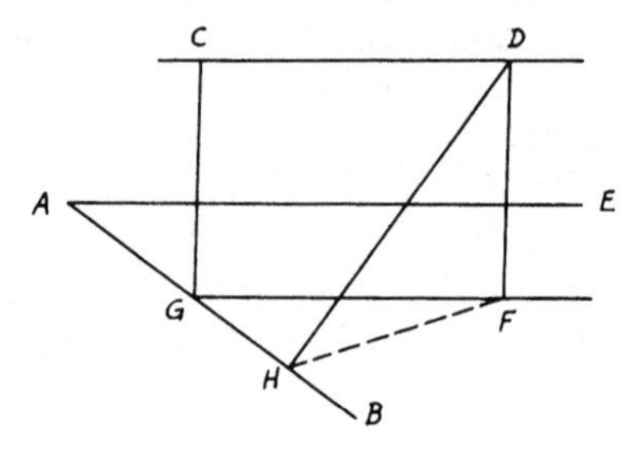

FIGURE 211

(ii) Draw any other straight line DH between AB and CD. Since DF is $\perp$ plane ABE, $\therefore$ DF is $\perp$ FH. $\therefore$ DH is $> DF$, i.e., CG.

8.4. *Two parallel planes are cut by a straight line ABCD in B, C such that AB = CD. Two more transversals are drawn from A and D to cut the planes with B and C in E, F and G, H. Show that BEG and CFH are two equal triangles* (*Fig.* 212).

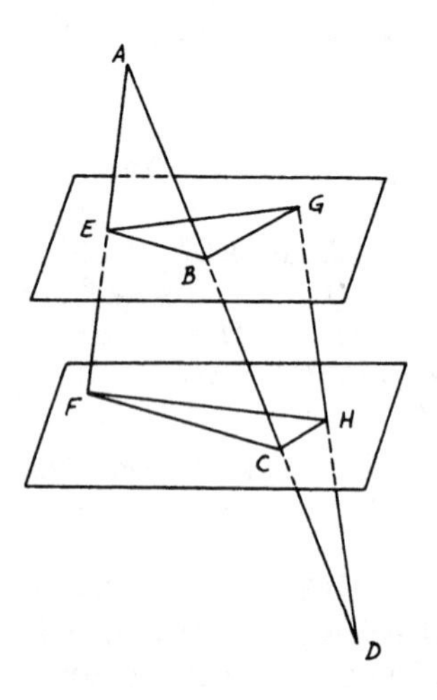

FIGURE 212

Proof: Plane ACF cuts the two $\|$ planes. $\therefore$ BE is $\|$ CF (Th. 8.16). Similarly, BG is $\|$ CH. Hence $\angle EBG = \angle FCH$. Again, $AB/AC = BE/CF = CD/BD = CH/BG$. $\therefore$ $BE/CF = CH/BG$. Therefore, the

$\triangle$s EBG, FCH have both $\angle$s EBG, FCH equal and the sides about these equal $\angle$s inversely proportional. $\therefore$ They are equal (Th. 4.94).

8.5. *DABC is a tetrahedron with D as vertex and ABC as base. E, F, G are the middle points of BC, CA, AB respectively. If DF, DG are perpendiculars to AC, AB and BAC is a right angle, show that DE is perpendicular to the base ABC.*

CONSTRUCTION: Join AE, EG (*Fig.* 213).

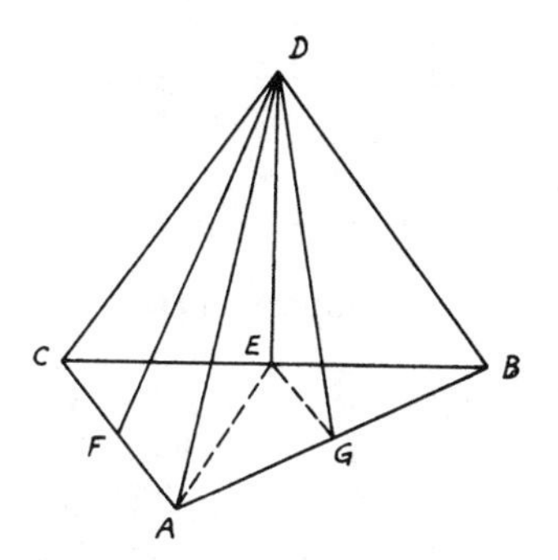

FIGURE 213

Proof: Since F, G are the mid-points of AC, AB and DF, DG are $\perp$s AC, AB, then $\triangle$s ADC, ADB are isosceles. Hence the edges DA, DB, DC of the tetrahedron are equal. $\because$ $\angle CAB$ = right angle and E is mid-point of CB, $\therefore$ $AE = EB$. Therefore, $\triangle$s DEB, DEA are congruent. $\therefore$ $\angle DEB = \angle DEA$ = right angle. Hence DE is $\perp$ the base plane ABC (Th. 8.4).

8.6. *ABCD is a square and a is the length of its side. From A, C two perpendiculars AE, CF are drawn to the plane ABCD such that AE = AC and CF = $\frac{3}{2}$ AC. Find the length of BE, EF, BF in terms of a. Then show that the triangle BEF is right-angled at E and that FE is perpendicular to the plane BDE.*

CONSTRUCTION: Join AC, FD and draw $EG \perp FC$ (*Fig.* 214).

Proof: Since EA is $\perp$ plane $ABCD$, $\therefore$ EA is $\perp AB$. $\therefore$ $BE^2 = AE^2 + AB^2$. But $AE^2 = AC^2 = 2a^2$. $\therefore$ $BE^2 = 3a^2$. Hence $BE = \sqrt{3}\,a$. Again, CF is $\perp$ plane $ABCD$, i.e., $\perp$ CA and $\because$ EG is $\perp$ FC, $\therefore$ $EGCA$ is a square. $\therefore$ EG = and $\parallel$ AC. $\because$ $CF = \frac{3}{2}AC$, $\therefore$ $GF = \frac{1}{2}AC$. $\therefore$ $EF^2 = FG^2 + GE^2 = \frac{1}{4}AC^2 + AC^2 = \frac{5}{4}AC^2 = \frac{5}{2}a^2$. $\therefore$ $EF = \sqrt{\frac{5}{2}}\,a$. Similarly, FC is $\perp BC$. $\therefore$ $BF^2 = \frac{9}{4}\cdot 2a^2 + a^2 = \frac{11}{2}a^2$. $\therefore$ $BF = \sqrt{\frac{11}{2}}\,a$. Again, $\because$ $BE^2 + EF^2 = 3a^2 + \frac{5}{2}a^2 = \frac{11}{2}a^2 = BF^2$.

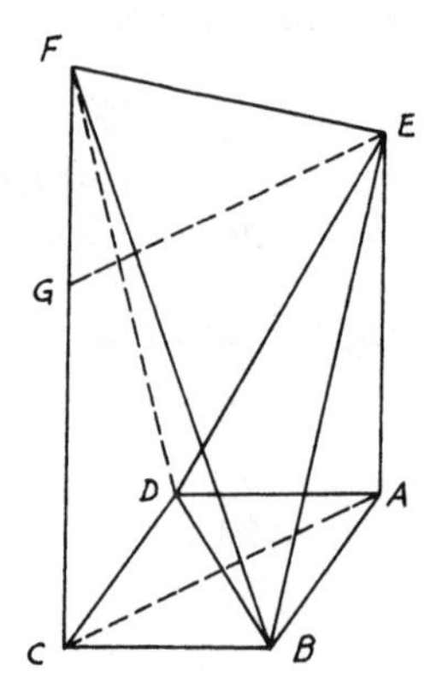

FIGURE 214

Therefore, $\triangle BEF$ is right-angled at E. Also, $\triangle$s EAB, EAD and FCB, FCD are congruent. $\therefore$ $EB = ED$ and $FB = FD$. $\therefore$ $\triangle$s EBF, EDF are both congruent. $\therefore$ $\angle BEF = \angle DEF =$ right. $\therefore$ $FE \perp ED$ and also $\perp EB$. $\therefore$ $FE \perp$ plane BDE (Th. 8.4).

8.7. *ABCD is a tetrahedron in which the angles ABC, ADC are right. If M, N are the middle points of BD and AC respectively and AM = MC, show that MN is the shortest distance between AC and BD.*

CONSTRUCTION: Join NB, ND (*Fig.* 215).

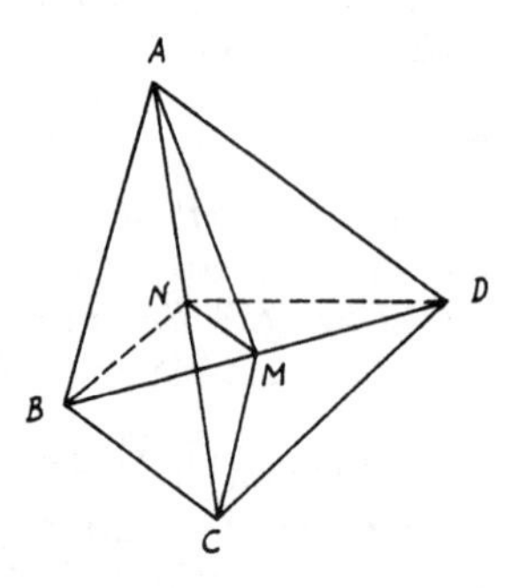

FIGURE 215

Proof: Since $AM = MC$ (hypothesis), $\therefore$ $\triangle$s AMN, CMN are congruent. $\therefore$ $\angle ANM = \angle CNM =$ right angle or MN is $\perp AC$. Again, since $\angle ABC = \angle ADC =$ right angle and N is the mid-point of AC (hypothesis), $\therefore BN = DN = \frac{1}{2} AC$. But M is the mid-point of BD. $\therefore \triangle$s BNM, DNM are congruent. $\therefore \angle NMB = \angle NMD$ = right angle. Hence MN is also $\perp BD$. $\therefore$ It is the shortest distance between AC, BD.

8.8. *From D, a fixed point outside plane P, a straight line DE is drawn to cut the plane in E. F is a point taken in DE such that $DE \cdot DF$ is constant. Find the locus of F.*

CONSTRUCTION: Draw $DA \perp$ plane P. Join EA. In the plane DAE, draw $FB \perp DE$ to meet DA in B. The locus of F is a sphere on DB as diameter (*Fig.* 216).

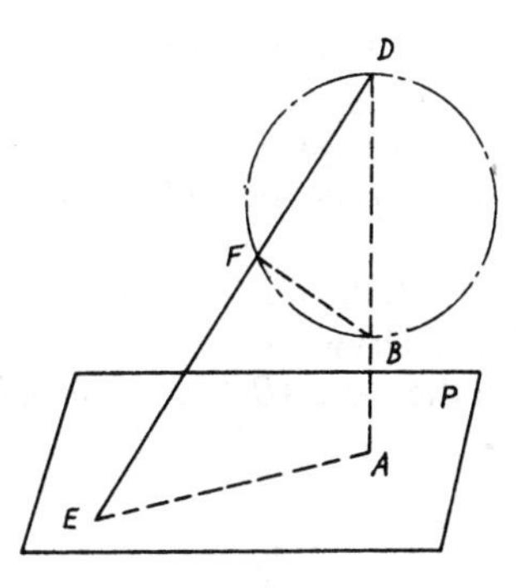

FIGURE 216

Proof: $\because \angle DAE = \angle DFB$ = right angle, $\therefore BAEF$ is cyclic. Hence $DE \cdot DF = DB \cdot DA$ = constant. Since D is a fixed point and P is a given plane, $\therefore$ the perpendicular DA is fixed. $\therefore DB$ is fixed in position and magnitude. But $\because \angle DFB$ = right, the locus of F is a sphere on DB as diameter.

8.9. *ABCD is a parallelogram of plane paper in which $AD = 2\,AB = 20$ inches and the $\angle A = 60°$. E, F, G are the middle points of AD, BC, DC respectively and AF, BE intersect in M; paper is folded about BE so that the plane ABE is perpendicular to the plane BCDE* (*Fig.* 217). *Find the area of $\triangle AMG$.*

Proof: Before paper is folded, $ABFE$ is a rhombus and AF is $\perp BE$ at M; i.e., AM is $\perp$ to the line of intersection of the perpendicular planes. Hence AM is $\perp$ the plane $BCDE$. $\therefore AM$ is $\perp MG$. $\therefore \angle AMG$ = right angle. Since $\triangle ABE$ is isosceles and $\angle A = 60°$, $\therefore \triangle ABE$ is equilateral. $\therefore AM = \sqrt{10^2 - 5^2} = 5\sqrt{3}$ inches. But $MG = \frac{1}{2}(DE + BC) = 15$ inches. $\therefore \triangle AMG = \frac{1}{2}(5\sqrt{3} \cdot 15) = 37.5\sqrt{3}$ inches^2.

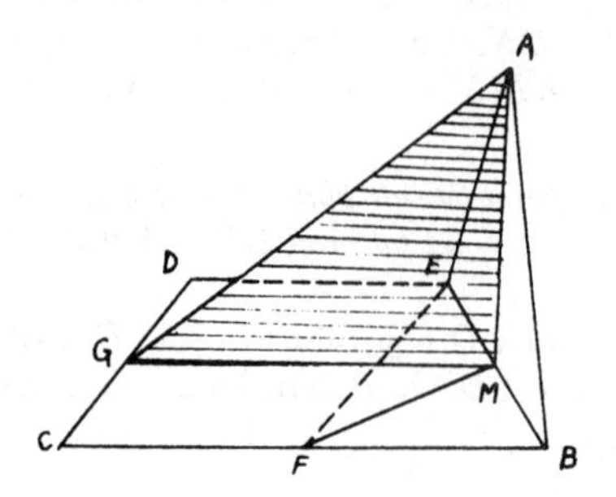

FIGURE 217

8.10. *Find the total surface and volume of a regular tetrahedron and if these are $\sqrt{2}$ ft^2 and $2\sqrt{2}/3$ ft^3 respectively, find the corresponding lengths of the side.*

CONSTRUCTION: Let $SABC$ be any regular tetrahedron having S as vertex and ABC as base. Draw $SH \perp CB$ in the face $\triangle CSB$. Draw also SD the altitude on the base ABC, then join CD and produce it to E on AB (*Fig.* 218).

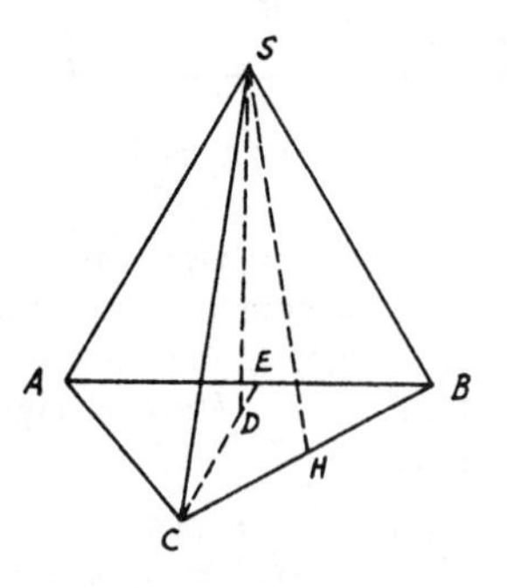

FIGURE 218

Proof: Total surface of tetrahedron $= 2\,BC \cdot SH = 2\,BC(\sqrt{3}/2\ BC) = \sqrt{3}\,BC^2$. If this surface area $= \sqrt{2}$ ft^2, $\therefore\ BC = \sqrt[4]{2}/3$ ft. Volume of tetrahedron $= \frac{1}{3}\ \triangle ABC \times SD$. Similar to Th. 8.29, $\because$ all sides are equal and CE is $\perp AB$, $\therefore\ \triangle ABC = \frac{1}{2}\,AB \cdot CE = a^2\sqrt{3}/4$ (a being the side of tetrahedron). But D is the centroid of $\triangle ABC$. $\therefore$ $CD = \frac{2}{3}CE = \frac{2}{3}\,a\sqrt{3}/2 = a/\sqrt{3}$. Again, $SD^2 = SC^2 - CD^2 = \frac{2}{3}\,a^2$. $\therefore$ $SD = a\sqrt{2}/3$. Hence volume of tetrahedron $= \frac{1}{3}\ a^2\sqrt{3}/4\ \ a\sqrt{2}/3$ $= a\sqrt{2}/12$. If this volume $= 2\sqrt{2}/3$ ft^2, $a = 2$ ft.

8.11. *ABC is a right-angled triangle at A. AD is drawn perpendicular to the hypotenuse BC. The triangle ADC is folded around AD to ADC′ such that the angle between the planes C′AD and BAD is* 60°. *Show that the volume of the tetrahedron* $C'BAD = \sqrt{3}/12$ *the volume of the cube with AD as side.*

CONSTRUCTION: Let C be C' in the folded triangle $C'AD$. Draw $C'E \perp$ plane ABD, to meet BD in E (*Fig.* 219).

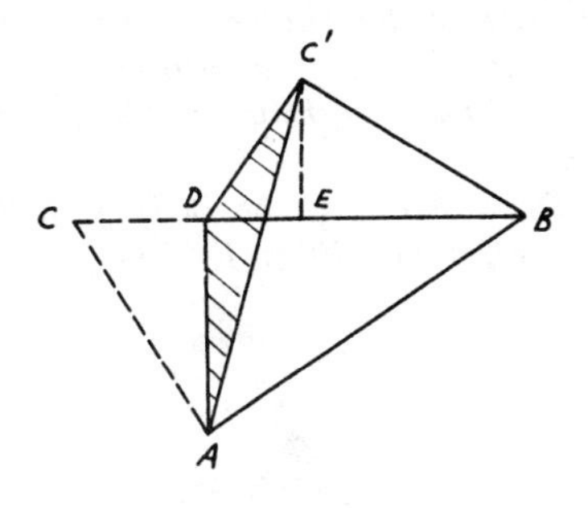

FIGURE 219

Proof: Before folding $\triangle ADC, AD^2 = CD \cdot DB$. After folding, $\because C'E$ is $\perp$ plane ABD, $\therefore C'E$ is $\perp DE$. $\because \angle ADB =$ right angle, $\therefore E$ is on DB. $\because$ The angle between planes $C'AD$, $ABD = 60°$, $\therefore \angle C'DE = 60°$. $\therefore C'E = CD \sin 60 = \sqrt{3}/2\, CD$.

Volume of terahedron $C'BAD = \frac{1}{3} \triangle DAB \times C'E = \frac{1}{3} (DB \cdot AD/2) (\sqrt{3}/2)\, CD = \sqrt{3}/12\, AD^3$.

8.12. *AB, BC, CD are three sides of a cube the diagonal of which is AD. Show that the angle between the planes ABD, ACD is* $\frac{2}{3}$ *of a right angle.*

CONSTRUCTION: Draw BE, BF $\perp$s to AD, AC. Join EF, BD (*Fig.* 220).

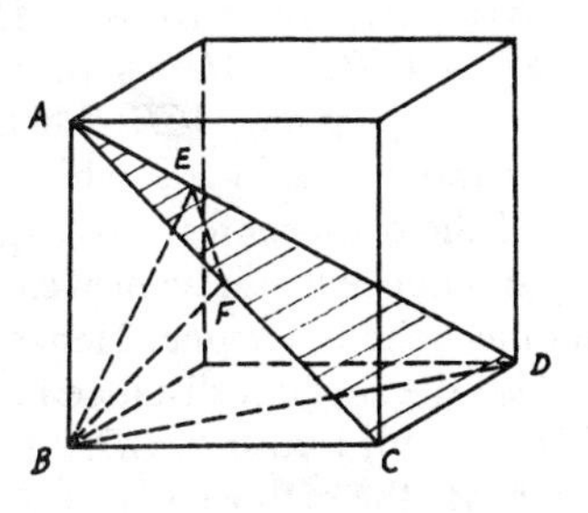

FIGURE 220

Proof: $\because$ DC is $\perp$ CB, $\therefore$ plane ADC is $\perp$ plane ABC. $\therefore$ BF is $\perp$ plane ADC. $\therefore$ BF is $\perp FE$ (Th. 8.4). Let a be the side of the cube. $\therefore$ AB, BD, AD are equal to a, $\sqrt{2}\,a$, $\sqrt{3}\,a$ respectively. $\because$ AB is $\perp$ BD, $\therefore$ $BE\cdot AD = AB\cdot BD = a^2\sqrt{2}$. $\therefore$ $BE = a\sqrt{2/3}$. Again, $BF = a/\sqrt{2}$. $\therefore$ $\sin FEB = BF/BE = (a/\sqrt{2})(\sqrt{3/2})(1/a) = \sqrt{3}/2$. $\therefore$ $\angle FEB = 60°$, which is the angle between the planes ABD, ACD.

8.13. *If in a tetrahedron ABCD the directions of AB, CD be at right angles and also those of AC, BD, so also will the directions of BC, AD be at right angles. Also the sum of the squares of each pair of opposite edges is the same and the four altitudes and the three shortest distances between opposite edges meet in the same point.*

CONSTRUCTION: Draw AO $\perp$ plane BCD. Join BO, CO, DO and produce them to meet CD, DB, BC in E, F, G. Join AE, AF, AG (*Fig.* 221).

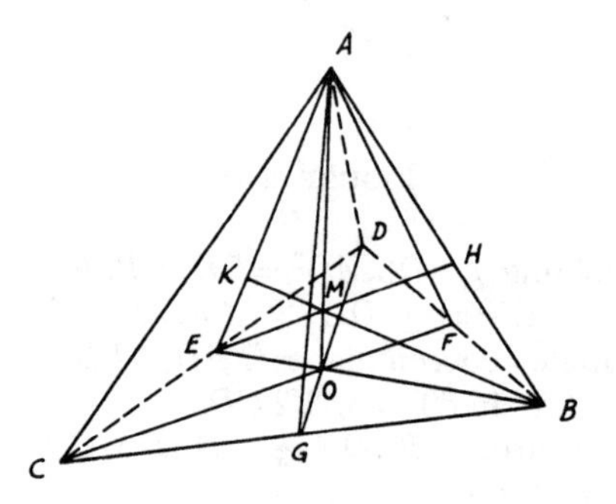

FIGURE 221

Proof: $\because$ BD is $\perp$ AC and AO in the plane ACF, $\therefore$ BD is $\perp$ AF, CF. Similarly, CD is $\perp$ plane ABE and $\perp$ AE, EB in that plane. $\therefore$ O is the orthocenter of $\triangle BCD$. $\therefore$ CB is $\perp$ DG and CB is also $\perp$ AO. $\therefore$ CB is $\perp$ plane AGD. $\therefore$ CB is $\perp$ AD. Since BD is $\perp$ AF, CF, $\therefore$ $AD^2 - AB^2 = DF^2 - BF^2 = CD^2 - CB^2$. $\therefore$ $AD^2 + BC^2 = AB^2 + CD^2$; also $= AC^2 + DB^2$. Draw BK $\perp$ plane ACD. Then, as before, K is the orthocenter of $\triangle ACD$ and $\therefore$ lies in AE, which is $\perp$ CD, as proved. $\therefore$ BK meets AO in M the orthocenter of $\triangle AEB$. Similarly, the $\perp$s from C, D meet AO, and hence it can be proven that any three of the four perpendiculars meet the other one. Hence they all intersect in the same point M. Also the shortest distance between CD, AB is the $\perp$ EH from E on AB, for this is also $\perp$ CD. $\because$ CD is $\perp$ plane AEB, $\therefore$ EH passes through M, the orthocenter of $\triangle AEB$. Similarly, the other shortest distances pass through M.

8.14. *M is the vertex of a pyramid with the parallelogram $ABCD$ as its base. In the face MBC, a line EF is drawn parallel to BC and cuts MB, MC in E, F. Show that AE, DF if produced will meet in a point the locus of which is a straight line parallel to the base $ABCD$ (Fig. 222).*

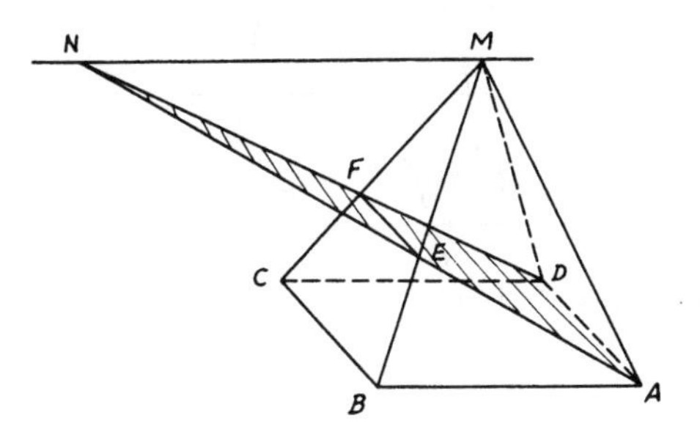

FIGURE 222

Proof: EF is $\parallel CB$ (hypothesis) and $ABCD$ is a $\square$. $\therefore$ EF is $\parallel AD$. $\because$ EF is $< BC$, $\therefore$ also $< AD$. Hence EF and AD form a plane of a trapezoid $AEFD$. $\therefore$ If AE, DF are produced they will meet in a point N. In $\triangle AND$, $NF/ND = FE/DA$. $\therefore$ $NF/ND = FE/CB = MF/MC$. $\therefore$ $NF/FD = MF/FC$. $\because$ $\angle MFN = \angle CFD$, $\triangle$s MFN, CFD are similar and MN is $\parallel CD$ and hence $\parallel$ to its plane $ABCD$. Since M is a fixed point and MN is $\parallel$ a fixed plane $ABCD$, $\therefore$ the locus of N is a straight line through the vertex M and $\parallel$ plane $ABCD$.

8.15. *A straight line AD cuts two parallel planes X, Y in B, C respectively such that $AB : BC : CD = p : q : r$. From A, D two other straight lines are drawn cutting planes X, Y in E, F and P, Q respectively. Show that $(rp + pq)CF \cdot CQ = (rp + rq)BE \cdot BP$ (Fig. 223).*

Proof: $AB/BC = p/q$. $\therefore$ $AB/AC = p(p + q)$. $\because$ Planes X, Y are parallel, $\therefore$ BE is $\parallel CF$ (Th. 8.16). $\therefore$ $AB/AC = BE/CF = p(p + q)$. $\therefore$ $CF = ((p + q)/p)BE$. Similarly, BP is $\parallel CQ$ and $DC/DB = r(r + q) = CQ/BP$. $\therefore$ $CQ = (r(r + q))BP$. Hence

$$CF \cdot CQ = \left(\frac{p + q}{p}\right)\left(\frac{r}{r + q}\right)BE \cdot BP$$

$$= \left(\frac{rp + rq}{rp + pq}\right)BE \cdot BP.$$

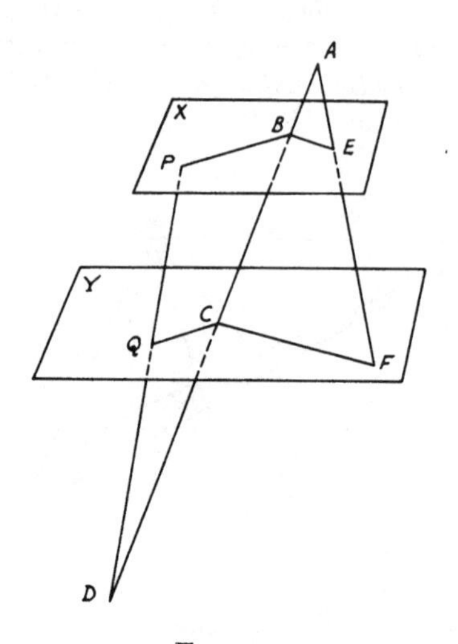

FIGURE 223

8.16. *A, B, C, D are four points in space. The straight lines AB, DC; EF, GH are divided by points E, F; G, H such that $AE{:}BE = DF{:}CF$; $AG{:}DG = BH{:}CH$. Prove that the straight lines EF, GH lie in one plane.*

CONSTRUCTION: From E draw the line $C'ED' \parallel CD$ and from C, D the lines CC', $DD' \parallel EF$. Divide EF in M such that $EM : MF = AG : GD = BH : CH$. From G, H draw GG', $HH' \parallel EF$ also meeting AD', BC' in G', H'. Join $G'E$, $H'E$, GM, HM (*Fig.* 224).

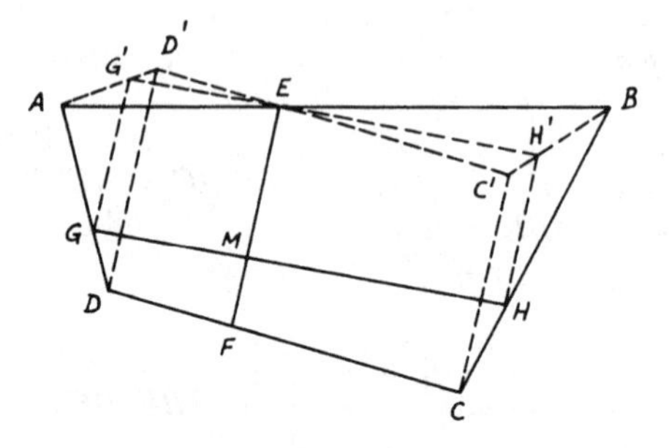

FIGURE 224

Proof: $\because EM/MF = AG/GD$, $\therefore EF/EM = AD/AG = DD'/GG'$. But $EF = DD'$. $\therefore EM = GG'$. $\because$ They are $\parallel$ lines, $\therefore EG'$ is $\parallel MG$.

Similarly, EH' is $=$ and $\parallel MH$. $\because DD', EF, CC'$ are $\parallel$, $\therefore ED'/EC' = DF/CF = AE/BE$ and $\angle D'EA = \angle BEC'$. $\therefore$ $\triangle$s $D'EA$, $C'EB$ are similar. $\therefore \angle D'AE = \angle C'BE$ and $AD'/BC' = AE/EB$. $\because AG'/AD' = AG/AD = BH/BC = BH'/BC'$, $\therefore AG'/BH' = AD'/BC' = AE/EB$. $\because \angle EAG' = \angle EBH'$, $\therefore$ $\triangle$s AEG', BEH' are similar. $\therefore \angle AEG' = \angle BEH'$. $\because AB$ is a straight line, $\therefore G'EH'$ is a straight line. Since MG is $\parallel EG'$ and MH is $\parallel EH'$, $\therefore GMH$ is also a straight line. Therefore, EF, GH intersect in M. $\therefore$ They lie in one plane.

8.17. *A prism has two equal and parallel base quadrilaterals $ABCD$, $A'B'C'D'$. A point E is taken on AD such that $DE = \frac{2}{3} AD$ and $D'C'$ is produced to F such that $C'F = \frac{1}{2} D'C'$. Show that $A'C$, EF intersect in a point and find the ratio in which $A'C$ is divided by this point.*

Construction: Take E' on $A'D'$ such that $D'E' = \frac{2}{3} A'D'$. Join $A'F$, CE, $C'E'$ (*Fig.* 225).

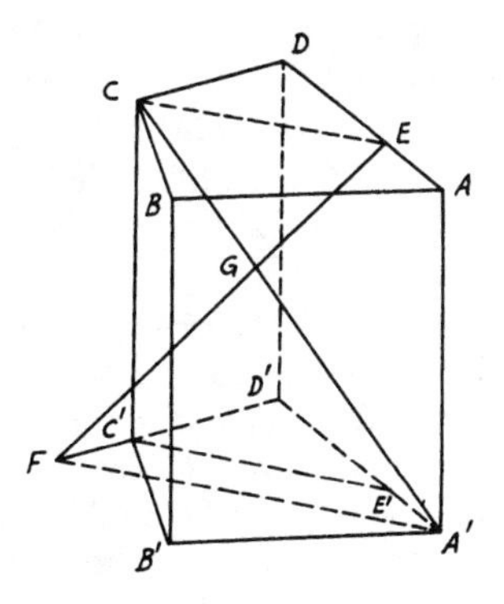

Figure 225

Proof: Since $DE = \frac{2}{3} DA$, also $D'E' = \frac{2}{3} A'D'$. $\therefore CE$ is $=$ and $\parallel$ $C'E'$ (since the bases of prism are equal and parallel). $\because D'C' = 2\,C'F$. $\therefore D'C'/D'F = \frac{2}{3} = D'E'/A'D'$. $\therefore C'E'$ is $\parallel A'F \parallel CE$. $\therefore A'C$ intersects EF in G. In $\triangle D'A'F$, $D'C'/D'F = C'E'/A'F = \frac{2}{3}$. $\because C'E' = CE$, $\therefore CE/A'F = \frac{2}{3}$. $\because CE$ is $\parallel A'F$, $\therefore CG/GA' = CE/A'F = \frac{2}{3}$.

8.18. *$ABCD$ is a tetrahedron with A as vertex and AD is perpendicular to the base BCD. If PQ is the shortest distance between the edges AC, DB and CR is drawn perpendicular to DB, show that $CP : PA = CR^2 : DA^2$.*

Construction: From R draw RS equal and $\parallel AD$. Join AS, CS and draw $PT \parallel AS$ to cut CS in T. Join RT (*Fig.* 226).

Proof: $\because QP$ is $\perp AC$ at P (hypothesis) and $\because ASRD$ is a parallelogram, i.e., AS is $\parallel DR \parallel PT$ and also PQ is $\perp BR$ at Q, $\therefore PQ$ is also $\perp PT$. $\therefore PQ$ is $\perp$ plane PTC (Th. 8.4). But BR is $\perp RS$ and RC,

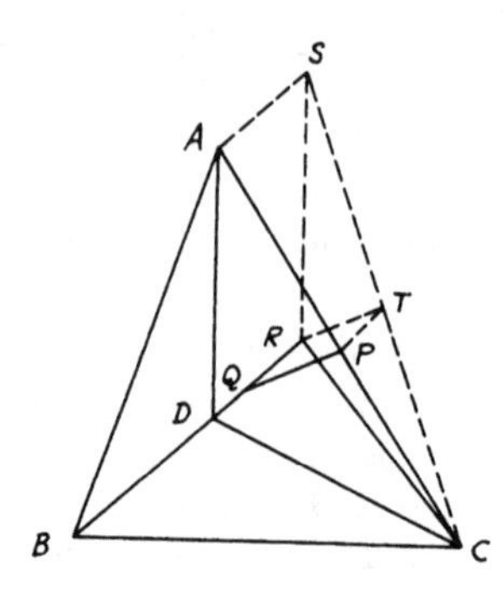

FIGURE 226

hence $\perp$ plane SRC. $\therefore BR$ is $\perp RT$. $\because BR$ is $\parallel PT$, $\therefore PT$ is $\perp RT$. $\therefore PQRT$ is a rectangle. Hence PQ is $\parallel RT$. $\because PQ$ is $\perp$ plane PTC, $\therefore$ also RT is $\perp$ plane PTC. $\therefore RT$ is $\perp TC$. $\because SR$ is $\parallel AD$ and hence $\perp$ plane BRC or $\perp RC$. $\therefore \triangle SRC$ is right-angled at R and RT is $\perp$ hypotenuse SC or TC. $\therefore CR^2 = CT \cdot CS$ and $SR^2 = ST \cdot CS = AD^2$. $\therefore CR^2/AD^2 = CT/ST = CP/PA$.

8.19. *A cube is constructed inside a right circular cone with vertex A such that four of its vertices lie on the cone surface and the other four vertices on the base of the cone. If the ratio between the altitude of the cone and its base radius is* $\sqrt{2} : 1$, *show that the side of the cube* $= \frac{1}{2}$ *the cone altitude.*

CONSTRUCTION: Let the cube be $DEFGHLMN$ of side a. Draw the altitude AO of the cone on the base cutting the upper face of the cube in O'. Join LN and produce it to meet the base circumference in B, C (*Fig.* 227).

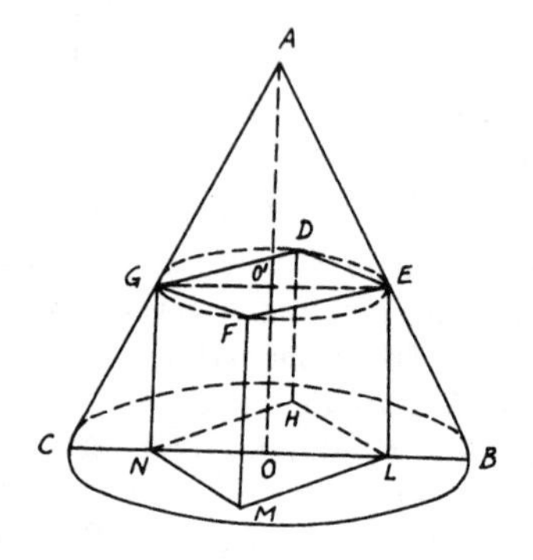

FIGURE 227

Proof: Since the cone is right and cube vertices lie on its surface and base, then O, O' are the centers of the faces $DEFG$, $HLMN$ of the cube and hence $EO'G$, LON are diameters of the upper circular section and respectively $\parallel$ and coplanar to the base. $\because AO : BO = \sqrt{2} : 1$ and EG is $\parallel BC$, $\therefore AO/BO = AO'/EO' = \sqrt{2}$ $\therefore AO' = \sqrt{2}\, EO'$. But $AO' = AO - OO' = AO - a$ and $EO' = \frac{1}{2} EG = \sqrt{2}/2\, a$. $\therefore AO' = \sqrt{2}\, EO' = a = AO - a$. $\therefore a = \frac{1}{2} AO$.

8.20. *ABCD is a tetrahedron and M, L, R are three points on the edges AB, AC, AD. If the lines MR, ML, LR are produced to meet the sides of the base BD, BC, CD or produced in X, Y, Z, show that XYZ is a straight line: Desargue's theorem* (*Fig.* 228).

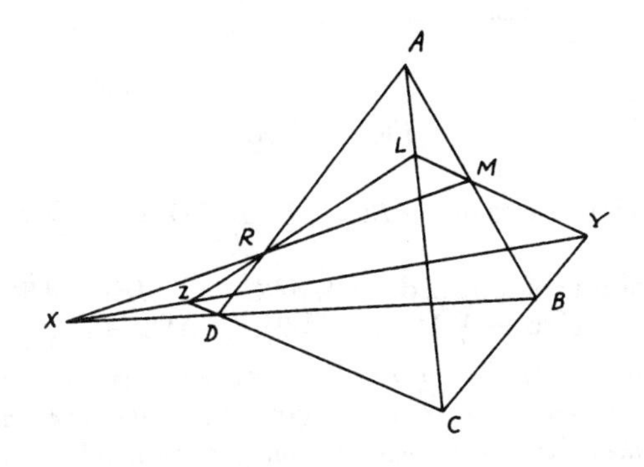

FIGURE 228

Proof: In $\triangle ABD$, $(BX/XD)(DR/RA)(AM/MB) = 1$. Also in $\triangle ABC$, $(CY/YB)(BM/MA)(AL/LC) = 1$. Also in $\triangle ADC$, $(DZ/ZC)(CL/LA)(AR/RD) = 1$ (Menelaus' Th. 5.13). Multiplying yields $(BX/DX)(YC/YB)(DZ/ZC) = 1$. Hence in the base triangle BCD, X, Y, Z are collinear.

8.21. *MA, MB, MC are three edges of a cube meeting in the vertex M. If a denotes the side of the cube, show that* (*i*) *volume of tetrahedron MABC* $= a^3/6$; (*ii*) *area of triangle ABC* $= a^2\sqrt{3}/2$; (*iii*) *the perpendicular from M on plane ABC* $= a\sqrt{3}/3$.

CONSTRUCTION: Draw $MD \perp AB$ and join DC (*Fig.* 229).

Proof: (i) Volume of tetrahedron $MABC$ is the same as volume of tetrahedron. $CMBA = \frac{1}{3} MC \cdot \triangle AMB = (a/3)(a^2/2) = a^3/6$.

(ii) $\because MD$ is $\perp AB$, $\therefore$ it bisects AB in D. Again $BC = AC =$ face diagonal. Hence ABC is an equilateral $\triangle$. $\therefore CD$ is $\perp AB$ (since D is the mid-point of AB). $\because AB = a\sqrt{2}$, $MD = a/\sqrt{2}$, and CM is $\perp MD$, $\therefore CD^2 = MC^2 + MD^2 = a^2 + (a^2/2) = \frac{3}{2} a^2$. $\therefore CD$

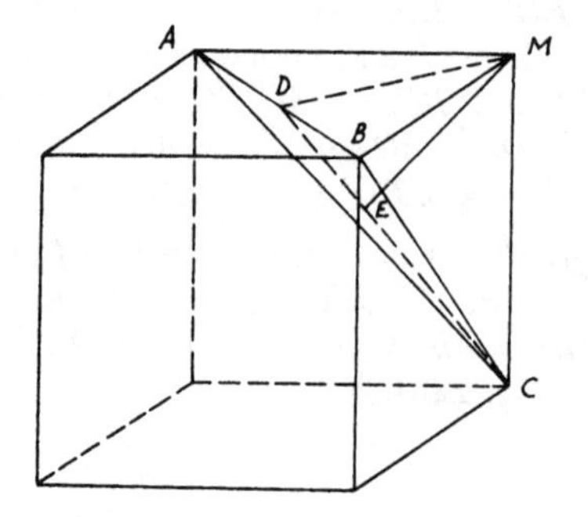

FIGURE 229

$= a(\sqrt{3/2})$. Hence area of $\triangle ABC = \frac{1}{2}\, AB \cdot CD = (a\sqrt{2}/2)(a\sqrt{3}/\sqrt{2})$
$= a^2\sqrt{3/2}$.

(iii) $\because$ Volume of tetrahedron $MABC = \frac{1}{3}\, ME \cdot \triangle ABC$ (ME is $\perp$ plane ABC), $\therefore\ a^3/6 = \frac{1}{3}\, ME(a^2\sqrt{3}/2)$. $\therefore\ ME = a/\sqrt{3} = a\sqrt{3}/3$.

8.22. *If the area of the side surface of a right circular cone is equal to* $\frac{1}{4}$ *of the area of the circle having as radius the generator of the cone, show that the ratio between the volume of the cone and that of a sphere which passes through the vertex and base circle of the cone is* 225 : 2048.

CONSTRUCTION: Let ABC be the right circular cone with A as vertex and BC a diameter in its base. Draw its altitude AD and produce it to meet the sphere in E. Join CD, CE (*Fig.* 230).

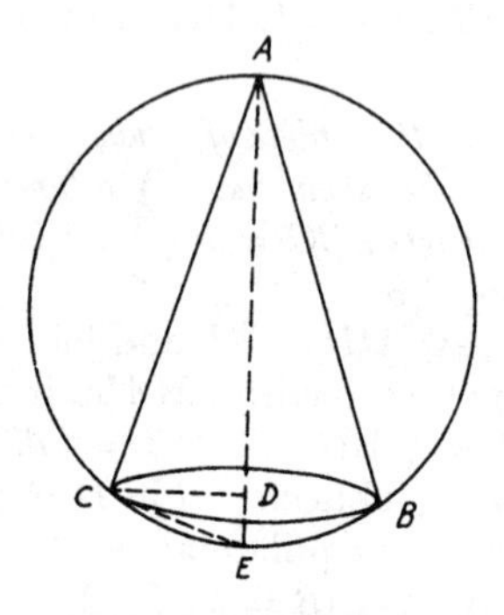

FIGURE 230

Proof: Let the altitude $AD = h$, generator $AB = l$, radius of base $\odot = r$ and radius of sphere $= R$. $\because$ Area of cone side surface $= \pi rl = \frac{1}{4}\pi l^2$ (hypothesis), $\therefore l = 4r$. $\because AD \perp$ base $\odot$, $\therefore \perp CD$. $\therefore AC^2 = AD^2 + CD^2$ or $l^2 = h^2 + r^2$. $\therefore h = r\sqrt{15}$, $\because$ volume of cone $= (\pi/3)r^2h = (\pi/3)r^3\sqrt{15}$. Since AE is a diameter of the sphere, $\therefore \angle ACE =$ right. $\because$ CD is $\perp AE$, $\therefore$ $l^2 = AD \cdot AE$. $\therefore$ $AE = l^2/h = 16r/\sqrt{15}$. $\therefore$ $R = 8r/\sqrt{15}$. Volume if sphere $= \frac{4}{3}\pi R^3 = \frac{4}{3}\pi(512\ r^3/15\sqrt{15}) = 2048\ \pi r^3/45\sqrt{15}$. $\therefore$ Volume of cone : volume of sphere $= (\pi/3)r^3\sqrt{15} : 2048\ \pi r^3/45\sqrt{15} = 225 : 2048$.

8.23. *MA, MB, MC are three straight lines in space; each is perpendicular to the other two lines. If x, y, z are the lengths of these lines respectively, show that (i) volume of tetrahedron MABC* $= \frac{1}{6}xyz$; *(ii) area of the triangle* $ABC = \frac{1}{2}\sqrt{x^2y^2 + y^2z^2 + z^2x^2}$; *(iii) the altitude from M on the plane* $ABC = xyz/\sqrt{x^2y^2 + y^2z^2 + z^2x^2}$.

Construction: Draw $MD \perp BC$ then join AD. Let MN the altitude from M on ABC be denoted l (*Fig.* 231).

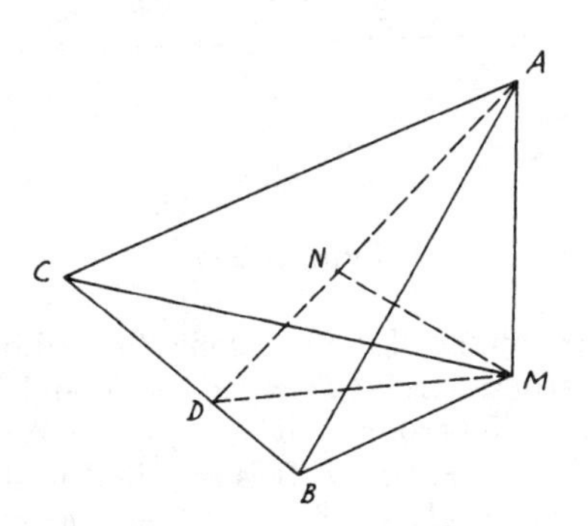

Figure 231

Proof: (i) Volume of tetrahedron $= \frac{1}{3} AM \cdot \triangle MBC = \frac{1}{6} AM \cdot MB \cdot MC = \frac{1}{6} xyz$.

(ii) $\because AM$ is $\perp$ plane MBC and $MD \perp BC$, $\therefore AD$ is $\perp BC$ also (Th. 8.4). In $\triangle MBC$, $\angle M$ is right. $\therefore$ $MB \cdot MC = 2\ \triangle MBC = MD \cdot BC$. $\therefore$ $MD = MB \cdot MC/BC = yz/BC$. $\because$ $AD^2 = AM^2 + MD^2$, $\therefore AD = \sqrt{AM^2 + MD^2} = \sqrt{x^2 + (y^2z^2/BC^2)} = (1/BC)\sqrt{x^2 \cdot BC + y^2z^2}$. But $\triangle ABC = \frac{1}{2} AD \cdot BC = \frac{1}{2}\sqrt{x^2 \cdot BC^2 + y^2z^2}$. $\because$ $BC^2 = y^2 + z^2$, $\therefore$ $\triangle ABC = \frac{1}{2}\sqrt{x^2y^2 + x^2z^2 + y^2z^2}$.

(iii) $\because$ Volume of tetrahedron $= (l/3)\triangle ABC$, $\therefore$ $\frac{1}{6}$ $xyz = l/6\sqrt{x^2y^2 + y^2z^2 + z^2x^2}$. $\therefore l = xyz/\sqrt{x^2y^2 + y^2z^2 + z^2x^2}$.

Also $\therefore 1/l^2 = (x^2y^2 + y^2z^2 + z^2x^2)/x^2y^2z^2 = 1/x^2 + 1/y^2 + 1/z^2$.

This result is also true in a two-dimensional problem. If l is the length of the altitude from the right vertex on the hypotenuse C and x, y are the lengths of the sides surrounding the right angle in any right-angled triangle, then $xy = cl$. $\therefore x^2y^2 = c^2l^2 = (x^2 + y^2)l^2$. $\therefore 1/l^2 = (x^2 + y^2)/x^2y^2 = 1/x^2 + 1/y^2$.

8.24. *If V denotes the volume formed by the rotation of a right-angled triangle around its hypotenuse and V_1, V_2 the volumes formed by the rotation of the triangle around the sides of the right angle, show that $1/V^2 = 1/V_1^2 + 1/V_2^2$.*

CONSTRUCTION: Draw from the right vertex B of $\triangle ABC$, $BD \perp AC$. Let a, b, c be the lengths of the sides BC, CA, AB, and $BD = h$ (*Fig.* 232).

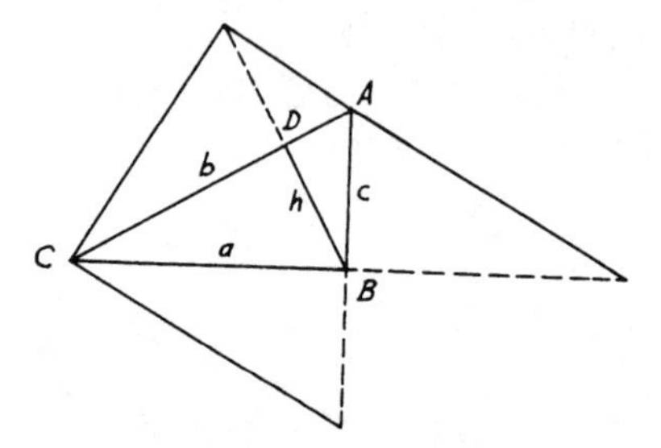

FIGURE 232

Proof: V = the volume of the two cones formed by the rotation of $\triangle$s ADB, CDB around AD, $CD = (\pi/3)h^2 \cdot b$ and V_1 = volume of cone formed by rotation around $AB = (\pi/3)a^2 \cdot c$. Also, V_2 = volume of cone around $BC = (\pi/3)c^2 \cdot a$. It is required to show that $V_1^2V_2^2 = V^2(V_1^2 + V_2^2)$ or $a^4c^2 \cdot c^4a^2 = a^6c^6 = h^4b^2(a^4c^2 + c^4a^2)$. But $h^4b^2(h^2b^2a^2 + h^2b^2c^2) = h^6b^4(a^2 + c^2) = h^6b^6 = a^6c^6$.

8.25. *On the same circular base of a hemisphere with center O, a frustum of a cone is constructed inside the hemisphere, the bases of which are parallel. If the ratio between the radii of the cone bases is* 1 : 2, *show that the ratio of the side surface area of the cone to the hemispherical surface area is* 3 : 4.

CONSTRUCTION: Let AB, CD be two $\parallel$ diameters in the bases of the frustum, AB being the common diameter with the hemisphere. Join DO and draw $CE \perp AB$. Let $OB = r$ (*Fig.* 233).

Proof: $\because CD = \frac{1}{2} AB = BO = OD$ (hypothesis) and CD is $\parallel BO$, $\therefore OBCD$ is a rhombus in which $CO = CB$. $\therefore CE$ bisects BO in E. Since side surface area of cone frustum $= \frac{1}{2} BC(\pi \cdot AB + \pi \cdot CD) = 3\pi r^2/2$ and hemispherical surface area $= 2\pi r^2$, the ratio between their surface areas = 3 : 4.

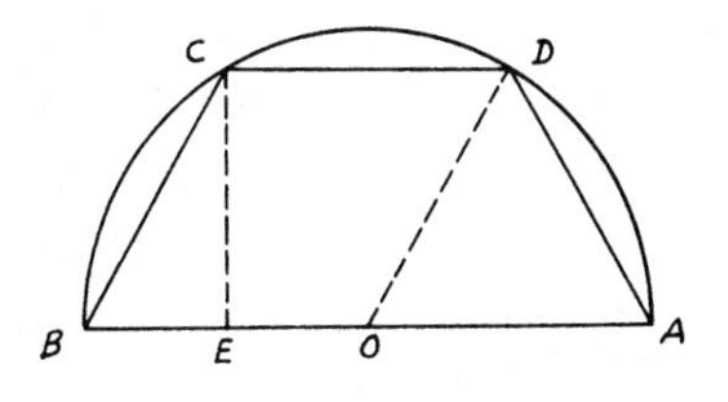

FIGURE 233

8.26. *If a sphere with radius r touches internally a right circular cone whose altitude and base radius are h, R, show that* $1/r^2 = 1/R^2 + 2/rh$. If V, S *and* V_1, S_1 *are the volume and surface of sphere and cone respectively, show also that* $V : V_1 = S : S_1$.

CONSTRUCTION: Let BC be a diameter in the base of that cone and A be the vertex. Draw $AO \perp BC$, O being the base center. If M is the center of the sphere which touches the cone at E, join ME (*Fig.* 234).

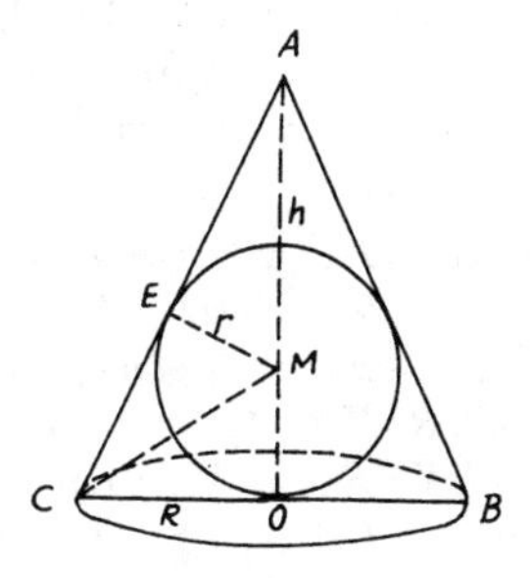

FIGURE 234

Proof: $\because$ $AE^2 = AM^2 - ME^2 = AM^2 - MO^2 = AO^2 - 2\,AO \cdot MO = h^2 - 2\,hr$ and $\because$ $\triangle$s AME, ACO are similar, $\therefore$ $AE/AO = EM/CO$. $\therefore$ $(h^2 - 2\,hr)/h^2 = r^2/R^2$ or $1 - (2r/h) = r^2/R^2$. Dividing by r^2 and rearranging, $\therefore$ $1/r^2 = 1/R^2 + 2/rh$. Join MC. Since MC bisects $\angle ACB$ in plane ACB, $\therefore$ $CO/AC = MO/AM$. Let the generator $AC = l$. $\therefore$ $R/l = r(h - r)$. $\therefore$ $R(h - r) = r \cdot l$. $\therefore$ $Rh = r(l + R)$. $\therefore$ $V : V_1 = \frac{4}{3}\pi r^3 : (\pi/3)R^2h = 4\pi r^2 : \pi R(l + R)$. $\therefore$ $S : S_1 = 4\,\pi r^2 : \pi R(l + R) = V : V_1$.

8.27. *If the inclined generator of a frustum of a cone is equal to the sum of the radii of the bases, show that the altitude = twice the geometric mean of these radii, and the volume = the entire surface of the frustum times* $\frac{1}{6}$ *its altitude.*

CONSTRUCTION: Take a projection of the cone frustum $ABCD$ in which $AB = 2R$ and $CD = 2r$. Draw $CM \perp AB$ and let the generator $AD = CB = l$, and the altitude $= h$ (*Fig.* 235).

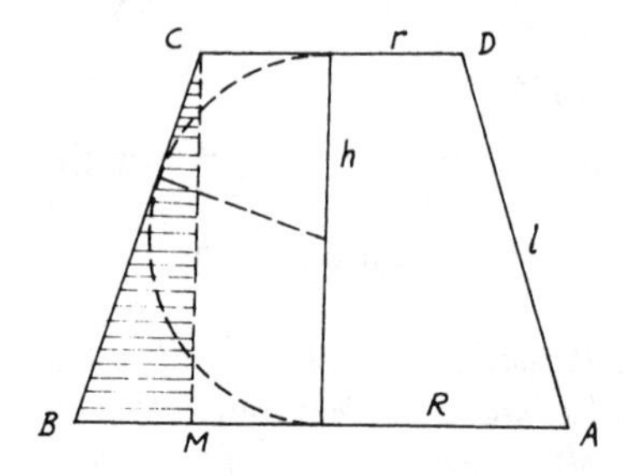

FIGURE 235

Proof: In $\triangle CBM$, $CM = h$. $\therefore CB^2 = CM^2 + BM^2$ or $h^2 = l^2 - (R - r)^2$. $\because l = R + r$ (hypothesis), $\therefore h^2 = (R + r)^2 - (R - r)^2 = 4Rr$. Therefore, $h = 2\sqrt{Rr}$. Again, entire surface of frustum $= (l/2)(2\pi R + 2\pi r) + \pi R^2 + \pi r^2$. Entire surface $\times (h/6) = (h/6)\pi[(R + r)^2 + R^2 + r^2] = (\pi h/3)(R^2 + r^2 + Rr)$. $\therefore$ The volume of this frustum $= (h/3)(\pi R^2 + \pi r^2 + \sqrt{\pi R^2 \cdot \pi r^2}) = (\pi h/3)(R^2 + r^2 + Rr) =$ entire surface $\times (h/6)$.

8.28. *On AOB as diameter and O as center a circle is described and two points D, E are taken on AB such that $OD \cdot OE = OB^2$. On DE as diameter a semi-circle is described such that its plane is perpendicular to the plane of the circle O. If any points R, L are taken on the circle O and X, Y on the semi-circle DE, show that $RX : RY = LX : LY$.*

CONSTRUCTION: Draw RP, LQ, XF, $YS \perp ABE$. Join LS. Let M be center of the semi-circle (*Fig.* 236).

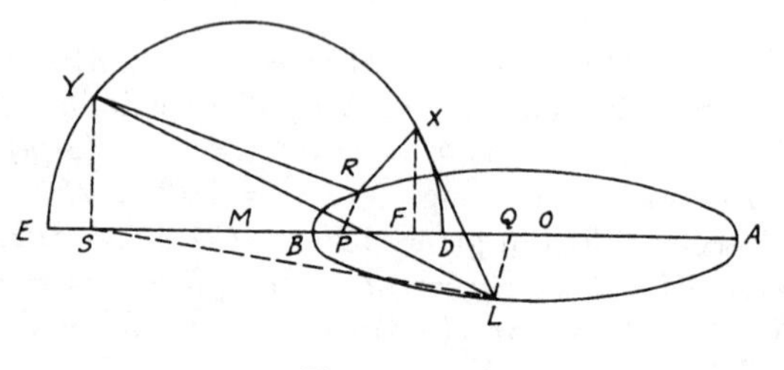

FIGURE 236

Proof: Since $OD \cdot OE = OB^2$, then, if the semi-circle plane is rotated a right angle to coincide with the plane of $\odot O$, they will cut orthogonally. If the radii of $\odot O$ and semi-circle are r, r_1, then $OM^2 = r^2 + r_1^2$. $\because$ YS is $\perp AB$, and planes of $\odot$s are $\perp$ each other. $\therefore$ YS is $\perp$ plane $\odot AOB$. $\therefore$ YS is $\perp$ SL which lies in the plane of $\odot O$. But $LY^2 = YS^2 + SL^2 = LQ^2 + QS^2 + YS^2 = r^2 - QO^2 + QS^2 + r_1^2 - MS^2 = OM^2 - QO^2 + QS^2 - MS^2 = 2\,QM \cdot OS$ and $LX^2 = 2\,QM \cdot OF$. $\therefore$ $LX^2 : LY^2 = OF : OS$. Similarly, $RX^2 : RY^2 = OF : OS$. Hence $RX : RY = LX : LY$.

8.29. *$ABCDA_1B_1C_1D_1$ is a given regular prism, the faces of which are parallelograms and P is a fixed plane. Find the locus of the point E in the plane P such that the sum of the squares on its distances from the vertices of the prism is constant.*

CONSTRUCTION: Join the diagonals AC_1, A_1C, BD_1, B_1D, which intersect in one point M (since the figure is a regular prism). Draw $MO \perp$ plane P and join ME, OE (*Fig.* 237).

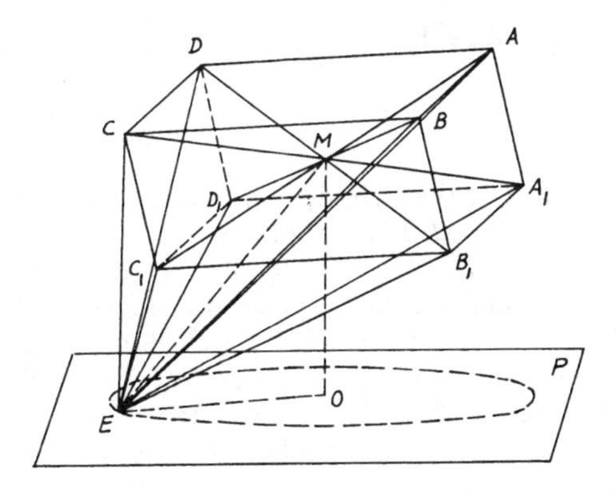

FIGURE 237

Proof: Since M is the middle point of all the diagonals of prism, $\therefore$ $EA^2 + EC_1^2 = 2\,EM^2 + 2\,MA^2$. Similarly with the other three pairs of diagonally opposite vertices. Adding up gives $8\,EM^2 = EA^2 + EB^2 + EC^2 + ED^2 + EA_1^2 + EB_1^2 + EC_1^2 + ED_1^2 - 2\,(MA^2 + MA_1^2 + MB^2 + MB_1^2)$. Assuming that the sum of the squares of the distances from E to the vertices $= S$ and since $EM^2 = MO^2 + OE^2$, $\therefore$ $8\,(MO^2 + OE^2) = S^2 - 2\,(MA^2 + MA_1^2 + MB^2 + MB_1^2)$. $\therefore$ $OE^2 = \frac{1}{8}\{S^2 - 2\,(MA^2 + MA_1^2 + MB^2 + MB_1^2 + 4\,MO^2)\} =$ constant. $\because$ The projection O of the fixed point M on the fixed plane P is fixed, the locus of E is a $\odot$ in plane P with OE as radius.

8.30. *The total surface area of a cylinder whose diameter is equal to its altitude and which is constructed outside (inside) a sphere is the mean proportional between the surface area of this sphere and the total surface area of the right circular cone whose generator is equal to the diameter of its base and which is also constructed outside (inside) the same sphere. Prove that this is also the case with their volumes.*

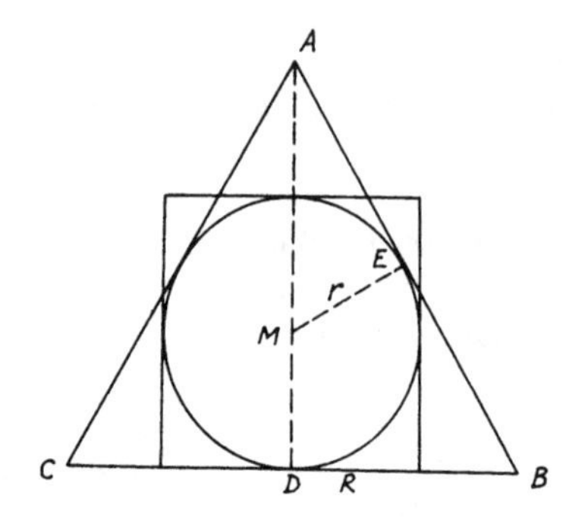

FIGURE 238

CONSTRUCTION: Let the cylinder and cone be described outside the sphere with center M. Let the sphere touch the base and surface of cone in D, E. Join AD, ME. R, r denote the radii of cone base and sphere (*Fig.* 238). Let the shown projection represent the solids in question. Total area of cone $= \pi Rl + \pi R^2$. But generator $l = 2R$. So, total area of cone $= 3\pi R^2$. Area of sphere $= 4\pi r^2$. Since cone ABC has $\angle A = 60°$, $\therefore ME/AE = r/R = \tan 30°$. $\therefore r = R/\sqrt{3}$. $\therefore$ Area of sphere $= \frac{4}{3}\pi R^2$. If cylinder altitude $= h$, total area of cylinder $= 2\pi rh + 2\pi r^2$. But $h = 2r$. So total area of cylinder $= 6\pi r^2 = 2\pi R^2$. Therefore area of sphere $\times$ total area of cone $= 3\pi R^2 \cdot \frac{4}{3}\pi R^2 = 4\pi^2 R^4 =$ (total area of cylinder)^2. Again, volume of cone $= \frac{1}{3}\pi R^2 H$. But H the altitude of cone $= R\sqrt{3}$. So, volume of cone $= (\sqrt{3}/3)\pi R^3$. Volume of sphere $= \frac{4}{3}\pi r^3 = 4\pi R^3/9\sqrt{3}$. Volume of cylinder $= \pi r^2 h = 2\pi r^3 = 2\pi R^3/3\sqrt{3}$. Hence volume of sphere $\times$ volume of cone $= 4\pi^2 R^6/27 =$ (volume of cylinder)^2. Similarly for the case when the cylinder and cone are described inside the sphere.

Miscellaneous Exercises

1. Given three non-intersecting straight lines in space, draw from any point in one a straight line intersecting both the others. Examine the cases in which more than one line can be so drawn, also the cases in which no line can be so drawn.

2. Draw a plane perpendicular to a given straight line through a given point either in the line or without it.
3. If a pyramid stands on a square base and has equilateral triangles for its faces, prove that the perpendicular from the vertex to the base is equal to half the diagonal of the base.
4. A piece of wire is bent into three parts, such that each of its exterior parts is double the middle one and perpendicular to the plane containing the other two of which the middle part is one. Show that the line joining the ends of the wire = three times the middle part.
5. Of all the straight lines which can be drawn to meet a given plane from a given point outside it, the least is the straight line perpendicular to the plane, and those which meet the plane in points equally distant from the foot of the perpendicular are equal, and of two straight lines the one which meets the plane at a point farther from the foot of the perpendicular than another is the greater and conversely.
6. AB is a straight line. From any point A on the line, AC, AD are drawn perpendicular to AB, and from another point B in the line BE, BF are drawn perpendicular to AB. Show that the planes ACD, BEF are parallel.
7. AB, CD are two parallel lines and from any point E outside their plane, EF is drawn perpendicular to AB meeting AB in F and FR drawn perpendicular to CD meeting CD in R. Show that ER is perpendicular to CD.
8. MA, MB are two straight lines intersecting in M and MC is another straight line not in their plane such that $\angle AMC = \angle BMC$. From any point C in MC, CD is drawn perpendicular to the plane MAB. Show that DM bisects the angle AMB. (From C draw CE, CF perpendiculars to MA, MB and join DE, DF.)
9. From the point P outside two intersecting planes in AB, two perpendiculars PQ, PR are drawn to the planes. If QS is also drawn perpendicular to the plane where R lies, then prove that RS is perpendicular to AB.
10. Find the locus of points equally distant from two given points.
11. AB, AC are two perpendiculars from any point A on two intersecting planes in DE. Show that BC is perpendicular to the parallel from C to DE.
12. ABC is a triangle and M is the orthocenter. If MP is drawn perpendicular to the plane of the triangle, show that the line joining P to any one of the vertices of the triangle will be perpendicular to the parallel from that vertex to the opposite side.
13. Find the locus of points equally distant from three given points.
14. Given a straight line and any two points, find a point in the straight line equally distant from the two points.

15. Prove that the sides of an isosceles triangle are equally inclined to any plane through the base.

16. A given line l is parallel to a given plane. x is a variable line in this plane. Show that unless x is parallel to l, the shortest distance between x and l is constant.

17. Draw a straight line to meet three given straight lines and be parallel to a given plane. (Let the lines be AB, CD, EF. Draw through CD a plane $\parallel$ to given plane. Let AB, EF cut this plane in P, Q. Then PQ is such a line.)

18. Find the locus of the foot of the perpendicular from a given point on a plane which passes through a given straight line.

19. From a point P, PA is drawn perpendicular to a given plane, and from A, AB is drawn perpendicular to a line in that plane. Prove that PB is also perpendicular to that line.

20. OA, OB, OC are three straight lines mutually perpendicular. From O perpendiculars OP, OQ, OR are let fall on BC, CA, AB. Prove that BC bisects the angle QPR externally.

21. OP is at right angles to the lines OA, OB, which are also at right angles to each other. OC, OD are drawn bisecting the angles POA, POB. Prove that COD is $60°$. (Make $OC = OD$, and draw CA, $DB \perp OA$, OB.)

22. From a point P, PA, PB are drawn perpendicular to two planes which intersect in CD meeting them in A and B. From A, AE is drawn perpendicular to CD. Prove that BE is also perpendicular to CD.

23. Prove that the intersection of two planes, each of which contains one of two parallel straight lines, is parallel to those lines.

24. Planes are drawn through a given point each containing one of a series of parallel straight lines. Prove that they intersect any plane which intersects their common line through the given point in three concurrent lines.

25. Given a plane and two points on the same side of it. Find the point in the plane the sum of the distances of which from the given points is a minimum. (Similar to Problem 1.16.)

26. If a straight line is equally inclined to each of three straight lines in a plane, it is perpendicular to the plane in which they lie.

27. Draw two parallel planes, one through each of two straight lines which do not meet and are not parallel. (Let AB, CD be the lines. Draw $AE \parallel CD$, $CF \parallel AB$. Hence plane AEB is $\parallel$ plane CFD.)

28. If two straight lines are parallel, each is parallel to every plane passing through the other.

29. Draw a line which will be parallel to one given line and intersect two other given lines, no two of the given lines meeting.

30. In a tetrahedron in which each edge is equal to the opposite edge (called an *isosceles tetrahedron*) prove that (a) the faces are congruent; (b) the lines joining the mid-points of opposite edges are perpendicular to the edges which they bisect and to one another. [(a) Each face has sides a, b, c. (b) Let L, L', M, M', N, N' bisect AB, CD, AC, BD, AD, BC of tetrahedron $ABCD$. Then $LM = \frac{1}{2} BC = M'L'$ and $LM' = \frac{1}{2} AD = ML'$. Hence $LML'M'$ is a rhombus. Hence LL', MM' are perpendicular, so MM', NN' and NN', LL'. Now NN' is perpendicular to LL' and MM' and therefore to LM and $L'M$, i.e., to BC and AD.]

31. Prove that a sphere can be inscribed in an isosceles tetrahedron and that the lines joining its corners with the centroids of the faces meet at the center of this sphere.

32. The six planes bisecting the edges of a tetrahedron perpendicularly meet in a point, which is the center of the circumscribing sphere.

33. If a pair of points is taken on each edge of a tetrahedron such that the pairs on any two adjacent edges are concyclic, show that the twelve points lie on the same sphere.

34. The section of a tetrahedron by a plane is a parallelogram if and only if the plane is parallel to a pair of opposite edges. (Let the parallelogram be $LML'M'$, where L, L', M, M' are on AB, CD, AC, BD of a tetrahedron $ABCD$. Then the three planes LML', ABC, DBC meet at a point; and this is at infinity since LM, $M'L'$ are $\parallel$. Hence LM is $\parallel$ BC and so LM' to AD.)

35. In a tetrahedron the lines (called the *medians*) joining the mid-points of opposite edges bisect one another at the centroid.

36. In Theorem 8.28 of a tetrahedron $ABCD$, the volumes subtended by the faces at G are equal, G being the centroid of the tetrahedron. (For $ABCD : GBCD = AA' : GA' = 4 : 1$, and so on.)

37. Show that the shortest distance between two opposite edges of a regular tetrahedron is equal to half the diagonal of the square described on the edge.

38. Two planes which are not parallel are cut by two parallel planes. Prove that the lines of section of the first two with the last two contain equal angles.

39. If two parallel planes are cut by three other planes which have no line common to all three, and no two of which are parallel, the triangles formed by the intersections of the parallel planes with the three other planes are similar to each other.

40. Two straight lines do not intersect and are not parallel. Find a plane upon which their projections will be parallel. (Refer to Exercise 27.)

41. Two similar polygons, not in the same plane, are placed with their homologous sides parallel. Prove that the straight lines which join corresponding vertices are either parallel or concurrent.

42. Show that a tetrahedron can be formed of any four equal congruent triangles, provided the triangles are acute-angled.

43. Of parallelograms which are parallel to two opposite edges of a tetrahedron, the one of greatest area bisects each edge. (For in a tetrahedron $ABCD$; $AL : AB = LM : BC$, i.e., $LM \propto AL$. So $LM' \propto BL$. See Exercise 34 for the parallelogram $LML'M'$ on AB, CD, AC, BD. Also the angle MLM' is constant, being the angle between BC and AD. Hence the area varies as $AL \cdot LB$.)

44. If a solid angle at O is contained by three plane angles AOB, BOC, COA and D is any point in the plane ABC, prove that the angles AOB, BOC, COA are together less than twice the angles AOD, BOD, COD.

45. Divide a straight line similarly to a given divided straight line lying in a different plane.

46. Two straight lines which intersect are inclined to each other at an angle equal to $\frac{2}{3}$ of a right angle and to a given plane, each at an angle equal to half a right angle. Prove that their projections on this plane are at right angles to each other.

47. If perpendiculars to two faces of a tetrahedron from the opposite vertices intersect, prove that the edge in which the faces intersect is perpendicular to the opposite edge. (Refer to Problem 8.13.)

48. A pyramid is described with a parallelogram as base. Show that if its four triangular faces are equal in area, their projections on the base are the triangles into which the parallelogram is divided by its diagonals; and the parallelogram is a rhombus.

49. Through one of the diagonals of a parallelogram a plane is drawn. Prove that the perpendiculars let fall from the end of the other diagonal on this plane are equal.

50. Prove that the greatest tetrahedron which can be inscribed in a given sphere is equilateral. (Take B, C, D fixed in a tetrahedron $ABCD$. Then the perpendicular from A on BCD must be greatest; i.e., A must be on the diameter of the sphere through the circumcenter of BCD. Hence $AB = AC = AD$.)

51. Prove that a sphere can be described through two circles in different planes, provided these circles have two common points (i.e., through the common points and a point on each circle).

52. Prove that a circle PQR and its inverse with respect to a point, not in its plane, lie on the same sphere (i.e., the sphere $PQRP'$).

53. A and B are two given points on two given planes which intersect in the line l. Find a point P on l such that $AP + PB$ may be least.

54. The points X and Y move on given lines which are not in the same plane. Show that the center of XY moves on a plane.

55. Draw a plane to cut the given lines OA, OB, OC, OD, no three of which are in the same plane, in the parallelogram $ABCD$.

56. Prove that in the tetrahedron of Problem 8.13, each altitude meets the opposite face at its orthocenter.

57. Prove that in the tetrahedron of Problem 8.13, a sphere can be drawn through the centers of its edges and the feet of the shortest distances between opposite edges.

58. If a plane cuts a sphere and the area of the circle of intersection is $\frac{1}{12}$ of the surface area of the sphere, find the distance of this plane from the center of the sphere in terms of its radius.

59. Prove that in an isosceles tetrahedron, where every two opposite edges are equal, the three angles at any vertex are together equal to two right angles.

60. In an isosceles tetrahedron, prove that the centers of the inscribed and escribed spheres coincide. (Refer to Exercise 31.)

61. In an isosceles tetrahedron, show that each plane angle of every face is acute.

62. In an equilateral tetrahedron $ABCD$, E bisects the perpendicular from A on BCD. Show that EB, EC, ED are mutually perpendicular. [Let AG be the $\perp$ from A on BCD. Then, by symmetry, G is the centroid of the equilateral $\triangle BCD$ and $AG \perp BG$. $\therefore$ $4\,BE^2 = 4\,BG^2 + 4\,GE^2$. But, $BG = \frac{2}{3}\,BL$ (if $BL \perp CD$) $= \frac{2}{3}\,BC(\sqrt{3}/2)$. $\therefore$ $BG = BC/\sqrt{3}$ and, $4\,GE^2 = AG^2 = AB^2 - BG^2$. $\therefore$ $4\,BE^2 = (4\,BC^2/3) + AB^2 - (BC^2/3) = AB^2 + BC^2 = 2\,BC^2 = 4\,CE^2 = 4\,DE^2$. Hence $BE^2 + CE^2 = 2\,BE^2 = BC^2$. Hence $BE \perp CE$, so $CE \perp DE$ and $DE \perp BE$.]

63. Show that all parallel sections of an isosceles tetrahedron which are parallelograms have the same perimeter.

64. A plane is drawn through an edge of a tetrahedron and the center of the opposite edge. Show that the six planes so drawn meet in a point.

65. Prove that the sum of the squares of the edges of a tetrahedron is equal to four times the sum of the squares of the lines joining the mid-points of opposite edges.

66. If the lines joining corresponding vertices of two tetrahedrons meet in a point, the intersections of corresponding edges lie on a plane.

67. If a sphere can be drawn to touch the six edges of a tetrahedron, the sum of each two opposite edges is the same.

68. If the section of a tetrahedron by a plane is a square, show that two opposite edges must be perpendicular and, in that case, show how to construct the square. (With the figure of Exercise 34, LM is $\parallel BC$ and $LM' \parallel AD$. Hence, if $LML'M'$ is a square so that $LM \perp LM'$, then $BC \perp AD$. Again, if $LML'M'$ is a square $LM = LM'$. But, $LM : BC = AL : AB$ and $LM' : AD = BL : AB$. Hence $BC \cdot AL = AD \cdot BL$. Hence to get L we divide AB so that $AL : BL = AD : BC$. Then, draw $LM \parallel BC$, $ML' \parallel AD$, and $L'M' \parallel BC$.)

69. Prove that the three lines joining the middle points of opposite edges of a tetrahedron are concurrent and that they form, two and two, the diagonals of three parallelograms, such that the angles between their sides are the angles between the opposite edges of the tetrahedron.

70. A pyramid stands on a quadrilateral base. Draw a plane such that its lines of intersection with the faces of the pyramid may form a parallelogram in which one angle is given.

71. Prove that the sections of a tetrahedron made by planes parallel to a pair of opposite edges are rectangles if the opposite edges are at right angles.

72. Prove that if the line joining one vertex of a tetrahedron to the orthocenter of the opposite face be perpendicular to that face, the same is true with regard to the other vertices.

73. Three planes intersect in a point, and a plane is drawn through the common section of each pair perpendicular to the third plane. Show that these three planes intersect in a straight line.

74. If through the edges AB, AC, AD of a tetrahedron $ABCD$ planes are drawn such that their intersections with the planes CAD, DAB, BAC respectively bisect the angles at A in those planes, these planes have a common line of intersection.

75. If BAC, CAD, DAB are three plane angles containing a solid angle, prove that the angle between AD and the straight line bisecting the angle BAC is less than half the sum of the angles BAD, CAD.

76. A solid angle is contained by three plane angles BOC, COA, AOB. If BOC, COA are together equal to two right angles, prove that CO is perpendicular to the line which bisects the angle AOB.

77. $ABCD$ is a tetrahedron; AD, BD, CD are cut in L, M, N by a plane parallel to ABC. BN, CM meet in P; CL, AN meet in Q; AM, BL meet in R. Show that the triangles LMN, PQR are similar.

78. $ABCD$ is a face and AE a diagonal of a cube. BG is drawn perpendicular to AE and DG is joined. Prove that DG is perpendicular to AE.

79. Given three straight lines not in the same plane. Draw through a given point a straight line equally inclined to the three. (Through given point O draw OA, OB, OC $\parallel$ to the given skew lines. Bisect $\angle$s AOB, BOC by OD, OE; through OD, OE draw planes ODF, OEF $\perp$ planes AOB, BOC. Hence any line through O in plane ODF makes equal angles with OA, OB and any line through O in plane OEF makes equal angles with OB, OC. Hence OF is equally inclined to given lines.)

80. On the same side of the plane of a parallelogram $ABCD$, three straight lines Bb, Cc, Dd are drawn in parallel directions, and such that Cc is equal to Bb and Dd together. Prove that the points A, b, c, d lie in one plane.

81. In the tetrahedron $OABC$, OA, OB, OC are equal. Prove that each of them is greater than the radius of the circle circumscribing the triangle ABC.

82. Prove that in any tetrahedron, the plane bisecting any dihedral angle divides the opposite edge into segments which are in the ratio of the areas of the adjacent faces. If each edge of a tetrahedron is equal to the opposite edge, the plane bisecting any dihedral angle bisects the opposite edge.

83. A pyramid stands on a triangular base. Show that if the circles inscribed in three of the faces touch each other, the circle inscribed in the fourth face will touch all the others.

84. If $ABCD$ be a tetrahedron and G the intersection of lines drawn from A, B, C to the middle points of BC, CA, AB respectively, prove that the squares on AD, BD, CD are together equal to the squares on AG, BG, CG with three times the square on DG.

85. In a tetrahedron the straight line joining the middle points of one pair of opposite edges is at right angles to both edges. Prove that of the other four edges any one is equal in length to the opposite one.

86. AB, AF, CB, CD are edges of a cube, AF, CD being opposite edges. Find the angles between the planes ABD, AFD, ABF. (Let $ABEF$ be a face of the cube. BD, BE are both $\perp AB$. $\therefore EBD$ is the angle between planes ABD, ABF and $\angle EBD = \frac{1}{2}$ right angle. Similarly, the angle between planes AFD, $ABF = \frac{1}{2}$ right angle. Draw $BG \perp AD$. $\therefore DA \cdot AG = AB^2 = AF^2$. $\therefore FG$ is $\perp AD$ and $BG = GF$. Draw $GH \perp BF$. $\therefore BH = HF$. $\therefore H$ is the intersection of AE, BF. Again, $GH : AH = DE : AD = AB : AD = BG : BD$ and $BD = AE = 2\,AH$. $\therefore BG. = 2\,GH$. $\therefore \angle BGH = \frac{2}{3}$ right angle. $\therefore \angle BGF = \frac{4}{3}$ right angle.)

87. If P be a point equally distant from the vertices A, B, C of a right-angled triangle, of which A is the right angle and D the middle point of BC, prove that PD is at right angles to the plane ABC. Prove also that the angle between the planes PAC, PBC and the angle between the planes PAB, PBC are together equal to the angle between the planes PAC, PAB.

88. $ABCD$ is a tetrahedron, $EFGH$ a plane section cutting AC in F. Prove that if the section is a rhombus in a plane parallel to BC and AD, then $BC : AD = CF : AF$.

89. Under what circumstances is it possible to draw a plane so as to cut the faces of a tetrahedron in four lines forming a square?

90. Prove that if each solid angle at the base of a tetrahedron is contained by three plane angles which make up two right angles, the tetrahedron will have all its faces equal and similar.

91. Prove that in an isosceles tetrahedron where every pair of opposite edges are equal, the inscribed and circumscribed spheres are concentric.

92. If a cone in which the generator equals the diameter of the base touches a sphere along the contour of the circle of the base, and a cylinder envelops the sphere and the cone so that the vertex of the cone lies in the upper face of cylinder, find the ratio of the total surface areas and the ratio of the volumes of these three sides.

93. O_1, O_2, O_3, O_4 are the centers of the circles inscribed in the faces BCD, ACD, ABD, ABC respectively of the tetrahedron $ABCD$. Prove that AO_1, BO_2, CO_3, DO_4 will meet in a point if the rectangles $AB \cdot CD$, $AC \cdot BD$ and $AD \cdot BC$ are all equal.

94. Prove that a sphere can be described through two circles which touch one another but are not in the same plane. (This is the limit of Exercise 51, when the two points coincide.)

95. Prove that the cone joining any point to a circular section of a sphere cuts the sphere again in a circle.

96. Prove that four planes can be drawn through the center of a cube, each of which cuts six edges of the cube in the corners of a regular hexagon and the other six edges produced in the corners of another regular hexagon. The area of the second hexagon is three times that of the first and the sides of the second are perpendicular to the central radii drawn to the corners of the first.

97. A cone is described such that its base circle touches internally the sides of a square base of a regular pyramid which has the same vertex as the cone. If the common altitude of cone and pyramid is $\sqrt{3}/2$, the side of the square base, find the ratio of the volume of the sphere inscribed in the cone to the volume of solid between pyramid and cone.

98. Prove that every two circular sections of a cone are either parallel or inverse with respect to the vertex.

99. Find the locus of a point in a given plane whose distances from the given points, which are not in the plane, are in a given ratio. (Let A, B be the given points and P the variable point. Then since $PA : PB$ is given, the locus of P is a sphere. The required locus is the circle which is the section of this sphere by the given plane.)

100. Draw a plane to touch three given spheres. (Let S be a center of similitude of the spheres 1, 2, and S' of 1, 3. Then a tangent plane through the line SS' to 1 will touch 2, 3.)

INDEX